Anyone Can Code

Anyone Can Code
The Art and Science of Logical Creativity

Ali Arya

CRC Press
Taylor & Francis Group
Boca Raton London New York

CRC Press is an imprint of the
Taylor & Francis Group, an **informa** business
A CHAPMAN & HALL BOOK

First Edition published 2021
by CRC Press
2 Park Square, Milton Park, Abingdon, Oxon, OX14 4RN

and by CRC Press
6000 Broken Sound Parkway NW, Suite 300, Boca Raton, FL 33487-2742

CRC Press is an imprint of Informa UK Limited

British Library Cataloguing-in-Publication Data
A catalogue record for this book is available from the British Library

Library of Congress Cataloging-in-Publication Data
Names: Arya, Ali, author.
Title: Anyone can code : the art and science of logical creativity / Ali Arya.
Description: First edition. | Boca Raton : CRC Press, 2020. | Includes
bibliographical references and index.
Identifiers: LCCN 2020030146 | ISBN 9780367199692 (pbk) |
ISBN 9780367199746 (hbk) | ISBN 9780429244421 (ebk)
Subjects: LCSH: Computer programming. | Programming languages (Electronic
computers) | Logic.
Classification: LCC QA76.6 .A78 2020 | DDC 005.13—dc23
LC record available at https://lccn.loc.gov/2020030146

ISBN: 978-0-367-19974-6 (hbk)
ISBN: 978-0-367-19969-2 (pbk)
ISBN: 978-0-429-24442-1 (ebk)

Typeset in Minion Pro
by codeMantra

Visit the companion website: http://ali-arya.com/anyonecancode

Contents

PART 1   Getting Started

PART 5   More about Objects and Classes

Chapter 11 ▪ Class Hierarchies   459

Chapter 12 ▪ Object Identities   497

PART 6   Moving Forward

List of Sidebars

List of Tables

List of Exhibits

Preface

Anyone Can Code is a tricky title for a book. It may give the impression that the book is one of those super-simplified resources that helps *anyone* make a nice-looking computer program, for example, a game or a mobile app. It is not, but there is a truth in that title, and it deserves an explanation. The title and the whole book are based on how I view the computing education. And my view is the result of a long journey.

I wrote my first computer program in 1985 when I was a university freshman studying Electrical Engineering. Michael Jackson's "Thriller," hair bands, and breakdance were the big things, and the PC era had just started. I loved the creative power that programming gave me and found the underlying problem-solving process fascinating. So, I moved, and sometimes sneaked and forced my way, through various computing platforms and went from batch processing punch cards for mainframe computers to terminals, early personal computers such as Commodore 64, ZX Spectrum, and MS-DOS PCs. I soon discovered computer graphics as the combination of my visual arts and computing passions, and eventually tried software development on Windows, Unix, Mac, and later web and mobile platforms, using various tools and languages, and for different types of applications from industrial automation to games. Throughout this process, I shared my experience with friends, colleagues, and students, and accumulated years of teaching experience in both industry and academia.

This journey taught me many things about computers, work, life, and myself, but it particularly shaped my belief about programming. I learned that despite many complex and seemingly unnatural forms, programming is not really about the complicated languages and strange names; it is about thinking. More precisely, it is a process that some people call *logical creativity*, as it combines logical thinking with creativity and creative problem-solving. And here is where the book title comes in: despite the scary looks, computer programming is an activity that we are all equipped to do. We all think and solve problems on a daily basis. We can all *Code*, but we need to approach it in the right way: instead of getting stuck with tools, we should learn the concepts; instead of memorizing this and that syntax, we should learn how to think, a process that involves both rigorous methodology and creative design. That is how the second part of the book title came in. The *Art and Science of Logical Creativity* part of the book title says that we can all design and write programs, but it is a process that we need to learn the right way.

A few years ago, as a random plan after a lunch date, my wife and I tried a dance lesson. Unlike any other dance classes I had attended before, the instructors at the Fred Astaire

Dance Studio combined multiple types of ballroom dances in each session. The logic was that there are so many common concepts among seemingly different dances such as Tango, Swing, and Foxtrot. Learning them together helps the learner to go deeper than the surface and get the fundamental ideas while playing with variations. This experience inspired the particular approach of this book that introduces programming concepts within the context of multiple languages. Such an approach emphasizes the concepts and the notion of logical creativity. It allows the learners to see how these concepts can be translated into different languages while maintaining the same meaning.

However, just like many other creative design tasks, software development is a hierarchical process that starts with basic and simple elements and creates more and more complex modules on top of each other. This *modularization* is the thread that connects many key concepts in programming from arrays and functions to classes, design patterns, and architectures. Modularization, logical thinking, and the multi-language approach are the key features of this book.

Anyone Can Code: The Art and Science of Logical Creativity introduces programming as a way of problem-solving through logical thinking. It uses the notion of modularization as a central lens through which we can make sense of many software concepts. This book takes the reader through fundamental concepts in programming by illustrating them in three different and distinct languages: C/C++, Python, and Javascript. This approach is not the simplest way to start programming and "build a nice-looking app." But, based on my many years of experience, it gives the beginner and intermediate learners a strong understanding of what they are doing so that they can do it better and with any other tool or language that they may end up using later. It is designed to cover material for the first two programming courses in a typical university program: an introductory course and an intermediate one with subjects such as Object-Oriented Programming. This book can also be used by individual learners at their own pace.

The popularity of computer games and the multi-disciplinary nature of game development projects that involve programmers, artists, designers, writers, and many other experts have prompted the idea of "learning by making games." This idea is particularly useful in the case of learning programming where almost all topics from basic concepts to advanced ones (such as graphics, networking, and database) will be used and practiced frequently. As such, this book uses the notion of "learning by making games" quite frequently. While many other examples are provided, you will see a clear focus on game development as a context for explaining various programming topics. This allows the reader to practice what is being said through a hopefully engaging experience that can result in programs similar to many popular desktop and mobile games. But no learning experience will be effective if it is not accompanied by sense-making that allows the learner to really understand what happened and why.

Reflection (making sense of the experience) is found to be necessary for learning as "doing" by itself will not help us learn if we don't understand what we did and how it has affected us. It is defined as a generic term for a variety of affective and intellectual activities in which an individual engages in order to gain new understandings and appreciations by exploring their experiences. While essential and quite helpful, reflection may not be a very natural and comfortable process for many people. Most of us do think about our experiences and the work we have done, but, usually, this is not in any organized fashion, it happens randomly, and we don't lead the thinking to proper results. I encourage the readers to think of programming, and reading this book, as reflective processes where they make sense of what they are doing by thinking about them during and after the experience. To help readers with the reflective aspects of this experience, and in addition to self-test questions and projects, this book includes reflective questions. They are grouped into three categories:

1. Explicit Story: Questions such as "What was this chapter about?" or "What was your task in this project?" encourage the learner to think about what they did and try to review it.

2. Implicit Story: Questions such as "How do you feel about this subject?" or "How do you evaluate your performance on this task?" encourage the learner to reflect on the effect of the tasks and their reactions to them in order to see the strengths and weaknesses.

3. Upcoming Story: Questions such as "Where do you see this concept used?" or "How will you approach the next task based on what you did this time?" encourage the learner to see the big picture and plan ahead.

Throughout this book, I provide various examples of the concepts I am discussing. They are almost always followed by tasks that expand on the examples or involve variations. In order to successfully perform these and any future projects that use those concepts, you need to (1) design and write the programs and (2) make sense of what you have done through reflection. You will not learn programming by reading this or any other text. Learning to program requires an experiential approach and thinking about what you have done, so you understand the concepts. Reflective questions are provided throughout this book to help with this learning approach. They are as important as writing the code itself. Feel free to consider other reflective questions you may have and answer as many as you can.

After decades of doing software development and trying many different roles, I still find programming a fascinating, rewarding, and enjoyable activity, which is a big part of my life. I hope reading and using this book helps you discover and establish your own partnership with it and develop skills that not only provide you with new professional options but also help you be a stronger and more creative and logical problem-solver.

In the words of the iconic Mr. Spock who symbolized logical thinking for many of us, "Live Long and Prosper."

Ali Arya
April 2020, Ottawa

Acknowledgments

I always tell my students, "You don't learn programming by listening to people or even by reading sample codes. You learn by writing a lot of code and making sense of it." Writing a book about programming is somewhat similar but needs more. First, there is a significant amount of listening, to people who are learning to see what they find hard or easy, and to other instructors to learn from their experiences. Then there is the need to read a lot of code by a lot of people to see what mistakes they make and what good solutions they use. Finally, and after you make sense of all these and come up with a way of presenting your thoughts, there must be many more conversations to make sure what you are saying makes sense to others.

Writing this book was the result of many years spent doing all the things I mentioned above, starting way before I thought about writing the book. From the first friend that I sat with to share my limited knowledge to those who reviewed the final version of this text, many were involved in shaping my understanding of programming and my approach to teaching and learning it. The list is too long to include here. I am grateful to all of them, but some I should mention.

Writing this book would not have been possible without the experience of working with many students who attended my classes since I was a PhD student and particularly those at the School of Information Technology, Carleton University. They helped me shape my ideas about computer programming and how to learn it, and for that and their patience, support, and advice, I am honored and grateful. Some of them shared my passion for programming, and some didn't (and occasionally got bored and fell sleep in class). Some ended up doing (or surprisingly teaching) software development, and some didn't write a single line of code after graduation. Regardless, I am proud of them all. I hope to have contributed to their lives and careers positively, not just through the practice of algorithmic thinking but with our human interactions. I know I learned a lot from them, enjoyed those interactions immensely, and will always strive to prepare a better learning environment where everyone feels valued, understood, and supported. Hopefully, this book is a small step toward such support. I would like to particularly thank Max Leeming, Marie-France Curtis, Geoffrey Datema, and Alex Larocque for their direct assistance in creating the content of this book.

Many colleagues and academic reviewers also provided invaluable suggestions to improve the content. I am grateful to all of them, especially Dr. David Sprague, whose comments were essential in resolving some of the problems with this book.

I would also like to thank the whole publishing team for their support and excellent work. My special gratitude goes to my editor, Randi Cohen, for believing in my book idea and guiding me through the process.

Last but certainly not least, for this book and many other good things in my life, I am grateful to my wife, Luciara; my son, Caio; and our cats, Aztec and Daphne. I cannot promise there won't be any more late-night or weekend working (just like many other flaws and weaknesses I have), but I know I can count on your support and forgiveness. Love you!

Definition of Key Terms

Abstract Class:	A class with missing implementation for one or more functions
Abstract Data Type:	An abstract description of a data type and its related operations
Algorithm:	A series of operations to perform a task
Class:	A collection of related code and data
Code:	The instructions (operations) in a program
Coding:	The process of creating a computer program based on a set of requirements. In this book, I use the terms "coding" and "programming" interchangeably. Others may use the term "coding" to mean the specific act of writing the code
Compile-Time:	The time of writing the code and translating it to machine language (a.k.a. Build Time)
Compiler:	A program that translate code to machine language
Data:	The information in a program
Data-Centered Design:	Designing a software based on what data it has
Dynamic:	Any information known and set at run-time
Encapsulation:	Combining code and data into classes
Function:	A module of code
Inheritance:	Extending a class by defining new "child" classes based on the "parent" class
Interface:	A class without any implementation of its function
Iteration:	Repeating part of the code through loops
Method:	Function in a class
Object:	An instance of a class (a variable created based on the class)
Object-Oriented Programming:	A programming paradigm based on objects
Polymorphism:	The ability of objects to behave using different identities

Programming: The process of creating a computer program based on a set of requirements. This includes design, implementation, and some testing. Others may use the term "programming" in a more specific way (the act of writing the program instructions) or a more broad one (from gathering the requirements to testing and releasing)

Property: A data item in a class

Requirements: The features that a product needs to have from the user's point of view

Run-Time: The time (and environment) associated with executing the program

Selection: Choosing different paths of code

Software Implementation: Creating the actual program (textual instruction or other form) based on a design

Software Design: The process of deciding on software modules based on requirements

Software Development: The full process of producing a software product, starting with requirement gathering to release and maintenance

Static: Any information known and set at compile-time

Structure: A collection of data items (in C) and data and code (in C++)

Structured Programming: A programing paradigm based on selection, iteration, and functions

Three HOW Questions: How to initialize, How to use, and How to change a data item

User-Defined Type: A new data type based on combining existing ones

Variable: A module of data that can change its value during run-time

Abbreviation

3HQ: Three HOW Questions
ADT: Abstract Data Type
API: Application Programming Interface
OO: Object-Oriented
OOD: Object-Oriented Design
OOP: Object-Oriented Programming
SP: Structured Programming
UDT: User-Defined Type

Companion Website

This book has a companion website: http://ali-arya.com/anyonecancode.
The website includes

- All sample code files, with more detailed documentation

- More examples

- Solution to selected tasks from the book

- Guides on how to install and use the software tools required for the examples

PART 1

Getting Started

A good programmer is someone who always looks both ways before crossing a one-way street.

GOAL

Programming is about logical creativity, solving problems through logical thinking and design, and also formal methods of presenting those thoughts and solutions.

In the first part of this book (Introduction, Chapters 1 and 2), I aim to provide some historical and background information on computer programming and introduce the basic concepts in logical creativity. This includes algorithmic thinking as the main approach to design and present solutions and programs that implement them.

In the next part of this book (Chapters 3–5), I will talk about actual programming languages as the tools we use to present the logic we have in mind as a set of computer instructions.

These two parts are necessary for the remaining parts of this book where I tackle real-world programming problems. After reading the Introduction, you may continue with Chapters 1 and 2 to prepare yourself for writing programs by understanding some basic concepts before proceeding to actual programming in Chapter 3. Alternatively, you may move to Chapters 3–5 after the Introduction, if you just can't wait to get some code written and executed. But make sure you get back to Chapters 1 and 2 at some time during the process, before moving to Chapter 6.

Introduction

```
┌─────────────────────────────────────────────────┐
│                    Topics                         │
│   • Getting started with programs and programming │
│        • Software development process             │
└─────────────────────────────────────────────────┘
```

```
┌─────────────────────────────────────────────────────┐
│  At the end of this chapter, you should be able to:   │
│ • Describe the main phases of a typical software       │
│   development process                                  │
│       • Define the concept of modularization           │
└─────────────────────────────────────────────────────┘
```

HELLO, WORLD!

Since Brian Kernighan and Dennis Ritchie wrote their seminal book *The C Programming Language* (Kernighan & Ritchie, 1978), the most common starting point for programmers has probably been to write a simple "Hello, World!" code; one that displays the famous greeting message on the screen. It makes sense. Computer programming is the process of creating programs that are instructions for computers to perform different tasks. Greetings are the most common things to say at introductions, and we are, in fact, being introduced to programming and creating new programs to interact with users and the "outside world." In a sense, we are "introducing" the computer to that world. So, we tell the computer to say hello to the world!

In order to tell computers to do things, we need to learn their language. And just like people, computers have multiple languages. Here is how to say "Hello, World!" in some of them:

- C: `printf("Hello, World!");`
- C++: `cout << "Hello, World!";`
- C#: `Console.WriteLine("Hello, World!");`

- Python: `print("Hello, World!")`

- Java: `System.out.println("Hello, World!");`

- Javascript: `document.write("Hello, World!");`

- HTML: `<p>Hello, World!</p>`

While using most of these languages requires setting up software on your computer or visiting particular websites, there are pretty easy options to try. Let's try the HTML for a quick start. **HyperText Markup Language (HTML)** is the primary language for creating web pages. HTML code describes the content of the page. We will talk about it more in Chapter 3, but for now, just open your favorite text editor and enter the HTML line above in it:

```
<p>Hello, World!</p>
```

This line of code defines a paragraph with the text "Hello, World!". You can see more details about the syntax of HTML later, but one of the key concepts in HTML is an **element** that is a part of the document that HTML code is describing. P (short for Paragraph) is an HTML element. Each element is identified with an **opening tag** (a name inside < > pair) and a **closing tag** (same as the opening but starting with a/symbol).

Save the code into a file with extension. html (for example, hello.html), and then, open that file with your browser by using the file manager program on your computer (File Explorer for Windows, Finder for Mac, or similar programs), locating the file you just created, and double-clicking on it. You should see an almost-blank page with your greeting text on it. Congratulations! You just wrote and executed your first program. Executing (or **running**) a program means having the computer perform the instructions.

☞ **Key Point:** Programs are instructions telling a computer what to do and how to do it. Programming is the process of designing and creating programs.[1]

Now let's make it a little more complicated. Go back to your text editor and add this line:

```
<p id="demo">?</p>
```

The second line creates a second paragraph with "?" as the text. It also gives a name (id, short for identifier) to the paragraph. You can use any name. I called it demo. HTML elements can have different information associated with them. A name or id is one such information that in HTML terminology, we call **attribute**. Each attribute is defined inside the opening tag for the element using name, an = sign, and a value inside " " (double quote).

---

[1] See Section I.1 for a discussion on the definition of programming and what it involves.

You can try saving the file and refreshing the browser to see the results (two paragraphs). After that, we are going to make things a little more interesting. Add the following text at the end of your HTML file:

```
<script>
demo.innerHTML = "What's up?";
</script>
```

The above lines are telling the computer to add a code written in **Javascript** language to the HTML page. Combining HTML that describes the web page and Javascript that adds interactivity and advanced operations to the page is a very common practice. The Javascript part is another element identified with <script> and </script>, just like a paragraph was identified with <p> and </p>. The Javascript code itself tells the computer to modify the text (called innerHTML) for the paragraph named demo. Save the file and refresh your browser to see both paragraphs.

If you have questions about details such as syntax and various elements in the program, just hold that thought. We will talk about all those details in the rest of the book. For now, let's wrap up our example. Add another line to your Javascript code to make it like this:

```
<script>
demo.innerHTML = "What's up?";
document.write("My name is Ali.");
</script>
```

Replace Ali with your own name and see the results. Voila! Now you have written a simple program using two popular languages. As complicated as it may seem, we all can write computer programs. That's why this book is titled *Anyone Can Code*.

☞ **Key Point:** Some programming languages like HTML "describe the output," while others like Javascript "describe the actions." Either way, a program instructs the computer to do something.

🖐 **Reflective Questions:** Do you feel overwhelmed with the sample program? Make sure you write the code and try it hands-on. Does it make any difference when you wrote and run the program yourself? As we move forward, things will be more clear. Embrace some uncertainty and explore the topic.

Now, let's try to spice up our program with some user interaction. Replace your HTML code with the following lines:

```
<p>Hello, World!</p>

<p id="demo">0</p>

<input id="userdata">

<button onclick="demo.innerHTML = userdata.value * 2;">Double</button>
```

The first two lines are the same as before, but the `script` element has been replaced with two new lines. They add an input box and a button to the web page. The input has an id, so we can refer to it later. We give names to things if we want to refer to them later. Keep that in mind. The button, on the other hand, has an operation defined for when we click on it. The operation is written in Javascript. It reads the value of the input box, multiplies it by two, and then changes our demo paragraph with the results.

☑ **Practice Task:** Try making a very simple calculator by expanding this example. Make two input boxes to enter two numbers and four buttons to select the arithmetic operation.

For now, don't worry about having a nice interface (for example, showing a keypad, memory, and other fancy things that a calculator has for the above task). Once uploaded to the web, your program can be a nice online application.

🖐 **Reflective Question:** Have I convinced you that anyone can write computer programs?

You can see the full code for the calculator on the Companion Website.[2] Now here is a tricky part. Many beginners think of programming as learning how to use the programming languages; their syntax, keywords, symbols, operators, and so on. That probably sounds right to you. After all, we just created a program by learning some simple rules about two languages, and I had said earlier that programming languages are like the languages people use. What is the first thing we do when we need to interact with other people? Learn their language, right? Not quite!

In her book, *Working in a Multicultural World*, Luciara Nardon says: "Knowledge about different cultures such as their language and customs is necessary but not enough (when interacting with them). The thoughts and processes matter the most." (Nardon, 2017) Interacting with other cultures is really about knowing how intercultural relations work and how people

---

[2] See the information at the start of the book before the Introduction.

(including ourselves) think. Working with different cultures involves a certain skill and competency that is beyond each language or culture. It provides us with a way of thinking and behaving that is based on proper observation, understanding both sides, reflecting and sense-making, and making decisions at the presence of uncertainty. And here is a secret about programming: It is not really about those languages. It is about a way of thinking.

Imagine that instead of making a simple calculator, you wanted to have a fully functional one. It needs a keypad to enter numbers digit by digit, be able to erase, remember the previous number and results, and other features. This is no longer a simple and casual problem that you can easily solve. It requires more complicated solutions and steps to follow to build that solution. Steps that are logically related and build on each other. Some of these steps are:

- Finding a method to update a number based on the new digit. For example, if you have 21 and enter 5, 21 should shift to the left so 5 can be inserted and make the number 215.

- Providing some memory to keep a previous number, if you start entering a new one. For example, pressing + means we are done with the first number, so we save it somewhere and reset the display to enter the second number.

- Performing operations with the number on display and what was saved.

Notice that these steps are all related and rely on each other. We need to not only think about a solution, but we also need to think about it in a logical manner so the computer can follow these steps and get to the result we want. Unlike what you see in many other books on programming, in this book, I am not focusing on a particular language. While learning the details of specific programming languages is important, there are a few basic and common concepts that help us establish that way of thinking we need in order to develop software programs. In the following chapters, I will discuss these concepts within the context of multiple languages and subject areas, which allows us to distinguish between what is special about a language and its syntax and the general concepts that are common among all of them.

If you are reading this book and plan to learn programming, there is good news for you, which I've already said: *Anyone can code!* Think about the arts, for example. Not everybody can become a Picasso, but everybody can learn to draw and paint some reasonably nice shapes. We all have the potential. We just need some support and a certain amount of practice. The same is true about programming. If you can calculate the tip at a restaurant or use Microsoft Office, then you can write a program. I know people who started with so much trouble that just listening to a short talk about programming (not even understanding it) was difficult for them, but they managed to do some nice programming after a couple of months. Our brain (at least a part of it) is made for thinking, and that's what programming is really all about. A way of thinking. But here comes a question: What type of thinking?

Programming is an act of problem-solving: we write programs to solve problems, and how to write them itself is a problem we solve. When solving a problem, whether it is casual or important, we sometimes make decisions that we can't justify rationally. We refer to our "gut feeling," rely on our intuition or preference, or just guess something. In none of these cases, it is quite possible to describe the process we go through so that someone else can repeat and get to the same result. There is a place for this way of thinking, but computers don't have our sense of intuition or aesthetics, and so require a well-defined and logical process (and so do we, in many cases). Of course, computers can also make art and do random things, but there is a logic underneath them all.

Programming is based on what we call **Algorithmic Thinking**. You may call it "thinking like a computer" or "thinking like Mr. Spock" if you are a Star Trek fan. It is a logical way of thinking based on well-defined rules and operations. A set of these rules and operations, allowing us to make a decision or solve a problem, is called an algorithm.[3] A simple example of an algorithm is addition. When we add two numbers, we add the digits on the right-most column, and if the result is more than 10, we leave the right digit and carry over a one to the next column of digits and add them. This process continues all the way to the last column of digits on the left side. Assembly instruction set for your new furniture is another example of an algorithm although it is less precise than mathematical operations. And so is your method of the piling and folding laundry based on whose clothes or what types they are.

We will discuss algorithms in Chapter 2, but for now, let's just say that they are well-defined sequence of rules and operations that depend on each other, flow logically, and together allow us to solve a problem. This definition is important because it shows the relationship between parts of an algorithm and also its purpose.

> ☞ **Key Point:** Algorithms are the basis of programming and are formed by a well-defined sequence of rules and operations that depend on each other, flow logically, and solve a problem.

Exhibit I.1 shows a simple algorithm that we can use in a program to calculate a restaurant bill. We usually show algorithms in two common methods:

- **Flowchart**, a graphical representation with various operations shown with shapes in Exhibit I.1

- **Pseudocode**, a plain text that is informal but somewhat structured into a clearly written text where each line corresponds to an operation.

The text version of the Exhibit I.1a will be something like this:

1. Add the price of all food items

2. Add tax

3. Ask the tip amount

4. Add tip

---

[3] There are multiple formal and informal definitions for algorithms. Such definitions are discussed in mathematics and computer science literature and are beyond the scope of this book that has a more elementary and practical purpose. See, for example, *Introduction to Algorithms*, by Cormen et al.

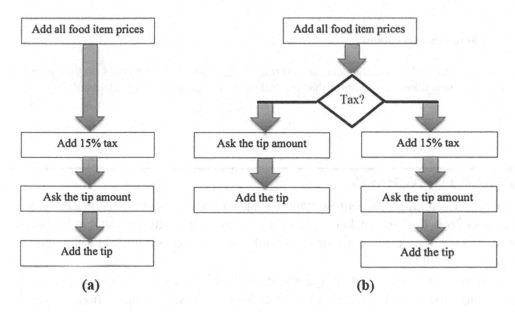

EXHIBIT I.1   Simple algorithm. (a) Fixed set of operations, (b) with a rule.

The logic we follow in this example is quite simple. We add the price of all food items, then add 15% service tax (assuming there is such tax in your location and it is 15%), ask the customer to enter the tip amount, and finally add the tip to have the final amount. This algorithm is well defined and has a fixed set of operations, so every time we use it, the same actions will happen. Algorithms can also have rules that control the operations based on certain actions. For example, we can have a rule that says, "if there is service tax, then add 15%." This rule changes the sequence of operations based on a certain condition, so in some locations, we will have four operations and in some others three,[4] as shown in Exhibit I.1a and b.

In flowcharts, checking a condition with true and false possibilities is usually shown with a diamond shape (see checking for Tax in Exhibit I.1b), while operations are represented by rectangles. Lines or arrows show the dependency and order of operations (the flow of algorithm). I don't use flowcharts in this book, but you can see a short description in the following sidebar. In Chapter 2, I will discuss pseudocode in detail.

☑ **Practice Task:** Write a short algorithm for a simple program you commonly use.

---

[4] Programming is problem-solving, but in most programming problems, there is more than one solution. Don't be surprised if, throughout this book, your solution is not the same as what I show. They both can be correct, although one may be more efficient, suitable for a certain condition, or just easier to implement. More on this later.

 Reflective Questions[5]

- Do you understand the concept and role of algorithms? In what cases do they work?
- Can you think of algorithms that you follow in non-programming activities?

SIDEBAR: FLOWCHARTS

A flowchart is a graphical representation (diagram) of a process. This process can be a workflow, decision-making, algorithm, or any other activity that can be broken down into separate tasks, with input, output, and decision points that follow each other sequentially.

It is common to show different elements with different geometrical shapes such as rectangle for activities, diamond for decisions (with two or three choices), parallelogram for input/output (data), and ovals for start and end. Arrows connect these elements to show the flow.

For example, the flowchart below shows a process that involves doing A, checking the condition B, and doing C as long as B is true.

For many people and because of the logical structure, programming seems a radical paradigm shift in thinking and problem-solving. This shift to thinking algorithmically is the most challenging part of programming, not the strangely named tools and complicated syntax of different programming languages. If you can get the thinking part right, tools and languages will be easy to learn afterward. Don't be too alarmed, though. Algorithmic (or logical) thinking does not belong to one side of your brain that you don't use! You use it every day, maybe without noticing, when you solve causal or important problems. And programming is really an act of problem-solving using computers.

Just like any other type of problem-solving, programming can also be a creative process[6]; this means we develop a novel solution and come up with a new idea or design. As such, it is not fundamentally different from any other type of creative process, such as visual thinking that artists and designers are used to. In this book, I refer to algorithmic thinking and programming as **Logical Creativity**, as they are activities aimed at creating novel designs and solutions following a logical thinking process. We discuss problem-solving, creativity, and algorithmic thinking in more detail in Chapter 2.

---

[5] Throughout the book, you will see "Key Points" and "Tasks" sections identified with the ☞ and ☑ symbols. Make sure you understand those points and perform the tasks. There may be "Reflective Questions" (identified with) following the key points and tasks. Even if there are no reflective questions, make sure that you reflect on the points and tasks to understand them.

[6] Not all problems require creativity as in "creating novel solutions." Some problems are routine and casual, and as such easy to solve based on existing experiences and knowledge. Similarly, many cases of programming, especially when starting with simple ones, are not particularly "creative." But we need to start with simple ones to learn the process.

☞ **Key Points**
- Logical creativity is a type of creativity (developing new solutions and designing new products) that follows an algorithmic and logical thinking process.
- Logical creativity is the foundation of programming.

Learning programming as a form of problem-solving and a creative process is the foundation of my approach in this book. This allows us to use known methods of problem-solving and creativity within the context of programming. Also, it helps us use the algorithmic thinking approach in non-programming tasks and solve our other problems more efficiently. But let's not get ahead of ourselves. We will talk about these in Chapter 2. But first, let's talk about a few important subjects that will help you explore the book and understand why it is structured in this particular way.

I.1 SOFTWARE DEVELOPMENT

The word "software" refers to the collection of information that a computer uses. The "soft" part is due to the fact that it can easily change, as opposed to the physical device, or hardware, that is less flexible.[7] The terms "computer programing," "coding," "software development," and "software engineering" are frequently used in the computer industry and literature. They are very much related but not quite the same. While there may be different definitions and interpretations, the most commonly used one defines computer programming **as the act of creating a program, a set of executable instructions in a formal language that can be (after some steps) executed by a computer to achieve a particular purpose or result**.[8] The key elements of this definition are (1) instructions or program, (2) language, (3) execution by a computer, and (4) purpose or results.

☞ **Key Point:** Programming is the process of creating a computer program (set of executable instructions) in a formal language for a particular purpose.

The term **Code** is frequently used as equivalent to **Program**, but it can also mean a part of the program (as opposed to data) and also a specific representation of a program's instruction (as in "machine code" vs. "human-readable code"). In this book, I make a distinction between code and data as two parts of a program: the information used by the program and the operations performed on that information. This separation is more conceptual because except for data files outside the program, every data used by the program is created through code and cannot be separated. Consider the following line from our first example:

```
<p id="demo">?</p>
```

---

[7] There is also "firmware" that means the program built into the hardware by manufacturer and generally not accessible to users, particularly, the start-up program in any computer, a.k.a. BIOS (Basic Input/Output System)

[8] This and the following definitions are what I use in this book and is similar to most common definitions in the literature, but not the only ones.

It creates a paragraph with a given name and text, so it is "data." But in order to make that data, we need to have an operation, so it is mixed with "code." Now, imagine the next instruction:

```
demo.innerHTML = userdata.value * 2;
```

This line is an operation performed on two existing data items, demo and userdata.

Coding and Programming, though, are used interchangeably in most contexts, including this book. They refer to creating a program for a particular purpose, i.e., based on a set of given requirements. **Software Development**, on the other hand, refers to a much broader process that involves understanding the user needs and requirements, designing a software product and its parts (modules), implementing the parts, integrating these parts, evaluating the product, and finally releasing and maintaining it. This is shown in Exhibit I.2 that demonstrates the **Waterfall Model** of software development. While it is not possible to do programming without some level of requirement analysis and testing, the term "programming" mostly, and in this book, refers to the design and implementation phases.

Waterfall Model is one of the oldest descriptions of the software development process. There are many alternatives, but Waterfall is still useful thanks to its simplicity (Sommerville, 2015). **Software Engineering** is about understanding these various models of the software development process and improving them. It allows us to think of software development as a rigorous and disciplined engineering process rather than an ad-hoc hacker-style activity.

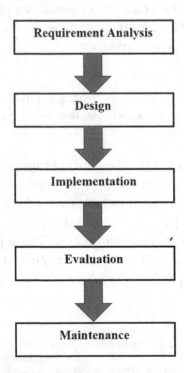

EXHIBIT I.2   Software development process.

TABLE I.1 Phases of the Software Development Process

Phase	Description	People	Output
Requirement analysis	Understanding what the software program is supposed to do	System analyst – domain experts, usually familiar with software development, so they have reasonable expectations of the software program	Requirements document, a.k.a. specification
Design	Defining the algorithms, modules, and their relation	Designer/programmer – software experts usually experienced programmers	Software modules description and design document
Implementation	Producing the actual program or code	Programmer – familiar with the programming language and platform to be used	Code and its documentation
Testing	Testing the final software product	Programmer/tester. Experienced in using test plans and finding errors	Test result
Maintenance	Installing and updating the program	All the above people plus field personnel who install and inspect the program	New versions of the software

Each of the phases in the software development process creates a different output and can potentially be done by different people, although in many small projects, one person may do all the work. Table I.1 shows the description, people, and output for each of these phases. You can see that a programmer's job can involve design, implementation, and testing of a software product. Designers and testers, on the other hand, have specific roles and don't even need to be programmers by training and experience (although they often are, especially designers).

In most software projects, these phases are overlapped and iterative. For example, designers may implement simple versions of the software to verify their design. Also, every programmer performs tests for various modules during the implementation before the full product is passed to testers for a final evaluation. They may even find design problems which result in going back and redesigning.

SIDEBAR: PROCESS MODELS FOR A SOFTWARE DEVELOPMENT PROJECT

In the early days of the computer industry, there was no standard process for software development. It took the industry many years and many failed projects to realize that creating a software product is an "engineering" process and requires clear guidelines, methods, and standards. The Waterfall Model was one of the earliest efforts to establish such a standard.

Later, software engineers realized that the process of software development is not quite linear and one directional; feedback mechanisms and repeating tasks based on the feedback are required in order to correct and improve the software. This resulted in a variety of non-linear and iterative models for software development. These modes still include tasks such as requirements analysis, design, implementation, and testing, but they are structured in a way that allows convenient feedback and cyclical development. Examples of such models are Spiral, Incremental, and Agile software development.

The book *Software Engineering* (Sommerville, 2015) is a great resource for learning about various aspects of the software development process and models.

This book is about programming and not the whole process of software development. Even though all five phases of the process are essential and require attention, in this book, I focus on design and implementation, which are more directly related to programming.

The initial experience of any new programmer usually involves learning the specific act of "writing code," the implementation phase, as we start with a very simple design. But as you can see in the process, programming is tightly related to software design that defines different parts and algorithms for the program. It is not really possible to completely separate design and implementation from each other. Almost always, the tasks are intertwined; programmers have to make design decisions, and designers need to know how their design is going to be implemented (in fact, they are generally experienced programmers). In this book, we start with the basic concepts in implementing simple programs, but we should always have an eye on the design side, and our focus on software design will ramp up as we get closer to the end of the book and deal with more complicated programs.

It is also not possible to develop any software without testing. Every programmer needs to test their program at some levels. Such testing includes at least running the program, noticing issues, and fixing the program to resolve the issues. The discussion of testing methods is outside the scope of this book, and so are various methods of collecting and managing software requirements.

☞ **Key Points**
- Software development involves various activities: requirement analysis, design, implementation, testing, and maintenance.
- Among those activities, design and implementation are the focus of this book, and I more specifically refer to them as "programming," even though programming overlaps with other activities.

✋ **Reflective Questions:** How does the software development process compare to other design and production projects you have seen? Do the activities listed above make sense to you based on what you may know of software development?

I.1.1 Design and Implementation

Design is the process of choosing and defining components and how they are related in order to achieve a high-level goal that none of those components individually could achieve. This may not be the best definition, but it works for many types of design. A simple example

EXHIBIT I.3 Sketch to painting. Van Gogh's *Potato Eaters*. Many details change at the implementation phase, but sketch gives a general idea. (Van Gogh, Vincent. *The Potato Eaters*. 1890. Van Gogh Museum, Amsterdam, The Netherlands.)

is the use of LEGO[9] pieces. Very complicated objects can be made from a few simple parts through a hierarchical approach where simple pieces are used to make complex parts, and these parts are then put together to make even more complex objects. Designers have to work with very few pieces that have limited colors and shapes. All complicated parts and objects have to be made of these small sets of pieces.

Sometimes the design process involves creating multiple versions of the final product. In engineering and manufacturing, this is called prototyping. For example, an industrial designer may create a cardboard version of a product to establish the concept prior to manufacturing. Another example of such a design process is an architect who designs a building and then passes it to engineers and workers to build or a visual artist who draws a sketch before painting.

Sketching is yet another design example that follows a similar process. The artist uses building blocks such as lines to create something more complicated such as a face. This sketch identifies the major elements of the final artwork, their size and location, the composition, and many other important aspects of the work we will eventually see. It will be used as an initial plan for a painting by the same person or another who knows how to use paint, brush, and canvas. The painting phase still involves many decisions that are not related to composition, for example, color and brush strokes, as illustrated in Exhibit I.3.

While no analogy is perfect, the examples above do resemble and illustrate the process of going from design to programming. Note that in all these cases, the second phase still involves significant work, expertise, and decision-making. An expert person (the designer or someone else) is still needed to finish the work. Similarly, software designers and programmers work together. In many cases, the programmer still makes lower-level and less critical design decisions, or as mentioned before in smaller projects, the programmer is the designer too. So even though there is no clear boundary, the task of design at any level involves dealing with basic components (a.k.a. parts or modules) while the programming involves translating ideas to an actual program using various tools, languages, and methods.

---

[9] LEGO is registered trademark of the LEGO Group.

I.1.2 Modularization

Discussing software design brings us to a fundamental concept in software development (and in this book), which is modularization. As I mentioned before, the design process involves choosing the right parts and establishing their relationship. In a LEGO project, that was choosing the pieces, making the parts using small pieces, and then assembling the object using the parts. In the artwork shown in Exhibit I.3, the sketch identified those parts (characters in that case) and the way they should be assembled (composition). Modules (sometimes called components) are the parts of a program, and modularization is the process of breaking down the program into modules and defining the software system in terms of these modules. We use modularization as a key concept throughout this book.

For example, to write a program based on the algorithm of Exhibit I.1 (calculating restaurant bill), we may notice that we need to perform three categories of operation:

- Getting data from the user (prices and tip)

- Performing various calculations

- Showing information to the user (total amounts).

This is a very common categorization in a program, input, process, and output. You may notice that the input and output are independent of the calculation. Regardless of what calculation we perform, we always need to get some data from users and show them something. So, it makes sense to divide our program into three modules, accordingly.

☞ **Key Point:** Modules are parts of a program that can be developed and used independently, similar to LEGO parts and pieces.

Let's consider a computer game as another example. If you think about a game as one piece, it is hard to know where to start creating it. Many of us have experienced that I-have-no-idea-what-on-earth-to-do feeling when starting a new project. Modularization helps us get over that hurdle. Breaking down the game into proper modules has the following advantages:

- We can divide the whole project into semi-independent parts that can be done by different people.

- The modules can be designed in a reusable way, so they can fit into other projects (same as LEGO pieces and parts).

- We can manage our time and group more efficiently as we know which parts need to be done first or which ones can be borrowed from other projects.

- We can get a better idea of what needs to be done as we will be dealing with smaller, more manageable tasks.

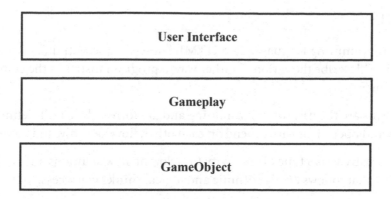

EXHIBIT I.4   Software modules for a simple computer game.

Exhibit I.4 demonstrates the concept of software modularization for a simple game. The design includes modules called gameplay (controlling how the game is played), GameObject (drawing and moving the objects in the game), and User Interface (interacting with the player). This is a good design because the modules are independent and reusable. For example, you may use the same gameplay but replace the objects with new ones or interact with the player differently.

Don't be alarmed if you couldn't guess what modules to use in these examples, or if the modules that I defined don't quite make sense yet. We will discuss the concept of modularization in detail throughout the book as it is essential in proper software design, and even small programs, as we will see soon, are made of modules. For now, all you need to remember is that we commonly divide a program into modules for more efficient programming.

You learn programming through designing and writing programs. The more you code, the more you learn. But you also need to reflect on what you read and do to make sense of it. As we move forward, remember to practice and reflect.

☑ **Practice Task:** Try to identify and visualize the main modules of some other software programs you use.

〰 **Reflective Questions**

- How comfortable are you identifying and visualizing modules?
- Will it help if you start by doing that for simple objects and programs?
- Do you see the value of modularization, and identifying the modules in advance?

HIGHLIGHTS

- Some programming languages like HTML "describe the output," while others like Javascript "describe the actions." Either way, a program instructs the computer to do something.

- Algorithms are the basis of programming and are formed by a well-defined sequence of rules and operations that depend on each other, flow logically, and solve a problem.

- Logical creativity is a type of creativity (developing new solutions and designing new products) that follows an algorithmic and logical thinking process.

- Logical creativity is the foundation of programming.

- Software development involves various activities; commonly grouped into requirement analysis, design, implementation, testing, and maintenance.

- Design and implementation are the focus of this book.

- Modules are parts of a program that can be developed and used independently, similar to LEGO parts and pieces.

END-OF-CHAPTER NOTES

A. Things I Should Mention

- In the movie *Ratatouille* (2007), Chef Gusteau's motto was **Anyone can cook**.

- All sample codes in this book, plus additional examples and software guides, can be found on the **Companion Website** (see the information at the start of the book before the Introduction).

B. Self-Test Questions

- What is an algorithm?

- What are the main tasks in software development?

- What is modularization?

C. Things You Should Do

- Learn to draw if you can. It helps you visualize your program design and follows a process that is not too different from programming, and it's fun!

- Take a look at these books:

  - *The Psychology of Computer Programming* by Gerald Weinberg

  - *Software engineering* by Ian Sommerville.

D. Reflect on the Experience of Reading This Chapter

- What did you expect from this chapter before reading it?

- What was it about, and what did you learn?

- What tasks did you perform, and what difficulties did you face?

- How did you feel about the material and tasks presented in this chapter?

- How can you improve your learning experience?

- How do you see this topic in relation to the goal of learning to develop programs?

CHAPTER 1

Computers, Programs, and Games

Topics

- History of computing
- Number systems
- Programs and programming languages
- Using games to learn about programs

At the end of this chapter, you should be able to:

- Understand the Von Neumann model of computing systems
- Perform binary operations
- Understand the basic structure of a computer program consisting of data and code
- Use GameMaker or similar tools to visually make simple programs

OVERVIEW

One of the first things that we do when starting a new relationship is getting to know each other. It is a reasonable thing to do because, in order to interact with people, we need to know some things about where they come from and how they live. Such knowledge complements our own personal characteristics and general interaction skills to build a good relationship. Programming is like a relationship with a computer, so it is helpful to start by collecting some general information about our new partner.

In this chapter, we will briefly review the history of modern computers and how they work. To understand computer operation better, I have also included a little bit about the mathematical system that is the underlying infrastructure for the operation of digital

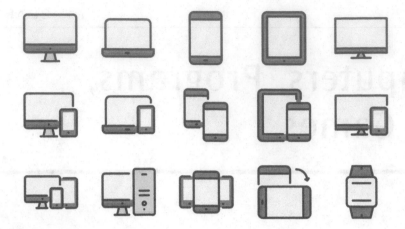

EXHIBIT 1.1   Computers come in various forms and shapes.

computers, and some definitions for the commonly used terms such as digital and computer. It's always good to know we are on the same page and mean the same thing when we use different words.

Once we know a little bit about computers, we should also make sure to have a general understanding of what a program is. **Programs** are the real subject of this book, and writing them is our goal. They are the means of communication with computers, and as we see very soon, they are the defining feature of computers. I use the example of a computer game to demonstrate what a program is and what are its important parts.

This brief introduction will prepare us for the next chapter that talks about how to design programs that can tell a computer what to do.

1.1 BRIEF HISTORY OF COMPUTING

Computer programming starts with the computer, not just literally with the word "computer" but also conceptually as we can't learn about programming without understanding computers. These days, we use computers everywhere. The presence of computers in almost any device we use is so prevalent that has resulted in terms such as Ambient Computing, Ubiquitous Computing, and the Internet of Things (IoT). Although these are not quite the same, they share a basic theme: they talk about how computers run most of our daily tools, from the phone in our pocket to the thermostat in our room (see Exhibit 1.1).

With such a diverse range of items that are or include a computer, it seems hard to define what a computer really is. The term "computer" literally means a person or object that determines a result through calculation. That definition will result in devices such as electronic or mechanical calculators or even an old abacus to be considered a computer. But a more modern definition is an electronic device that is programmable, receives data, processes the data, and provides output.[1] A program itself is a series of instructions to be executed by the computer. This is illustrated through the model presented by the mathematician John von Neumann in 1945 and illustrated in Exhibit 1.2.

---

[1] Dictionary.com

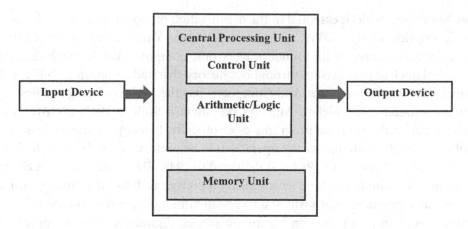

EXHIBIT 1.2   Von Neumann model of a computing device.

A detailed discussion of this model and computer organization in general is beyond the scope of this book. You can refer to books such as *Computer Organization and Design*, by Patterson and Hennessy. Here, I briefly talk about some of the most important things you need to know. According to von Neumann model, every computer has the following parts:

- Central Processing Unit (CPU or Processor) that itself includes arithmetic/logic and control units. This is really the brain of the computer that performs all the operations.

- Input such as keyboard and mouse.

- Output such as screen and printer.

- Memory where programs and other information reside. These are read and written by the CPU.

- Permanent Storage is not technically in the von Neumann model, but it is usually a part of any computer, and the memory part is considered to be not persistent. Loading is copying information from storage to memory. Saving is the opposite.

☞ **Key Points**
- A computer is a device that can execute ("run") programs and has input, processor, output, and memory.
- A program is a series of instructions (commands or operations).

🖐 **Reflective Questions:** How did you define a computer before reading the above text? Does this definition make sense to you? Does it help you understand computers and programs better?

The von Neumann model is essential in the organization of modern. The model was based on the description of the EDVAC (Electronic Discrete Variable Automatic Computer) system. EDVAC was not the first modern electronic computer. While many mechanical and electrical devices have existed throughout history that had computing ability, ENIAC (Electronic Numerical Integrator And Computer) has the honor of being the first digital electronic computer (see "Sidebar: Analog Computers"). Built in 1945, ENIAC was programmable but could not store programs electronically. It was programmed using hundreds of switches that determined the operations to be performed on the data. It also used a decimal system. Proposed in 1944 and delivered in 1949, EDVAC had built-in electronic memory and used a binary system, which meant it pretty much had the same organization as current-day computers. That's why the von Neumann model still works for us.

Despite complying with the von Neumann model, digital computer computers have changed significantly since the EDVAC. Throughout the 60s and 70s, the dominant form of computers is what we call **mainframes**. The basic electronic element that computers use is a **transistor**, a solid-state device[2] that can be used as an on-off switch or amplifier for electronic signals. The state of digital electronics did not allow manufacturers to have many of these transistors in small spaces, and so a mainframe computer occupied large rooms. Even the storage units were devices such as tape recorders that would take much space. Users would interact with mainframe computers through **terminals**, devices with keyboard, and screen that looked like today's desktop computers. Terminals didn't have any processing ability themselves and simply exchanged information between the user and the computer.

Gordon Moore, one of the industry pioneers, noticed that the number of transistors in electronic chips (integrated circuits) was doubling every two years. He stated this as a prediction, which is known as Moore's Law and is still applicable to the digital computers, although the trend seems to be slowing down.[3] The ability to fit many transistors in a small space eventually led to the development of "microprocessors" (full CPU on a single small chip) and therefore smaller computers.[4] The ability to create these smaller computers was accompanied by increasing their speed of operation and decreasing production costs. These all resulted in the introduction and popularity of personal desktop computers in the 1980s. The Mac computers by Apple and Personal Computers (PC, PC XT, PC AT, etc.) by IBM and later other companies started to be used in every office, school, and home. With the exceeding number and types of users, the number of application programs, and the interest in computer programming saw a significant and accelerating growth.

The exponential growth of computing power of modern computers was matched by improvements in storage and also the connectivity of computers. Magnetic and optical disks provided mass storage for computers that allowed them to have much more space for data and programs. Later, fully electronic storage devices (Solid-State Drive, or SSD, a.k.a. Flash Drive) helped reduce the size and speed of computers as they no longer need to use

---

[2] Earlier computers used vacuum tubes as basic electronic component which was much bigger and consumed more power.
[3] See Waldrop (2016) and Schaller (1997).
[4] The number of transistors in the first Intel microprocessor (4004, released in 1971) was 2300. Intel 80386 (1985) had 275,000 transistors, and the recent Intel Core i7 (2016) has 3,200,000,000 (http://www.intel.com).

mechanical parts for disk drives. At the same time, wired computer networks or LANs (Local Area Networks) and later the Internet and wireless networks made it possible for computers to share files and other resources remotely. The introduction of the World Wide Web, email, and other Internet technologies drastically changed the way we use computers and especially how we write computer programs. The web-based application programs needed to work with different user computers that resulted in new ways and languages for software programming.

Wireless connections and smaller devices also resulted in the development of mobile computers. Laptops and later smartphones and tablets became the popular computing devices of the 90s and the 21st century. This was followed by the smart connected devices (everyday devices with built-in Internet-ready computers) and what is called the Internet of Things (IoT).

1.1.1 Properties of Digital Media

Computers have changed over time and vary drastically in size, shape, speed, abilities, and many other aspects, but despite these variations, they still share some essential properties. In her book *Hamlet on Holodeck*, Janet Murray identifies the following important properties for digital media, i.e., any computer-based device (Murray, 1998):

- Rule-based or programmable. This is the key and defining property of computers, as shown in the von Neumann model.

- Participatory or interactive. Computer programs are usually (not always) made for interaction with the user. They can potentially allow the participation of many users and in many forms.

- Multi-dimensional. This can be a simple case of navigating through three-dimensional space in a game, but is actually far more than that and refers to the ability of computers to serve multiple purposes.

- Massive storage. Computers not only have access to a large amount of local data but also can use the almost indefinite amount of data online (through the Internet).

And I Believe We Can Add

- Connected and mobile. Computers can now go anywhere and still be connected. This means that (1) we can do the same task no matter we are, and (2) we can get localized services.

☞ **Key Point:** Computer-based systems (digital media) are programmable, interactive, and multi-dimensional, with massive storage and mobility.

When designing computer programs, it is important to pay attention to these properties to make sure our program takes advantage of them. For example, early websites were mainly static or at most had videos or simple visual effects. The introduction of Web 2.0 (O'Reilly, 2009) was based on the understanding that a website can be programmed to have dynamic and complex behavior similar to a desktop application, and it can be interactive or serve multiple purposes. Social media sites are good cases to demonstrate this as they allow various types of interaction and also can be used for messaging, content sharing, playing games, and many other purposes.

Another example that I recently saw was in an Automated Teller Machine (ATM). The cash dispenser provided the ability for users to select which bills they wanted. For a digital system, it makes sense perfect sense to program such interactivity, and it adds a significant value that was not easily possible in a non-digital system. While many ATM designers did not consider such a feature, it clearly shows a good understanding of digital media properties and affordances.

☑ **Practice Task:** Think of a digital system you frequently use and see if they use the properties of digital media effectively.

1.2 CIRCUITS AND NUMBERS, BITS, AND BYTES

You have probably heard that "all information in computers is ones and zeros." This may sound rather negative as if complaining that digital computers "don't have other things," but it is true and speaks of their power. Modern computers are electronic devices that use a binary system. To understand that, we first need to have a very basic understanding of electronics.

Electronic devices are generally distinguished from "electrical devices" based on the level of electricity involved. They work with a small amount of electricity (used as a signal), as opposed to electrical devices that usually use stronger electricity (used as energy). For example, a typical electronic device has 5–12-V (Volts) electricity, and that is why they have power adaptors to connect to electrical sockets. On the other hand, electrical devices use 110 or 220 V or even higher. Electronic devices are a combination of parts (that perform an operation on electronic signals) within circuits (that connect parts). We commonly refer to the input or output of any electronic part, or in general, any point on the circuit as a signal because we don't use it as a source of energy but a source of information.

In "analog" electronic devices, the amount of electricity at any point in the circuit, i.e., a signal, can be anything within a range. For example, if the power source is 5 V, then any value between 0 and 5 can potentially exist on the circuit. One of the main problems with analog systems that the value of signals can change a little due to electromagnetic interference that always exists around us. So, a signal that is supposed to be 2.5 V may end up being 2.6 instead. If the circuit is used for any type of data processing, this effect, which we call "noise," can cause serious issues as the results may be different from what we plan or not be repeatable. To avoid noise and increase reliability, "digital" electronic devices use only two values for a signal[5]: low and high, a.k.a. zero and one (or false and true). In the case of the 5-V system, these can mean 0–1 V for low and 4–5 V for high. The values between 1 and 4 are considered invalid. In this way, some small levels of noise cannot change the meaning of a signal.

SIDEBAR: ANALOG COMPUTERS

Before the widespread use of digital electronic computers that are based on digital (discrete value, such as binary) electronic signals, electronic computers did exist that use analog (continuous) electronics or mechanical systems to perform computation. While mechanical computers used various mechanisms that would act through controlling and processing force and movement of different parts, the electronic analog computers utilized switches, connections, amplifiers, and variable resistors, to define transformations that would control and modify an input signal. They were able to perform operations such as addition, subtraction, multiplication, division, integration, differentiation, logarithm, and inversion.

Due to the use of analog signals, these computers were sensitive to noise, and small changes in signal values could result in the wrong output or inability to reproduce the same output. The rise of digital electronic computers in the 1950s–1970s resulted in analog computers to be obsolete. With the recent advances in electronic technology, there is some new attention to analog computers as they may be able to help with quantization error, a problem that happens when we convert the naturally analog values to computer-usable digital. More information about analog vs. digital electronics is beyond the scope of this book but can be found in many textbooks (for example, Agarwal, 2005). The new attention to "analog" products as a reaction to the spread of digital devices goes beyond analog computers and has been discussed by many authors (Sax, 2016).

1.2.1 Decimal vs. Binary

Using a digital computer, it makes sense to use a mathematical notation that is better suited. The typical mathematical system to represent numbers (values) that we use on a day-to-day basis is called the decimal system or base 10. It is a positional notation, which means it has a limited number of symbols, and the value of each symbol depends on its position. If symbol 1 is the first digit on the right, then it means one, but if it is the second digit, then it means ten. In other terms, the value of each digit is the digit multiplied by its positional value, and the value of the whole number is the sum of all the values of all digits.

$$357 = 3*100 + 5*10 + 7*1$$

---

[5] Theoretically, a digital system can use more than two values by dividing the electrical range to multiple areas, but the binary (two-value) format is the common one. For the remaining of this book, we assume a binary format when talking about digital systems.

The Roman numbers, on the other hand, don't use the concept of digit and are not positional because the symbol I always mean one and V always means five. The decimal system gets its name from the fact that each digit can have ten possible values (0–9).

Other positional notations can be imagined numbers are not where digits do not have to be 0–9. The binary system allows only two possible values, 0 and 1. In order to convert from decimal to binary, you need to remember that when increasing a digit, if we exceed the maximum value, then the digit will turn to 0, and the next digit will be increased by 1. For example, in decimal, 9 plus 1 will become 10. In binary, it is 1 plus 1 that becomes 10. The positional value of the second digit in decimal is ten, but in binary, it is two. Similarly, the third digit has a positional value of a hundred (ten times ten) in decimal and four (two times two) in binary.

Two other systems commonly used in programming are octal (base 8, where each digit can be 0–7) and hexadecimal (base 16). In hexadecimal (or simply Hex) system, since the value of each digit can be up to 15, but we need to show it with only one digit, symbols A–F represent 10–15. As such, the number 10 in Hex means 16 in decimal. The octal system is not used very commonly in computer systems. The most common use of the Hex system is to represent binary data with fewer digits. Every Hex digit can be from 0 to 15 (F), so representing 4 binary digits. F (Hex) is the equivalent of 1111 (B), and FF (Hex) is 1111 1111(B). This is particularly important because computer memory is organized as units with 8 binary digits, which will be a value of 0–FF(H).

Before talking about computer programs, I will spend a little more time discussing these memory units and binary digits in the next section.

☑ **Practice Task:** Practice converting between number system, and make sure you understand how the digits increase. Try adding two numbers in binary or Hex.

✋ **Reflective Question:** The number systems in Table 1.1 are based on the concept of a digit. Does that concept make sense to you, as opposed to number systems that don't use digits, such as Roman numerals, I, II, III, IV, etc.?

1.2.2 Bits and Bytes

Information in digital computers is stored in a binary (two-valued) format. This means any point in the electronic circuit can have one of the two allowed values of 0 and 1. This matches a single digit in the binary number system. In computer terminology, we refer to this single-digit binary value as a **Bit**. As we saw in the discussion of number systems, to represent any number greater than 2, we need more than one digit. Since one bit is hardly enough to hold useful information, digital computers use a collection of bits to create a unit of information. Modern computers access data in 8, 16, 32, and 64 digits.

TABLE 1.1 Number Systems

Decimal	Binary	Octal	Hexadecimal (Hex)
0	0	0	0
1	1	1	1
2	10	2	2
3	11	3	3
4	100	4	4
5	101	5	5
6	110	6	6
7	111	7	7
8	1000	10	8
9	1001	11	9
10	1010	12	A
11	1011	13	B
12	1100	14	C
13	1101	15	D
14	1110	16	E
15	1111	17	F
16	1 0000	20	10

☞ **Key Point:** Bit is the binary digit. A single bit has only two possible values: low vs. high, or 0 vs. 1, or false vs. true.

A collection of 8 bits is called **a Byte**, which is the common unit of size for computer memories. Back in the early 80s, one byte was the amount of data that a CPU could read from or write to memory in a single instruction. Those were called 8-bit computers (or CPUs), such as the original Personal Computer (PC) released by IBM in 1981. IBM PC used an Intel 8088 CPU, which was 16-bit but, for manufacturing reasons, did 8-bit data transfer with the memory and other components.

Later, computers move towards 16, 32, and 64 bits. Most current computers use CPUs capable of memory access at 64 bits (8 bytes) at a time, which makes them much faster compared to original PCs. Some other factors that determine the computer speed are the frequency of executing instructions, or the "clock speed," the operations that can happen by each instruction, and the number of parallel instructions that can run at the same time (Patterson & Hennessy, 2013).

CPU is capable of performing arithmetic and logic operations on the data. While arithmetic operation (addition, subtraction, multiplication, and division) are performed on the whole number (byte or more), the logic operations are usually referred to as bitwise operations, as they operate on two bits. They include AND and OR and are illustrated in Exhibit 1.2. They have two inputs and one output. The result of AND operation is true (1) if both bits are true. Otherwise, it is false. The result of OR operation is true if at least one

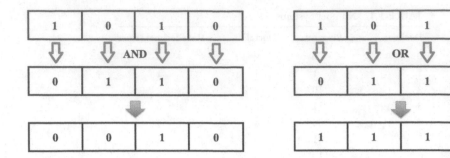

EXHIBIT 1.3   Bitwise operation for two bytes.

of the bits is true. It is false only if both bits are false. The operation acts on bytes or larger data in a bit-by-bit fashion. Every bit in the result is calculated by performing the operation on the corresponding bits in two input numbers, as shown in Exhibit 1.3.

1.3 WHAT IS A PROGRAM, ANYWAY?

Programs are instructions for the computer to execute. As you should know by now, these instructions have to be stored in binary form and as a series of bytes in computer memory (bits are too small to be a unit of memory). The CPU is the part of the computer that reads the instructions from memory and performs them. While there are standards for how to store text and multimedia content as binary, CPU manufacturers like Intel and AMD have their own proprietary system for the meaning of each binary instruction in programs. A binary program that is created to work on a particular CPU is called **Machine Code**.

Binary values are designed to be used by the CPU and are not conveniently readable by people. So, writing a program in the original machine code is very difficult and not practical for large applications. Programmers use special languages to write human-readable instruction, which will then be translated to machine code. The code written in these languages is referred to as **Source Code,** and the languages fall into two general categories:

- Low-level, written for a specific CPU

- High-level, independent of CPU that will run the program.

Low-level programming languages use either binary (Machine Code) or human-readable text (**Assembly languages**). Exhibit 1.4 shows an Assembly program and its equivalent

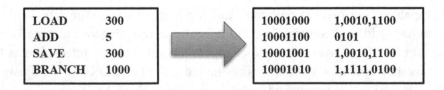

EXHIBIT 1.4   Assembly and Machine Code. Operations LOAD, ADD, and SAVE are common machine code instructions, but their name and equivalent binary values are hypothetical.

machine code. A special program called Assembler translates this text to machine code. The lines of code in Assembly programs have a one-to-one correspondence to the machine code and are CPU-dependent. Low-level programming languages are outside the scope of this book.

High-level programming languages are also human-readable plain text, but they are not CPU-dependent, and there is no one-to-one correspondence between their instructions and the machine code. This means that they can be used for any CPU, but they have to be translated to machine code for target CPU.

```
Sample high-level program
firstData = 9;
secondData = 5;
sum = firstData + secondData;
```

We will discuss the details of the compiling process in Chapter 3 when we start our focus on high-level programming languages, which are the main focus of this book. But before doing that, there are two subjects that will help us establish a better foundation for programming: **Visual Programming** and **Algorithms**.

Visual programming involves creating programs using easy-to-use graphical tools. It is an efficient way of developing programs and also a good starting point to learn the basic concepts of programming. Algorithms, on the other hand, are the common way of describing the logic of a program regardless of which language or method we use to develop the program.

We discuss visual programming in the remaining sections of this chapter and algorithms in the next chapter.

☞ **Key Points**
- Programs are instructions for computers, telling them what to do and how.
- Machine Code is the program in the binary language that the CPU understands.
- Source Code is the program in a human-readable language (high-level or low-level).

1.3.1 Visual Programming

Considering the complex nature of programming languages that require a significant learning curve, special software tools have been developed that allow beginner users to create programs without writing any code. This method is called visual programming and uses graphical user interfaces to define a program. The tool then translates the graphical representation into machine code. Many visual programming tools are easy to use and designed for making simple applications, but some of them are quite complicated and dedicated to scientific simulation or other professional purposes.

The concept of a visual programming tool, more formally known as **Visual Programming Language (VPL)**, has been around for quite a while. SketchPad and Pygmalion were some early examples from the 60s and 70s that allowed graphically

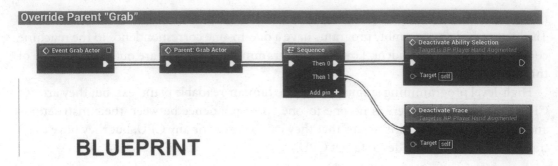

EXHIBIT 1.5   Unreal Engine's Blueprint in action.

representing and controlling the flow of instructions. These were later followed by efforts such as Executable Flowcharts and fully graphical simulation tools such as Simulink[6] that could represent a program as a network of connected modules with clear input and output that the user/programmer could arrange in order to achieve a desired final result. A commonly used example of visual programming in game industry is the Blueprint tool in Unreal Engine (see "Sidebar: Game Engines"). While Unreal allows programmers to write code for their games using C++ programming language (Chapter 3), it also has a visual tool, called Blueprint, to graphically design a game and its elements (Exhibit 1.5). Using Blueprint, programmers can choose game elements, connect them to each other, and define their properties and behavior.

☞ **Key Point:** Visual programming tools allow us to make programs through a combination of graphical elements.

Programs are sets of instructions, but no instruction works without the proper information. For example, if the instruction is to "move," we probably need to provide information such as where, in what direction, or how fast. In general, we can consider a program a combination of code and data, or operation and information. I should mention here that the terms "data" and "information" are not equivalent, in general. Information is commonly defined as data that can be used in or applied to a particular case. For the purpose of this book, though, I use them interchangeably as our data is what we are using in our program. Also, as I mentioned in Section I.1, code and data are not clearly separated in many programs. Inside a program, data items have to be created through some code, and most CPU instructions include operation and data at the same time. The distinction between code and data is more conceptual to distinguish between an operation and the information required for it. Some parts of the program are more about defining information elements.

---

[6] https://www.mathworks.com/products/simulink.html

I refer to these as data. Some other parts are about operations performed on the information, and I refer to them as code. This distinction will help us design and organize our program more effectively, as discussed later.

This book focuses on text-based high-level programming languages, but in the next section, I am going to use GameMaker,[7] a visual programming tool, to illustrate how a program combines these two elements. While in visual programming tools, the programmer does not see the code and data explicitly as in text-based languages, the elements of the program are still there. In the case of GameMaker, these two elements are very clearly distinguishable from each other. As such, GameMaker provides a fun and efficient way to start understanding and visualize the structure of a program.

> ☞ **Key Point:** Programs are made of information and operations. While these two are not separate, in this book, I consider them conceptually distinct and refer to them as data and code.

1.4 GAMES AS PROGRAMS: DATA VS. CODE

Computer games are perfect examples of programs and offer a good place to start learning about programming. The popularity of computer games and general interest in game development and game-based learning (Prensky, 2007) resulted in the development of many tools that use a graphical interface to build games. While some of these tools are aimed at beginners and even children (for example, Scratch illustrated in Exhibit 1.6), there are more advanced tools that are used professionally, such as Unity and Unreal Engines and their graphical editors. Many of these tools combine the convenience of visual programming with the advanced features offered by more traditional high-level text-based programming languages. The user can build basic features using a visual interface and add advanced features by writing a textual code.

GameMaker is an example of visual programming tools made specifically for making games. It is limited to Two-Dimensional (2D) games and does not support some of the advanced features of modern games, but it is commonly used to create all sorts of games, some of them quite impressive (https://www.yoyogames.com/showcase).

Computer games come in a variety of styles, gameplay, interaction, and backstory, but they all share some basic elements[8]:

1. An environment in which the game happens. We can call it game world, level, or any other name, but it generally has a visual setting and possibly audio addition. It can be animated and dynamic too.

---

[7] https://www.yoyogames.com/gamemaker

[8] This is my simplified and practical description and by no means an official and comprehensive definition. Games can, of course, include more elements, but these are the common ones that we see in almost any game.

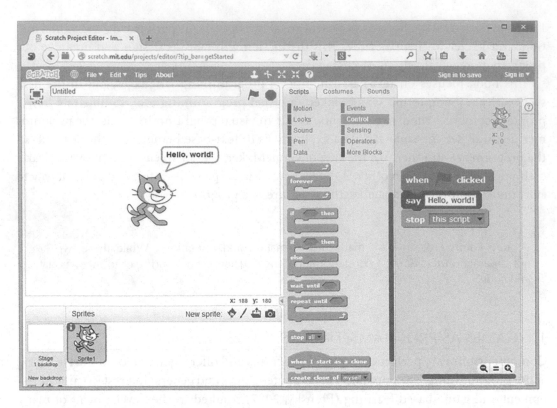

EXHIBIT 1.6   Hello, World! using Scratch.

2. A series of objects that can be of the following types:

   2.1. Controlled and moved by the player

   2.2. Controlled and moved by the game

   2.3. Mostly static

3. A logic that controls the gameplay.

If we think about computer games as programs, items 1 and 2 above are the main data (information), while item 3 represents the code (operations).[9] Of course, a typical game has other information and operations. For example, we can have a multiplayer game and, as such, will need information about users or operations related to connecting to the network. While these can and frequently do exist, for our simple illustration and getting started, we don't need to bother with them for now.

For example, imagine a game that includes an object that player moves (for example, with arrow keys on the keyboard), a few randomly moving objects that player has to avoid, and a few static objects that the player has to pick up to get a score. The information necessary

---

[9] Keep in mind that data elements are created using code. The first operation we perform on data is to create it. The distinction between code and data is more conceptual, not explicit. But it helps us design and organize the program.

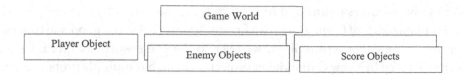

EXHIBIT 1.7   Modules of a simple game.

for this game is the size and the image of the background, and the number, shape, and location of each object. We can visualize the modules of this game, as illustrated in Exhibit 1.7.

This visualization is our first step in developing the game. In this step, we define the data in our program. The next step will be defining what happens to each of these entities (background, objects, and anything else we may have) and how. In GameMaker, the process of creating a game starts with such visualization of the data. By picturing what our game looks like, we identify the major data, consisting of our game world and the objects in it. The initial interface for GameMaker allows us to create these entities, as shown in Exhibit 1.7.

SIDEBAR: GAME DEVELOPMENT PROCESS

The development of computer games started as an almost purely software programming tasks. Back in the old days of video games, the majority of work required for making a game was in the code as the computers were not strong enough to have rich multimedia content. Things have changed now, and so the production of games is not just about programming anymore.

There are multiple groups of experts involved in developing a computer game:

- Designers who design the gameplay, write the story, and set up the levels
- Artists who create multimedia content
- Programmers who design and write the code
- Testers who evaluate and verify the game
- Management staff such as producers, project managers, and assistants.

The production model is somewhat similar to software projects and also movies. It generally involves:

- Concept development (creating a pitch document)
- Pre-production (creating a proof of concept and the game design document)
- Production (creating art and code)
  - This involves both software design and development.
  - The game design usually refers to the design of gameplay and story (what the game does), not the internal software design (how the game does it).
- Test (internal, alpha for selected users, and beta for public)
- Maintenance and upgrade.

There are many resources online and in print to learn more about game design and production. *Fundamentals of Game Design* by Ernest Adams (2014) is a good starting point.

Using the GameMaker terminology, we refer to the game world as room. Creating a room is the first step in a new GameMaker project. A room has multiple properties, but the most important ones are the size and background image. GameMaker makes a distinction between an entity such as room or object and its shape. For a room, the shape is referred to as a background. Your game can have multiple background images and multiple rooms. At any time, one of the rooms is used, and one of the background images is displayed for it.

Similarly, we can define objects and shapes. In computer graphics, a small 2D shape is usually called a Sprite. GameMaker allows you to define multiple sprites for your program. Once you create an object, you can assign one of your game sprites to it.

Finally, you can place the objects in the room to finish the appearance of your game. The game shown in Exhibit 1.8 is a GameMaker implementation of the one shown in Exhibit 1.7. It includes a room with no background image (only a plain color) that holds four groups of objects:

- Player object (smiley). These will be moved by the player using arrow keys.

- Enemy objects (skull). These will be moved by the game automatically, and play has to avoid them.

- Score objects (triangle). The player has to pick these static objects up.

- Wall objects (block). These static objects form a wall around the room.

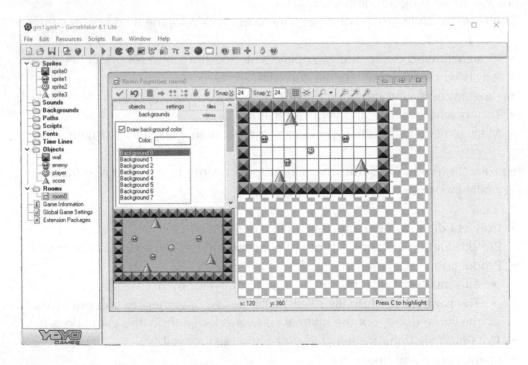

EXHIBIT 1.8   Defining objects for the simple game.

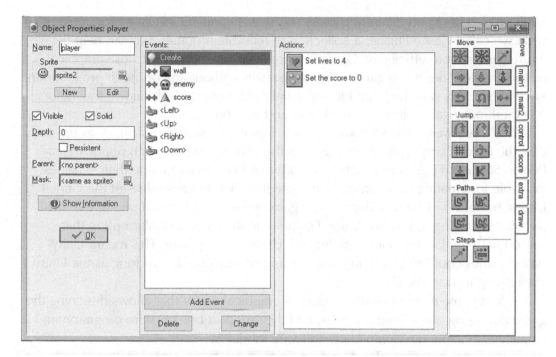

EXHIBIT 1.9 Defining Game Actions (operations) using object properties.

At this stage, we have defined most data elements that our program needs, but if you run the game, you will not see any action, and all those objects are static. This is because we have not defined any operations yet. That is the code part of the program, which allows us to treat objects differently. **Each program is made of data and code** [10] In order to define the code in GameMaker, we need to open the Properties window for any object and define any operation as an action in response to an event (see Exhibit 1.9). GameMaker uses a style of programming called **event-based**. We will see more about this in later chapters. For now, all you need to know is that the program defines a series of "things that can happen" as an event, and then selects actions that will be performed when the events happen.

GameMaker interface allows programmers to use from a set of predefined events such as Create (when the object is created), Collision (when the object hits another object), Key Press (when a key is pressed), and Mouse Click. GameMaker also has a large set of actions that can be chosen as a response to the event, such as moving, changing shape, and adding the score. All these are done graphically and without the need to write any text code.

SIDEBAR: GAME ENGINES

The popularity of computer games in the 80s and 90s resulted in many companies who were developing multiple games to consider reusing their code. For example, the software modules that implemented networking or user interaction could be

---

[10] As I mentioned earlier, these two may not be separated explicitly. The distinction can be more conceptual.

easily shared between different games. This was the initial motivation for creating what we call a Game Engine, a collection of reusable modules that can help develop future games more efficiently. Game Engines later included easy-to-use graphical editors to allow designing games and game levels without any or much programming knowledge. This was the basic idea of visual programming tools, as we saw before. It helped make game development easier and so more popular.

While there were "software kits" for developing 2D games as far back as the early 80s, the term "Game Engine" was used in the mid-90s and through popular First-Person Shooter (FPS) games such as id Software's Doom that made their "engines" available for creating new games. In the late 90s, Epic Games released the Unreal Engine based on its Unreal Tournament game series, which is still one of the most popular game engines. Unity Game Engine is another very popular option that started with a relatively unique ability to include a web player. This meant Unity-based games could run with any browser and on any platform as long as the Unity Player plug-in was installed.

Modern game engines usually include a graphical editor that allows designing the game and using predefined modules, and they support one or more programming languages to allow developing new software modules.

☑ **Practice Task:** Develop a simple game with GameMaker or a similar tool. If you don't have the software, document the design of it without actually making the game.

☝ **Reflective Questions**
- What was your game design and development process?
- Did you the separation of code and data make sense in this case?
- Where you able to see the program as a set of data (information) and code (operations performed on the data)?
- Do you think this view of the program can help you understand programming better?

In addition to easy and visual programming to make a game, we can use GameMaker to learn some basic concepts in programming:

- Define the general behavior of the program (main goal)

- Visualize the program so you can identify the main data elements

- Create the data items (information)

- Write the code (behavior or operation performed on the data).

> ☞ **Key Points**
> - While data and information are not, in general, equivalent, in this book we use the terms interchangeably to refer to the information the program needs.
> - While computer instructions frequently combine operations and information, in this book, I use the terms code vs. data to conceptually separate information from the operations.
> - Programs are made of data (information) and code (operations performed on the data).
> - It is usually easier to start designing a program by identifying and visualizing its data items.

HIGHLIGHTS

- A computer is a device that can execute ("run") programs and has input, processor, output, and memory.

- A program is a set of executable instructions in a formal language that can be (after some steps) executed by a computer to achieve a particular purpose or result.

- Programming is the act of creating a program.

- Software Development refers to a much broader process that involves requirement analysis, design, implementation, testing, and maintenance.

- In this book, I focus on design and implementation which I refer to as programming, even though it overlaps with the other activities.

- Coding and programming are used interchangeably in many contexts, including this book. Others may have broader or narrower definitions for these terms.

- Software engineering is about understanding these various models of the software development process and improving them.

- Computer-based systems (digital media) are programmable, interactive, and multi-dimensional, with massive storage and mobility.

- Bit is the binary digit. A single bit has only two possible values: low vs. high, 0 vs. 1, or false vs. true.

- Machine Code is the program in the binary language that the CPU understands.

- Source Code is the program in a human-readable language (high-level or low-level).

- Programs are made of data (information) and code (operations performed on the data).

- It is usually easier to start designing a program by identifying and visualizing its data items.

END-OF-CHAPTER NOTES

A. Things I Should Mention

- GameMaker software used to be available with a free version, but that is no longer supported. You can use a trial version and pay for it if you find it useful.

- Gamification is the idea of using game concepts and feature in non-game applications such as education and business. In her book, *Reality Is Broken*, Jane McGonigal talks about how we can use games to make world a better place.

B. Self-Test Questions

- What is the equivalent of 23 (decimal) in binary and Hex?

- How do we perform bitwise AND and OR operations on two bytes?

- What is a program?

- What are the high-level and low-level programming languages?

- What is the game development process in GameMaker?

C. Things You Should Do

- Practice making games with any visual tool you can access.

- Check out these web resources:

  - https://www.britannica.com/technology/computer/History-of-computing

  - https://www.computerhistory.org/timeline/

  - https://www.mathsisfun.com/binary-number-system.html

- Take a look at these books:

  - *A History of Modern Computing* by Paul Ceruzzi

  - *Fundamentals of Game Design* by Ernest Adams

  - *Visual Programming Languages* by Boshernitsan and Downes.

D. Reflect on the Experience of Reading This Chapter

- What did you expect from this chapter before reading it?

- What was it about, and what did you learn?

- What tasks did you perform, and what difficulties did you face?

- How did you feel about the material and tasks presented in this chapter?

- How can you improve your learning experience?
- How do you see this topic in relation to the goal of learning to develop programs?
- Can you see the code/data concepts in the programs you use?
- How do you think the concept of code vs. data can help designing programs?

CHAPTER 2

Logical Creativity

At the end of this chapter, you should be able to:

- Describe some general approaches to problem-solving
- Apply skills from other forms of the creative process to logical creativity
- Use algorithms to describe the problem-solving process
- Understand and use the data-centered algorithm design approach

OVERVIEW

In the previous chapter, we took the first step towards becoming a computer programmer. We learned what computers and programs are. Our next step is to see how to design a computer program. We do so by looking at programs as a means of problem-solving. We will see that despite variations in languages and applications, all programs are basically our attempts to solve problems or sometimes guide the computer to solve them on its own. The requirement analysis phase that I discussed in Section I.1 involves the proper definition of the problem, while design and implementation deal with producing the solution.

The problem-solving approach to programming has the main advantage of allowing us to use existing methods to solve problems in other areas. Psychologists, cognitive scientists, and many other researchers and professionals have long studied the act of problem-solving and the techniques we use for it. Once we know those techniques, we can try to apply them to our special case of computer programming.

We can also learn from other specific areas of problem-solving such as visual design. Despite looking completely different, an artistic process has many similarities to programming: they both solve problems, are creative, and in fact, follow similar steps. Equipped with this basic information, we will be ready to learn about algorithmic thinking as the main topic of this chapter.

Algorithms are one of the primary ways to describe the solution to a problem. They are clear and logically related steps we take to solve problems. While computer programming can be done in different languages, algorithms represent their common "way of thinking" that I mentioned in "Introduction." After we learn how to think algorithmically, we can present our thought in any programming language. Of course, in order to do that, we need to learn about the rules and syntax of these languages, which is the subject of the next three chapters of this book.

2.1 PROBLEM-SOLVING AND CREATIVITY

To many people, especially those who are starting to learn programming, the design and implementation of software programs seem like un-intuitive tasks that deal with complicated languages with strangely named elements and confusing syntax. At first interaction with computer programs, it is normal to feel that way, but the truth is computer programming is really about a basic skill that we all have: problem-solving.

Every day, we face many problems, and we solve them even without noticing that our brains have been involved in the quite complicated task of problem-solving. If you are like me, this starts by figuring out how much longer you can stay in bed. After that, the problems get more complicated (but maybe not harder). From deciding what to eat for breakfast to what path to take when walking to work or school, we are constantly using our abilities as problem-solvers. This continues throughout the day as we go to work, deal with relationship issues, or plan our finances. We are all expert problem-solvers, one way or another. That is why this book is called *Anyone Can Code*. There are many aspects to the design and development of good computer programs, but we have the basic skill.

Software design and programming involve two different types of problems. The first type involves what the user needs to be solved, for example, calculating the average of all grades for a student, or the total monthly profit of a business. As programmers, we usually receive a solution for that from the user (or someone we call a "domain expert"). But sometimes, we have to find that out first ourselves. The second type of problems is faced by the programmers only, and that is how to make a computer perform the solution for the first type of problem. These are the problems that the users don't see. There are also problems that the computer itself solves using what we call Artificial Intelligence (AI). In those cases, the programmer is still facing a type 2 problem, but the objective is not to make a computer program perform a given solution but to reach a new solution that the programmer is not aware of in advance.

☞ **Key Point:** Software design and programming are primarily problem-solving tasks. So, we can use established problem-solving methods for them.

Considering the role of problem-solving in programming, it is natural to rely on what we know about problem-solving when we approach programming. The problems we solve vary in complexity and significance. Some of them are well defined, meaning that they have clear starting and ending points. For example, we plan what to take for lunch. That can be extremely well defined when we have only one option in the fridge. On the other hand, some problems are ill-defined. It means we cannot be sure where to start or what the ending is going to be. We all want to be happier, may find our life not satisfying, or not be comfortable with our work situation. Regardless of the problem and its type, there are common concepts and methods that we use to solve it. That's why it is good to start by learning a few things about problem-solving as the first step to learn software development.

It is common to think of problem-solving as an act of navigation through a "problem space" where we start from an "initial state" and move to a "target state." (Sternberg, 1994) Consider, for example, taking public transit from location A to location B on a map (see Exhibit 2.1). The problem is to find a series of bus routes that take you to your

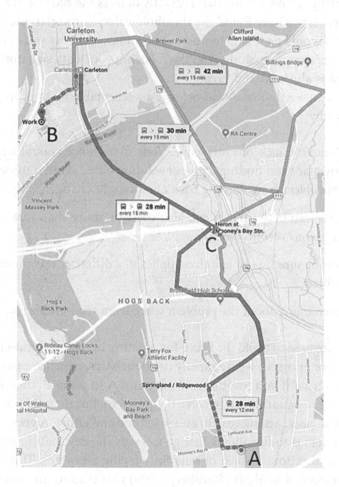

EXHIBIT 2.1   Directions from A to B. It may require taking different buses and transferring at different stops such as C.

destination if no single bus goes to both locations. This is a problem that we all may face if we take transit. It is also a common problem for map programs that give directions. So, while we see how humans solve this problem, we can see how that can be applied to computer programs.

One of the most common problem-solving methods that we all have used is trial-and-error. If you have no clue, you just try different solutions and hope that one works. This works fairly well if you want to buy a pair of pants and are not sure about your size. You try a few and see which one works. We may not want to admit, but computer programmers use this method many times, more than they probably should.

Imagine you are trying to automatically remove "red eye" from a photo by detecting it and then reducing the redness. Later in the book, we will talk about graphics programming and how these things are done, but for now, assume that you have a way of detecting the color of each point as a series of red, green, and blue values.[1] You can define a red eye as a small area with high red value surrounded by a bigger non-red area. But how red is "high red"? You don't know, and so you try different threshold values in your code and run it multiple times until you find a value that correctly detects the red eye in a few test images you have. In this case, you have found the threshold value using trial-and-error.

Another example could be trying to find out how many records you can have in a database without your program crashing. You try reducing your data size until your program doesn't crash. That is also trial-and-error.

Hopefully, you can see that trial-and-error is an acceptable method in some cases, but not often. Here is why:

- It generally shows a lack of knowledge. Your first choice should be understanding your problem space and finding a proper solution based on that understanding. That prevents the problem from coming back after a slight change.

- Trial-and-error can take time. It can also cost you more than just time if each trial involves using some resources.

- You can never be sure if your solution applies to a different version of the problem or a similar problem.

- You don't learn much about the problem space from trial-and-error.

Various methods are used to avoid the brute-force nature of the trial-and-error approach to problem-solving. For example, we frequently use **heuristics**, which are the rule-of-thumb-style guidelines such as "if you want to go from point A to point B, then follow a straight line." Heuristics are not guaranteed to solve the problem, but they are often good starting points and, in many cases, are successful and save time and effort. A good heuristic for our public transit problem is to use routes that go towards the destination or to a point in its vicinity or general direction, instead of randomly trying all routes. This is an example of what we call a means-end analysis (Sternberg, 1994) that tries to minimize the difference

---

[1] This is called RGB color system.

between the current state and target state. We can do it in a forward way (where can we go from A that takes us closer to B) or backward (which point goes to B that is closer to us).

Another common approach to problem-solving involves pattern-action rules (Sternberg, 1994). This approach relies on finding familiar patterns in the problem space and then taking actions that have been successful in previous cases with those patterns. For example, we can say, "If the destination is on the east side of the current location, then choose an intermediate point on your east side that has a direct route." The human brain is wired to be good at pattern recognition. Our ancestors, who were quick at recognizing the pattern of "Animal with scary horn running towards me" as a threat and ran away, were more likely to survive and pass their genes to the next generations. Pattern recognition helps us recognize a situation (such as a classroom) quickly without processing all the details, and then act properly (go sit on a chair and listen, instead of standing up front and dancing).

A successful pattern-action rule relies on the problem-solver creating an effective model of the problem with only relevant and important information and having a good set of previous cases to compare to. Presenting a problem or situation, as we can see, has an important role in our ability to deal with it. For example, in the case of going from A to B, paying attention to details such as what is going on at each point doesn't help find a transportation solution. Visualization (creating a clear and easy to use picture) and abstractions (hiding details that are not important to the main problem) are key skills in developing a helpful mental model of a problem. This is particularly true in software development, as we will soon see.

Last but not least, the topic of problem-solving is closely related to creativity. Creativity is a hard topic to define. A common definition is the ability to come up with new and useful ideas to solve problems (Sternberg, 1994). It has been studied by many researchers and is generally associated with knowledge, intellectual capacity, personality traits, and thinking style. Creative people are usually known for insight, openness to new experiences, perseverance, toleration of ambiguity, and perceptual process.

Not all the problems we face are at the same level of complexity and difficulty. Some are fairly simple or have been solved before in slightly different situations. Such "routine problem-solving" requires a small amount of effort and skill. On the other hand, dealing with new problems requires "creative problem-solving" are more complicated as they require solutions that are novel and appropriate for the new problem. Expected quality, significance, and other factors also affect the level of creative skills required.

Some aspects of creative problem-solving that are helpful in software development are

- *Divergent thinking* is our ability to "think outside the box," to consider various options and see things from different points of view.

- *Problem definition skills* are also important as they help us establish a representation of the problem that allows us to see the important aspects of it better. Two of the most important problem definition and representation skills are visualization (creating a visual description so we can understand better) and abstraction (removing details and specifics so we can focus on important aspects), which we talked about earlier.

- *Analogy* is another important aspect of the creative process which allows creative problem-solver to see similarities and, as such, use appropriate solutions.

A detailed discussion of theories on problem-solving and creativity is beyond the scope of this book. I encourage you to look at the book *Thinking and problem solving*, which is a collected set of articles edited by Robert Sternberg (1994) for a good review, although there are many other sources available. My goal here is to review some of the important things that will help us in our programming (logical creativity) process. Trial-and-error, heuristics, pattern-action rules, and the creative problem-solving techniques discussed above are commonly used in programming.

☞ **Key Points**
- Some common problem-solving methods are trial-and-error, heuristics, and pattern-action rules.
- Creative problem-solvers rely on skills such as divergent thinking, visualization, abstraction, and analogy.

Another creative process that shares many characteristics with logical creativity is visual design. In the next section, we will take a look at the visual thinking process and see how we can learn from it to improve logical creativity.

2.2 VISUAL DESIGN AND VISUAL THINKING

While visual artists and designers intuitively or through experience have developed a large body of knowledge and expertise about visual design, our scientific understanding of the way human brain processes visual input has recently changed significantly. This has been possible through research work in neurology, psychology, and even optics that have given us better insight into our visual thinking process. The key to this new understanding is the concept of visual query, which is a small act of attention that we make when looking at a scene (Ware, 2010). These visual queries can be detected using eye-tracking devices and show how we go through the details of a scene we are visually processing.

Visual thinking is a series of these queries that allows us to inspect and process a scene, recognize patterns, and make decisions. It involves a top-down task-driven process where our brain controls the attention in order to find what we need. While we usually deal with our visual stimuli with a task in mind, there is also a bottom-up process when we face any new scene. The bottom-up process is also aimed at recognizing objects to give us a proper understanding of our situation. It starts with an image being formed in our retina. Then, a hugely parallel processing machine made of billions of neurons goes through what we call feature extraction. These features are visual information that can help us find the important parts of the image. Size, orientation, and color are among these features that allow us to detect things such as a rectangular shape or a bloodstain on our shirt after shaving. The top-down and bottom-up processes are not really separate, and our brain is almost always sending signals in both directions.

Knowing the process of visual thinking helps us visualize our thoughts more effectively. As we saw in the previous section, visualization is a powerful tool in problem-solving as it allows us to present the problem and see it in different ways. We will see soon that visualizing the software program that we want to develop is a great starting point for any software project.

Visual thinking is a suitable tool not only for describing and starting your software project but also throughout the process. In fact, the process that a visual artist or designer follows is not particularly different from what a software developer does. If you have done some drawing or painting, you have probably noticed that artists don't just sit in front of a canvas and paint, for example, a great portrait, by getting the eyes completely right, then eyebrows, nose, and so on. The process is more cyclical and involves various activities. Exhibit 2.2 demonstrates this process.

The artist's visual thinking process, in the context of drawing, generally involves these steps:

1. Observing the scene

2. Identifying objects and their relations

   a. Choosing the right scene (angle, compositions, etc.)

3. Sketching basic shapes

4. Adding details through cyclical refinement.

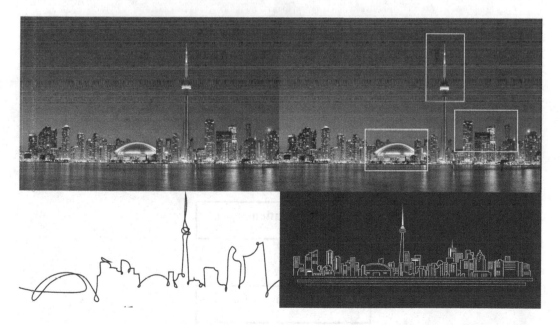

EXHIBIT 2.2 Visual design process. Observing, identifying the major entities, drawing the first sketch, and adding details.

Logical creativity and artistic creativity are not that different from the process point of view. Software development follows similar steps, and unlike what many believe, artists can do very well in software development if they can see these similarities and follow the process they are familiar with. The reason that many artists and individuals with more visual minds find programming strange and difficult is that they approach it incorrectly, and assume the point is to memorize textual syntax of different languages and tasks like that. Those are necessary but come much later once the core of your program is already designed.

Just like artists who use various tools such as models and sketches to visualize their design and refine it before doing the final work, software developers use different visualization tools and particularly diagrams and algorithms (programming sketches) to describe the software they need to develop (not just to others but to themselves).

Their process (Exhibit 2.3) also starts with proper observation of the "scene"; that is, the environment the software needs to be deployed. Then, we need to identify the major components of the software, which include information and operations. The next step is to decide how these components are related, which operation needs which information, and so on. This is what I refer to as "visualizing the program." It is an essential step in any software development. You can't make anything if you can't imagine it. Finally, we will sketch the program through algorithms, and then translate that into actual code and cycle through refinements until the work is presentable.

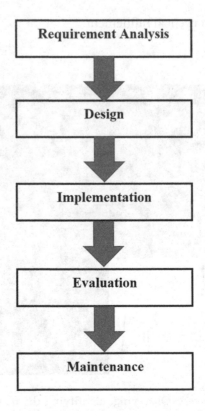

EXHIBIT 2.3   Software development process.

> ☞ **Key Point:** The following visual design process can help with programming:
> - Observe the scene (problem space)
> - Identify major entities
> - Sketch the basic structure
> - Add details and refine.

Before we can start with actual coding, we should feel comfortable with this process and particularly algorithms. In the next section, we will review some simple examples of algorithm design. They may look trivial and non-creative but are essential in establishing an understanding of algorithmic thinking, which our primary tool in software design and programming.

2.3 ALGORITHMS AND ALGORITHMIC THINKING

2.3.1 What Is an Algorithm?

As we saw in "Introduction," programming involves using different languages with different syntax and rules. But, more importantly, it relies on logical thinking. While there are various tools that help us understand the language rules or find syntax errors in our program, such as the improper use of symbols, logical errors may be much harder to detect. A sentence in natural languages may have correct spelling and grammar but make no sense (or be completely wrong). Similarly, a program may have no syntax error and look correct but fail to achieve its result due to incorrect logic. **Algorithms** are the tool we use to develop and describe the proper logic to solve a problem.

> ☞ **Key Point:** Programs have two general aspects: syntax and logic.
> - Syntax is language-dependent and easier to automatically correct.
> - Logic is task-dependent and requires proper thinking.

The word "algorithm" is derived from the name of Al-Khwarizmi, the great ninth-century Persian mathematician and astronomer who is considered the founder of algebra. An algorithm is a problem-solving method consisting of clearly and exactly defined instructions. In software terms, it starts with a clear definition of the problem (requirements) and results in a clear solution (program). The beauty of algorithms compared to other methods such as heuristics is the exact nature and the certainty of achieving a solution. Let me repeat these two as they are key characteristics:

- Well-defined: Algorithms are exactly defined, so the computer (or the person) following them knows what to do and performs the same action every time.

- Solution-oriented: Algorithms are written for a specific solution and are guaranteed to reach their expected result.

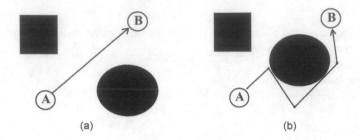

(a)   (b)

EXHIBIT 2.4   Point A to Point B. (a) Straight line heuristic (b) go-around-obstacle algorithm.

Compared to algorithms, heuristics are not clear and make no promises, but are generally easier to follow and take less time. Think about the problem of going from point A to point B. A simple heuristic will tell us to follow a straight line. This is simple and mathematically the shortest path between the points, as shown in Exhibit 2.4a.

If there is no obstacle, then this simple rule will get us to the destination quite simply and quickly. But there may be something in front of us, and that is when we need a more sophisticated rule that can guarantee we reach the destination. We can modify our rule like this: go straight, and if there is an obstacle, then move to the right a little, then try to go straight again, as shown in Exhibit 2.4b. This is what we call an algorithm, and we can describe it more formally as follows:

SAMPLE CODE #1[2]

```
1. Start from A

2. Assume destination is B

3. Go straight towards the destination

4. If you encounter an obstacle,

5.    Then move a little to the right

6. Back to step 3
```

While the description of algorithms does not have to follow any specific language or syntax, it is common to use a semi-formal style. This is called **Pseudocode**. Another method of showing algorithms is using a graphical representation called **Flowcharts** (see the Introduction, Exhibit 2.1, and Sidebar on flowcharts). In this book, I primarily use Pseudocode as it looks more like actual code, but some people prefer flowcharts due to their visual approach. I will include flowcharts for some of the examples, especially in the earlier chapters.

There are no standards for writing pseudocode, but generally, the common structure of programming languages is followed without specific details. For example, to have an

---

[2] Throughout the book, I identify example programs with the phrase Sample Code and a number. While algorithms are not exactly program or code, sometimes I refer to them with those words as they are the basis of a future program.

algorithm that can take us from any starting point to any destination, we can modify the code above as follows:

SAMPLE CODE #2

```
1. Get Start (for example, A)

2. Get End (for example, B)

3. MoveToward End (MoveToward is an operation)

4. If obstacle,

5.    MoveToward right

6. GoTo 3 (GoTo is another operation)
```

Our second version is still not a program, as it is not written in any standard programming language and, as such, is not machine-readable. Algorithms are written to be human-readable and help us understand what we (or computers) are supposed to do. But this version is based on three specific and clear operations:

- Get receives the value of something (in this case, Start and End).

- MoveToward changes the location in the given direction with a certain amount that is not specified (unimportant detail at this time). It applies to the object we are controlling.

- GoTo also changes the location but by moving to (not towards) the given point. It applies to the algorithm itself.

The above algorithm is fairly well defined and does, in fact, exactly tell us (or a computer) what to do. But you may notice that it is not error-proof. We can imagine situations that the obstacle doesn't end on the right or has a curve that may take us back to where we started. Algorithms may be designed incorrectly or need refinements. The idea is not to claim that any algorithm we write is guaranteed to solve its problem but that algorithms are written with a solution in mind. We will talk about how to design and refine algorithms in the following sections, but first, let's review more examples and through them, learn how to write algorithms that are **well defined** (consisting of clear steps) and **solution oriented** (leading to a clear solution).

Many operations that we perform in our minds seem fairly simple to the point that we don't think about how they are done and what steps we take. Consider writing a program that receives two numbers and adds them. If we are doing this ourselves, we ask for or think of a number (say 5), then another one (say 7), and then add them up and decide what the result is (in this case 12). If we are asked what algorithm we used, we probably say, "I just added them."

```
Add two numbers
Say the result
```

But for a computer, this is not well defined because it "implies" that the three pieces of information are stored somewhere without explicitly saying where. A more appropriate algorithm would be

SAMPLE CODE #3

1. Get A

2. Get B

3. SUM = A + B

4. Print SUM

The main job of most programs is to process information. Pieces of information such as input numbers and the result are key to our algorithms, and in longer programs, they may be used multiple times. We generally refer to them as **variable** as they can change throughout the program, as opposed to constant numbers such as 5 and 7. In math, when we write an equation such as $X + Y = 5$, X and Y are variables, while 5 is a constant. Equality applies only for a certain values of those variables. The user name for a computer is a variable because its value changes based on who has logged in. The clothes you wear are variables, while some of your personal characteristics may (arguably) be constant. Variables are identified with a *name* and have *values* (that can change).

☞ **Key Point:** Information (data) in a program or algorithm is generally stored in "variables" that have name and value.

☑ **Practice Task:** Modify the Sample Code #3 to calculate a more complicated mathematical formula.

✋ **Reflective Questions:** Do you see the advantage of breaking down the addition operation into the four steps of Sample Code #3? Can you do that for other tasks?

Note that algorithms (and programs) are instructions to the computer to do something and not a statement of facts. As such, the = sign in algorithms (and later, in most programming languages) does not mean that two sides "are equal." It is called "the assignment operator" as it assigns the value of the right side to the data item on the left side. So, SUM = A + B means "set the value of SUM to A+B." Other common operators in algorithms and programs are the arithmetic operators + (addition), − (subtraction), * (multiplication), and / (division). Also, the names used here (A, B, and SUM) are arbitrary. We usually use names that make sense to us and others who may read our code.

> ☞ **Key Point:** The = sign is called the "assignment operator." It assigns the value of the right side (r-value) to the variable on the left side (l-value).

The above algorithm is well defined because it is made of clear operations: input, addition, assignment, and output. It does not imply anything about the sources of data and what we do with the result. It also uses names for all the different information we have. This makes it easier to refer to them later without confusion. Imagine making the above example more complicated by adding three numbers. In our mind, this is done by adding the first two numbers and then adding the third one to the result of the first addition. While we imply this process, we hardly think of it – particularly the need to store the result of that first addition. Having the name SUM helps us refer to the proper information in a way that is well defined for a reader and later for a computer. Remember that algorithms are (at least in the context of this book) written to be later translated into computer programs.

SAMPLE CODE #4

```
1. Get A
2. Get B
3. SUM = A + B
4. C = 8
5. SUM = SUM + C
6. Print SUM
```

The above algorithm has a couple of interesting features that we should pay attention to:

- It can easily be extended by repeating lines 4 and 5 to add more numbers (D, E, F, etc.).

- It demonstrates that the assignment operation is not the same as mathematical equality.

Keep in mind that algorithms are not mathematical equations, even if they look the same sometimes. Saying X = Y in math means that X and Y "are" equal. Algorithms are operations to perform and not statements of the facts. As such X = Y as an operation means "set X to the value of Y." That is why a statement such as "SUM=SUM+C" does not make any sense in math (unless C is zero), but in algorithms (and later in computer programs), it simply means "set SUM to the current value of SUM plus C" or "increase SUM by C."

2.3.2 Program Flow

Thinking about algorithms as operations, there is another important point that we should notice in all the above examples. Algorithms perform a series of operations one by one, from top to bottom or beginning to end, in a linear form. This is the default way that algorithms and programs run, and we refer to it as **sequential execution**. The logical nature of programming and the dependency of operations on each other as cause and effect are based on this key property of algorithms and programs. It is important to understand that what we do depends on what came before, the order matters, and unless we specifically control the flow of execution, it moves linearly forward, one operation to the next.

☞ **Key Point:** The sequential execution is the basis for the logical flow of programs and dependency of each operation on what came before.

The above statement about programs moving linearly forward has a key point to it that I hope you noticed: "unless we specifically control the flow." Program flow control is an essential part of any program or algorithm and allows us to deviate from the sequential execution. It happens in almost any program we write, and you have already seen it in our first example, going from Point A to Point B. There are two main mechanisms for flow control: selection (decision-making) and iteration (repeating). Both of these two mechanisms are illustrated in the example of moving from Point A to Point B (Sample Code #2), although we didn't discuss flow control at that time. The sequential execution after Step 4 is conditional. Step 5 happens only if a condition is met. Otherwise, we move to Step 6, skipping Step 5. The decision to execute 4a or not depends on the condition of having an obstacle. Following that, Step 6 changes the direction of execution and goes back to Step 3, and this process continues indefinitely.

☑ **Practice Task:** How can we modify the algorithm of Sample Code #2 so that it will end if we reach Point B?
   Hint: Use another if statement, and make the GoTo conditional.

✍ **Reflective Questions:** A conditional GoTo is the base of what we call a selection or branching, commonly used in programming. It allows the program to "jump" to any point. Can you think of a more organized way to do this, so, in a program like Sample Code #2, it is clear how we may get to line 3?
   The next section will discuss such a "structured" way of doing selection and branching.

2.3.3 Selection

Consider the example of calculating the sum of two numbers. Now imagine the problem where A and B are two parts of the final grade in a course, and we have to calculate the total marks of a student and see if she passes the course (sum is more than 50). This is a very simple example of using Selection (decision-making) with two options. The selection process in software is similar to reaching a fork while moving along a road. You need to decide which way to go based on some logic, i.e., a condition.

SAMPLE CODE #5

```
1. A = 5

2. B = 7

3. SUM = A + B

4. If SUM < 50

5.    Print "Sorry. You Failed."

6. Else

7.    Print "Congrats. You passed."

8. Print "Goodbye"
```

At line 4, the algorithm checks a condition statement (SUM < 50) and selects which one of those two options will be executed. As you can see, conditional operations have multiple parts:

- An "If" statement shows that we are setting a condition.
- The "If" is followed by a "condition," another statement that can be true or false (in this case SUM < 50).
- Then, there are one or more lines of code that will be executed if the condition is true.
- Optionally, we may have an "Else" statement that is followed by the code that will run if the condition is false.

In some cases, we only do something when the condition is true, and we just skip and move on if it is false. In some other cases, we choose between two operations and then move on. To identify which code belongs to If or Else, we may use indenting (as in this example) or another method:

```
Using brackets:

  1. If SUM < 50

  2. {
```

    3. Print "Sorry. You Failed."

    4. }

Using special words:

    1. If SUM < 50

    2. BEGIN

    3. Print "Sorry. You Failed."

    4. END

☞ **Key Points**
- In programming, a language construct is a syntactically allowed combination of language elements that performs a certain task.
- Conditional statements are constructs that implement selection. "If" is the most commonly used conditional statement.
- The If statement has a condition and a body of code that runs if the condition is true. In Pseudocode, we can identify the code with indenting, or brackets, or BEGIN/END words.

🖐 **Reflective Questions:** Do you see any advantage in using if/else compared to go the GoTo? Using selection, you can choose one of the options to execute, but they are all moving forward. The GoTo command is less structured but also allows going backward. Iteration is another programming method that provides that ability. We will discuss them shortly.[3]

2.3.3.1 Detecting Odd and Even Numbers[3]

SAMPLE CODE #5A

  1. Get NUMBER

  2. R = Remainder of NUMBER divided by 2

  3. If REMAINDER == 0

  4.   Print "EVEN"

  5. Else

  6.   Print "ODD"

---

[3] In his famous 1968 article, the computer scientist Edsger Dijkstra argues about "GoTo Statement Considered Harmful." This argument is credited for starting what we now call structured or "Goto-less" programming.

This code detects odd and even numbers by checking the remainder of division by 2. Even numbers have a remainder of 0. You notice that line 2 does not look well defined as it doesn't say how to calculate the remainder and is very informal. In Pseudocode, we try to get closer to the formal structure of programming languages and avoid loose natural language descriptions. There are different ways to make that line look more formal and clear. Here is one:

R = NUMBER % 2

In the above line, we have defined a new operator (similar to +, −, *, and /) that performs the division but returns the remainder.[4] This operator doesn't exist in all programming languages, but in our algorithm, we can just assume it will exist in the language we are going to use or that the required code will be written to perform this operation.

You may have noticed the == in line 3. We talked earlier about the = (assignment operation). When you want to say "compare A and B" if you write "if A=B," it may be confused with an assignment operation. Remember that in an assignment operation, the left-side variable will receive the value of the right-side statement (can be more than one variable), but when we compare two items (to check a condition), no value changes. So, it makes sense to use a different sign for the comparison operation. It is common in programming languages to use == for this purpose.

☑ **Practice Task:** Write an algorithm to receive three values and check if they can be the angles of a triangle.
Hint: They have to add up to 180.

☑ **Practice Task:** Modify Sample Code #5A to show/detect factors of 5
Factors of 5 are 5, 10, 15, etc.

2.3.4 Iteration

Now imagine a case when we want to calculate the Grade Point Average (GPA) of a student and decide if he receives a medal with the condition being a GPA of 95 or more. This program introduces a new variable to our data, which is GPA. You notice that the If statement no longer needs an Else.

SAMPLE CODE #6

```
1. A = 5
2. B = 7
```

---

[4] Another common method for showing such operations is to use "functions" which we will introduce in Chapter 3.

```
3. SUM = A + B

4. GPA = SUM / 2

5. If GPA >= 95

6.    Print "Congrats! You receive a medal."

7. Print "Goodbye"
```

In addition to the new line 4 that calculates the GPA (assuming A and B are the grades from two courses), the other thing you may notice is the >= operator. This is a common notation to say "greater or equal." The algorithm seems pretty straight-forward, but what if we need to receive 3, 5, 10, or 50 grades and then calculate the GPA? We have already seen how to add three numbers but if we have a large number of data, then adding them line by line has two problems:

- The code becomes too long.

- We have to create many names.

As we saw in the A-to-B example, the code can be repeated as many times as we want until we get to where we want. Computers are pretty fast and can repeat a similar operation many times very quickly. This is a task that is quite annoying or even impossible for us to do. So, we frequently use computers to do that. In our sample code, we can start by using the GoTo operation that we saw earlier.

SAMPLE CODE #7

```
1. SUM = 0

2. COUNT = 0

3. Get A

4. SUM = SUM + A

5. Increment COUNT (or COUNT = COUNT + 1 or COUNT++)

6. If COUNT < 10 (assume we need 10 grades)

7.    GoTo 3

8. GPA = SUM / 10

9. If GPA >= 95

10.    Print "Congrats! You receive a medal."

11. Print "Goodbye"
```

This code introduces a couple of new things. The first one is COUNT. Remember that your algorithm (or code) has to be well defined, so any piece of information that is needed has to be explicitly defined. If the goal is to enter 10 grades, then as we keep entering new grades, we need to count how many we have so far, and that information has to be stored somewhere. That is the job of COUNT variable. You notice that the number 10 itself is not stored in a variable. In this program, it is **constant** (or **hard-coded**). If there is data that you are sure will never change, you may hard-code that value (use a constant value in the code), although I recommend you avoid hard-coding as much as possible.

The other new thing here is giving initial values to SUM and COUNT. Variables cannot be used in any operation if they don't have a value. The operation "Increment COUNT" is meaningless if we don't know what the value of COUNT is. Similarly, "SUM = SUM + A" cannot be performed if we don't know what SUM is. In general, the first time you use any variable in your program, it should be getting a value. In Sample Code #6, this happens at line 3 for SUM. But in Sample Code #7, both for Lines 4 and 5, SUM and COUNT need to have a value. Lines 1 and 2 are added with initial values for these two variables. This is called initialization. You can see that before we enter any grade, both these two variables have to be zero.

One problem with this algorithm is that when we are reading the Pseudocode, it is hard to know which parts are repeated. That is because we can have GoTo operations at any point in the program. Creating iterations using GoTo is not recommended even though it used to be common. In more modern programming, the use of GoTo is replaced with "loops," which are well defined and clear constructs that repeat a part of the code.

SAMPLE CODE #8

```
1. SUM = 0

2. COUNT = 0

3. While COUNT < 50

4.      Get A

5.      SUM = SUM + A

6.      COUNT++

7. GPA = SUM / 50

8. If GPA >= 95

9.      Print "Congrats! You receive a medal."

10.     Print "Goodbye"
```

You notice that While works almost exactly like If. It has a condition and a series of operations that will be executed conditionally. The only difference is that the operations will be repeated as long as the condition is true.

☞ **Key Point:** Loops are constructs that implement iteration. A "while" loop has a similar structure to an "if" statement (condition plus body of the code), but repeats the code as long as the condition is true.

✍ **Reflective Question:** Selection and iteration are the bases of what is called structured programming. Do you see how their combination can achieve what GoTo could do but in a more structured way?

The above code can be shortened by removing the details about COUNT (see below), but I personally prefer to keep the variable COUNT. It is an important piece of information that later we will use frequently.

SAMPLE CODE #9

```
1. SUM = 0
2. Loop 50 times
3.     Get A
4.     SUM = SUM + A
5. GPA = SUM / 50
6. If GPA >= 95
7.     Print "Congrats! You receive a medal."
8. Print "Goodbye"
```

Alternatively, we can combine all the lines related to COUNT into one:

SAMPLE CODE #10

```
1. SUM = 0
2. For COUNT=0; COUNT < 50; COUNT++
3.     Get A
4.     SUM = SUM + A
5. GPA = SUM / 50
6. If GPA >= 95
```

7.   Print "Congrats! You receive a medal."

8. Print "Goodbye"

This is an example of what we call a "for" loop, as opposed to #8, that used a "while" loop. We usually refer to "while" and "for" loops as indefinite and definite, respectively. This is because they are usually used in two different situations: when we have a condition (which results in repeating an indefinite number of times) and when we have a certain number of times to run the loop (with no other condition to check). A "for" loop in the above example is simply a "while" loop with a counter. It is common to use a "while" loop when we have a general condition but no idea how many times to run the loop and use "for" when we have a certain number of times. But conceptually, the two constructs are equal, and anything that can be written using one can also be written using the other. Both "if" and "while" have only one condition that is checked before the code executes, but "for" has that condition plus two other parts:

- An initial action (in the above example, COUNT = 0) that happens only once
- And another action (COUNT++) that happens at the end of each iteration.

Theoretically, these actions can be anything, including blank, which makes a "for" exactly the same as a "while" loop. On the other hand, these actions can be added in other lines as in Sample Code #9, which makes the "while" loop functionally similar to a "for" loop.

> **Key Point:** for loops are similar to while loops, but in addition to the condition, they have an initial statement (runs only once) and an ending statement (runs at the end of each iteration).

☑ **Practice Tasks**
- Use a while loop, and write an algorithm to move step by step to get to a destination point.
- Modify the algorithm to use a for loop, and take 20 steps in the direction of the destination.

✋ **Reflective Questions**

- Do you see the difference and similarity between while and for loops?
- Can you think of cases that can be done with one of them but not the other?
- Do you think it makes sense to have both?

2.3.5 Variables and the First Golden Rule of Programming

Remember that the computer cannot work with implied information. Anything we want to be done has to be explicitly stated. While in our mind, we simply add numbers and count how many numbers we have added, the computer needs to "know" this information:

- Before we get any grades, SUM is zero.

- There is a piece of information that says how many grades we have added (COUNT), and it has to start with zero too.

- Every time we add a new grade to SUM, we have to tell the program explicitly to increment COUNT.

Notice how we added new variables for each piece of information used in the program: A, SUM, GPA, and COUNT. It may not be the most efficient way of programming, but for those starting to code, it is a very helpful practice to define a new variable for every piece of information that the program needs or uses. I call this the **Golden Rule of Programming #1**. We will see more about this rule and my other (made-up but super helpful) rules in the following sections of this chapter and the rest of the book. But this one is probably the most important (hence #1) as it defines our data in a very prominent way.

☞ **Key Point:** Golden Rule of Programming #1. For any important piece of information, define a variable in your program.

Let's see another example of a simple yet common algorithm. Searching is a very common task in many programs. We all have used Google or another search engine to find websites that have certain information, just like we have searched for a word inside a document. Intuitively, we know how to do this: we go through the text and compare the words with what we are looking for. But how do we describe this to a computer so that it is well defined and solution oriented?

In order to search, first, we need to know what we are looking for. As you can see in the sample code below, we create a variable called KEYWORD and give it a value (for example, entered by the user). Then, we need a loop to read a word from the file and compare it to the keyword (lines 3–4). The tricky part of this process is that we don't know how many times we have to repeat. In such cases, our condition will be something like "repeat until we reach the end of the file," or "repeat while we have not reached the end of the file," as shown in line 2 below.

SAMPLE CODE #11

1. Get KEYWORD

2. While Not End-of-file

```
3.      Read WORD

4.      If WORD == KEYWORD

5.            Print "FOUND"
```

Note the use of == (comparison operator) in line 4. Make sure you don't confuse this with the assignment operator (=).

☑ **Practice Task:** Modify the search sample code so we can look for multiple keywords?

✋ **Reflective Questions:** Do you feel this example is giving you an idea of how search engines work? What is missing? How would you expand the example to get a more advanced search process? What other topics do you think you need to learn about before you can have a real web search program?

2.3.6 Loop Counter as a Variable

In the Sample Code #10, we have a variable as part of the "IF" statement that counts how many times we perform the operation. For now, it is used only for the purpose of counting. But this information can be used in many different ways. Imagine a program that wants to calculate the sum of all numbers from 0 to N (any given value). This will be very similar to the Code #10, except that we don't have the Get operation because we know that the numbers are 0, 1, 2, ..., N. Instead what we need is a value that starts at 0 and at each iteration increases by 1, and that is COUNT.

SAMPLE CODE #12

```
1. SUM = 0

2. For COUNT=0; COUNT < 50; COUNT++

3.      SUM = SUM + COUNT
```

☑ **Practice Task:** Modify sample code #12 so that it adds all the number from 10 to 50.

☑ **Practice Task:** Modify sample code #12 to do factorial. The factorial of N is the multiplication of all numbers from 1 to N.
   Hint: Note that the initial value can no longer be 0.

✋ **Reflective Questions:** How can you expand these examples to do other tasks in a similar way?

If you look at the examples we discussed in this chapter, you will notice that most of them start by identifying the variables we will use in the algorithm. This makes sense because a program is usually a set of operations that process certain information. Next, I will discuss this data-centered approach to problem-solving, which we will use in most examples of this book.

2.4 DATA-CENTERED APPROACH TO PROBLEM-SOLVING AND ALGORITHM DESIGN

Algorithms for programming are like sketches for artists. They are essential design tools for establishing what you want to do before you start writing code. The software design process is very similar to the visual design process we discussed earlier. Just like we start drawing a scene by observing and identifying the main objects, the algorithm design starts by "visualizing" the program and identifying the main elements.

These are both information and operations, but since most programs are created to process information, it is a good practice to start your design by identifying the main data elements and then move to what operations you need to perform on that data. This is what I call the **Data-Centered Approach** and is inspired by the visual design process we discussed in the previous section.

Remember from the Introduction that each software project has to start by understanding what the program needs to do, or what we called requirement analysis. The Data-Centered Approach is a simple way of doing requirement analysis for a program. It is based on the assumption that a program is a means of processing information, and as such, its requirements can be defined using the information it uses and the operations it performs on the information. The Data-Centered Approach has the following main steps:

1. Visualize your program. Use drawing, doodles, charts, or any other method that works for you, but try to "see" the code as a combination of elements.

2. Identify the main data items and how they are related.

3. Create a variable for each of your data items.

4. Describe the operation performed on your data through a "rough" algorithm. A good starting point for knowing what your code should do is to answer these **variable-related questions** or **Three HOW Questions (3HQ)**:

   1. How is each variable initialized?

   2. How does the value of each variable change?

   3. How is each variable used in the program/algorithm?

5. Refine the program by adding details and fixing issues

The concept of a "rough" algorithm in Step 4 is the same as an artist's sketch. Its main purpose is to establish the basic structure not to define all the details. It will be later refined to have a more detailed algorithm, followed by the actual program, which itself will go through refinement. We will talk about this refinement process in the next section, but first, we look at some examples of the data-centered approach.

> ☞ **Key Point:** Data-Centered Approach to programming defines the program in terms of its main data items and the operations performed on them. It is a simple way of doing requirement analysis.

To start, let's revisit our GPA code, Sample Code #8. This simple program is easy to visualize (Exhibit 2.5). The main data elements are grades, COUNT, SUM, and GPA. The relationship between these is very straight-forward: COUNT is related to the total number of grades; SUM is related to the added value of the grades, and GPA is related to SUM and COUNT. We can simply create a variable for each one of them. Later, we will talk about naming variables, but for now, let's just use something intuitive.

Steps 1–3 of the above process deal mainly with "what we have," that is, our information and entities. Step 4, on the other hand, deals with "what we do," the operations. The questions 4.1, 4.2, and 4.3 are good starting points because the operations we perform are generally what needs to be done with our data. Answering those questions creates the basic skeleton of the program, if not all, of it.

The code initializes (gives initial value to) SUM and COUNT. A simple rule is that if the first time we use a variable, it is getting a value, then we don't need to initialize it. For example, grade (variable A) and GPA don't need an initial value. But if the first time the variable is used, it will be part of any other operation, then it needs a value. For example, if COUNT has no initial value, the operation "COUNT = COUNT + 1" cannot be performed because it cannot calculate COUNT + 1.

Our next question is about when the variables change. The answer to that was in a loop where the grade is received from the user or read from a database, and SUM is updated. This results in line 3 and its group (3a-3c). GPA is changed (or set) in line 4.

Finally, we need to see where each variable is used. Grades are already used in calculating SUM. COUNT is used in line 3 for the loop. SUM will be used later in line 4 to calculate GPA, which itself is used in line 5 to make the final decision of the algorithm and perform the main task.

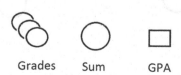

Grades   Sum   GPA

EXHIBIT 2.5   Visualization of the GPA program.

☞ **Key Point:** For each variable in your program, ask the three variable-related or "three HOW questions" (3HQ): how to initialize, change, and use.

✍ **Reflective Questions**

- Can you identify the answers to the 3HQ in the sample codes we have in this chapter?
- Do you think by answering these questions, most of the code will be in place?
- Does that show you the value of these questions and the data-centered approach?

2.4.1 Finding Min and Max

In many programs, we deal with a series of numbers and are required to find the minimum and maximum values. Again, while intuitively quite simple, it is important to have an algorithm that is clear and leads to a solution all the time. We start by visualizing our program. It consists of a set of numbers (values) but also includes two explicitly separate numbers that represent the min and max values. When we talked about creative problem-solving, I mentioned that problem definition and presentation are important skills. Exhibit 2.6 shows two different representations of this problem.

While conceptually the same, 6a hides max (or min) as a number within a series of data. As such, we need to know how to have all the data and access any particular number within that series. 4b, on the other hand, defines max as a separate data item. If we have max or min separately, the problem of how to keep and access the series of data no longer exists. Assuming that we read the sequence of numbers one by one, we only need to get the next number and update min and max if needed. Once used, the input data can be discarded, and the variable can get a new value from the next data.

First, let's see how we can do this for only two numbers. We have three pieces of information: two input data items and one to hold the max value. The logic is very simple; we assume max is the first data; then, if the second number is bigger, then we update max.

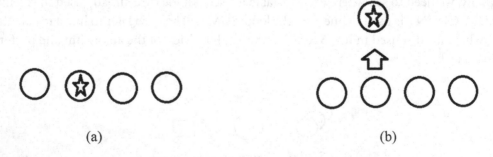

(a)   (b)

EXHIBIT 2.6 Identifying the maximum value in a series of data (item with a star). (a) It is shown as one of the data items. (b) It is a separate data and gets its value from one of the data items.

SAMPLE CODE #13

```
1. Get DATA1

2. MAX = DATA1

3. Get DATA2

4. If DATA2 > DATA1

5.     MAX = DATA2
```

It makes sense to initialize max with the first number because when we only have one number (first item), max has to be that.

Now, let' see what happens if we have a series of numbers and also how we can do both max and min.

SAMPLE CODE #14

```
1. Get DATA

2. MIN = DATA

3. MAX = DATA

4. While MORE DATA

5.         GET DATA

6.             If DATA > MAX

7.                 MAX = DATA

8.             If DATA < MIN

9.                 MIN = DATA
```

Note that the name of variables is arbitrary. We talk about naming in more detail later. For now, assume that we try to use something simple and meaningful.

☑ **Practice Task:** How can we modify Sample Code #14, so we only need to have one "Get DATA" instruction?
　　Hint: Remove the "Get DATA" outside the loop, and find out what should be the initial value of MIN and MAX.

🖐 **Reflective Questions:** How does the MIN/MAX algorithm compare to the search algorithm that we saw earlier? What is common between them? What other variations can you imagine?

2.5 STEPWISE REFINEMENT FOR ALGORITHMS

Problem-solving is a cyclical and iterative process. We hardly reach our final solution in one attempt. Think about the problem of "what to make for dinner?". It is quite common in my home as my wife and I and our child can be quite particular about our interests but don't necessarily have the same ideas. When dealing with this problem, you probably start by some basic requirements that guide the solution: "We should use what we have in the fridge." Or "It should include some protein and some vegetables." Combining your requirements, you will end up with your first rough solution: Chicken with mushroom and eggplant.

This gives you a general idea, but then you would need to know more details, so you decide on stir-frying them, and then, you choose and add spices and cooking sauces to the solution, and possibly a side of rice. And voila, you have your final dinner. It would have been counterproductive to spend too much time choosing spices at first when you don't know your ingredients yet. A fish spice may not work well on your chicken, or the sauce you end up using may already be too spicy to add more pepper.

Similarly, an artist follows an iterative process to finish a design. She first establishes the basics, parts and their relation, and then, details are added in multiple cycles, as shown in Exhibit 2.7. Final details of the face can only be added once we know how the whole body looks like as the expressions should match the body.

Algorithm and program design follow a similar process. There is no point thinking about how to format the print-out in Sample Code #11 until we know what the output is. In the Sample Code 12, our program relies on getting data from a file (or another source). Anyone using a computer knows that sometimes reading data from a file encounters problems such as a corrupted file or access restrictions (you must have seen the infamous "Error 404: File Not Found" online). The program should know what to do if something like that happens, but that detail is less important than the main flow of instructions, as shown in the sample code. Finally, a task may be made of sub-tasks. Getting data from a file may include opening the file, reading the data, and then closing the file.

In general, you can think of the program as a series of important "what" questions (as in "what to do?") plus some "how" (as in "how to do it?") and "what-if" (as in "what if this happens?") questions. There is no point dealing with how and what-if questions if we have not answered our "what" questions yet, and these questions all come in priorities based

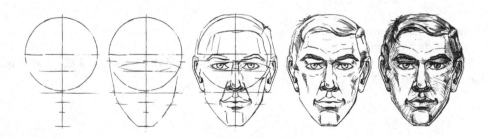

EXHIBIT 2.7   Stepwise refinement.

on their dependence. The process of iteratively going through the code and adding details based on our questions is called "stepwise refinement." Let's learn about it more through some examples.

Consider the relatively simple task of calculating the class average after a quiz. Our first attempt at solving the problem is to establish the general process:

```
Initialize variables

Input, add and count

Calculate and print
```

This clearly doesn't say much but gives us the general idea of what to do. Next, we can add details to each part.

```
Initialize sum to zero

Initialize counter to zero

Input the first grade

While input valid

        Add to sum

        Increment counter

        Input next grade

Set average equal to sum divided by counter

Print
```

This looks a lot more detailed but still is missing some things in Step 3.

```
If count is not zero

    Set average to total divided by count

    Print average

Else

    Print "no grades entered"
```

☞ **Key Point:** Start with a rough "big picture" or "basic" version of your algorithm. This should include the main data and operations. Then, add details when needed.

2.5.1 Restaurant Bill Calculation

You've probably heard the saying, "if you can calculate the tip in a restaurant, then you can do this." I said it myself at the beginning to convince you that you can code. However, calculating the tip still requires some thinking and problem-solving. And there is more to calculating the whole bill. Let's use our stepwise refinement approach to design a program for calculating the total bill in a restaurant.

How do we find the total amount we owe after a meal? Adding up all the item prices sounds like a good starting point.

```
1. Get P1 (price for meal item 1)

2. Get P2 (price for meal item 2)

3. SUM = P1 + P2

4. Print SUM
```

Unfortunately for us, there is always tax. So, we modify this to include that, assuming an 8% service tax:

```
1. Get P1

2. Get P2

3. SUM = P1 + P2

4. TOTAL = SUM * 1.08

5. Print TOTAL
```

We seem to have forgotten about the tip, didn't we?

```
1. Get P1

2. Get P2

3. SUM = P1 + P2

4. TOTAL = SUM * 1.08

5. Get TIP

6. TOTAL = TOTAL + TIP

7. Print TOTAL
```

Now the program looks complete as it answers all the "what" questions we had. However, no software development project is ever completely finished. Just think about the apps on your smartphone and how often they update. We may always find bugs (existing behaviors that were not intended) or new features (desired behaviors that were not planned). For example, there are some rules for calculating the tip. We could use them and simplify the program for the user. That is a "how" question.

```
1. Get  P1

2. Get  P2

3. SUM  =  P1  +  P2

4. TOTAL  =  SUM  *  1.08

5. TIP  =  TOTAL  *  0.15

6. TOTAL  =  TOTAL  +  TIP

7. Print  TOTAL
```

We can continue refining this by asking more "how" questions. This time, we ask the user for the tip percentage and then do the calculation.

```
1. Get  P1

2. Get  P2

3. SUM  =  P1  +  P2

4. TOTAL  =  SUM  *  1.08

5. Get  TIP _ PERCENT

6. TIP  =  TOTAL  *  TIP _ PERCENT

7. TOTAL  =  TOTAL  +  TIP

8. Print  TOTAL
```

Now, we can have a "what-if" question. What if we want this to be repeated for all customers?

```
1. While  Customers

2.      Get  P1

3.      Get  P2

4.      SUM  =  P1  +  P2

5.      TOTAL  =  SUM  *  1.08

6.      Get  TIP _ PERCENT

7.      TIP  =  TOTAL  *  TIP _ PERCENT
```

```
 8.      TOTAL = TOTAL + TIP
 9.      Print TOTAL
```

Now, let's ask, "what if we have more than two items?"

```
 1. While Customers
 2.      Get N (number of items for this customer)
 3.      SUM = 0
 4.      For COUNT=0; COUNT < N; COUNT++
 5.           Get Price (price for meal items)
 6.           SUM = SUM + Price
 7.      TOTAL = SUM * 1.08
 8.      Get TIP _ PERCENT
 9.      TIP = TOTAL * TIP _ PERCENT
10.      TOTAL = TOTAL + TIP
11.      Print TOTAL
```

You can even imagine more "what" questions to significantly change what the program does. Each one of these questions (and their answers) will refine our algorithm, add more detail, and get us closer to what we consider satisfactory. At that point, the algorithm can be passed to a software programmer for actual coding. Note that the programmer can be the designer or someone else. Either way, programming involves translating this algorithm into a formal language. The process of refinement doesn't end, though. The programmer (and the designer) will continue adding details, fixing errors, and possibly adding features. We will talk about the software development process in later chapters, but our next step is to learn more about computers and how real programs work.

☑ **Practice Task:** Consider the task of creating a simple calculator.

- Start by a simple case of asking the user to choose an operation and two operands. Then, perform the operation and show the result.
- Then, add a loop to keep doing this until the user chooses to end.
- Finally, add error checking if the user enters zero for the second operand of a division.

> ✋ **Reflective Questions:** Think about the stepwise refinement process.
>
> - How easy is it to identify the "basic" parts to do first?
> - Do you feel the stepwise refinement adds value to algorithm design?
> - Can you consider some existing programs and imagine the stepwise refinement process for them?
> - Do you see the relationship between stepwise refinement and incremental release of features that we see in modern apps? All the refinements don't need to happen in one version.

HIGHLIGHTS

- Software design and programming are primarily problem-solving tasks. So, we can use established problem-solving methods for them.

- The following visual design process can help with programming:

  - Observe the problem space

  - Identify major entities

  - Sketch the basic structure

  - Add details and refine.

- Programs have two general aspects: syntax and logic.

- Information (a.k.a. data) in a program or algorithm is generally stored in "variables" that have name and value.

- The = sign is called the assignment operator. It assigns the value of the right side (r-value) to the variable on the left side (l-value).

- The sequential execution is the basis for the logical flow of programs and dependency of each operation on what came before.

- In programming, a language construct is a syntactically allowed combination of language elements that performs a certain task.

- Conditional statements are constructs that implement selection. If is the most commonly used conditional statement that has a condition and a body of code that runs if the condition is true.

- Loops are constructs that implement iteration. A while loop has a similar structure to an if statement (condition plus body of the code), but repeats the code as long as the condition is true.

- `for` loops are similar to while loops, but in addition to the condition, they have an initial statement (runs only once) and an ending statement (runs at the end of each iteration).

- Golden Rule of Programming #1. For any important piece of information, define a variable in your program.

- For each variable in your program, ask the three HOW questions: how to initialize, change, and use.

END-OF-CHAPTER NOTES

A. Things I Should Mention

- I took a course on Urban Sketching in Ottawa School of Art a few years ago, and my instructor helped me establish the analogy between the visual design and algorithm design processes.

- In a sense, software design is like an art. To learn it, we first need to learn the craft, and how to use tools and perform basic tasks. Pure programming is more like a craft. This doesn't mean that craft people don't make design decisions or have no artistic talent. But the real beauty comes from a good design. The line between design and coding is not very clear. Most of the time, the programmers have to revise and refine the design and write algorithms. The tasks are quite overlapped or even done by the same person.

- Divergent thinking and presenting a problem in different forms are essential skills for software developers. They have to be coupled with recalling previously experienced patterns to allow creative problem-solving. Good software designers rely on these skills all the time.

- Experienced programmers may not explicitly write algorithms, but they do that in their minds or rely on well-established algorithms that they are familiar with.

- I sometimes summarize the data-centered approach in two steps: what we have (data) and what we do (operations). Feel free to use this short version if it works better for you.

B. Self-test Questions

- What are the common problem-solving methods?

- What is an algorithm?

- How does the assignment operation work?

- How do we control the linear flow of a program?

- What is a data-centered algorithm design?

- What is the Golden Rule of Programming #1?

- What questions do we ask for each variable?

- How do what-if and how questions help refine an algorithm?

C. Things You Should Do

- Check out these websites:

  - https://teachinglondoncomputing.org/resources/developing-computational-thinking/algorithmic-thinking/

  - https://www.coursera.org/learn/algorithmic-thinking-1

    - Coursera is a great place to take online courses.

- Take a look at these books:

  - *Think Like a Programmer* by V. Anton Spraul

  - *Thinking and Problem Solving* by Robert Sternberg

D. Reflect on the Experience of Reading This Chapter

- What did you expect from this chapter before reading it?

- What was it about, and what did you learn?

- What tasks did you perform, and what difficulties did you face?

- How did you feel about the material and tasks presented in this chapter?

- How can you improve your learning experience?

- How do you see this topic in relation to the goal of learning to develop programs?

- What is your approach to problem-solving?

- How comfortable are you with defining an algorithm for your everyday tasks?

- Do you see the point of visualizing a problem?

- Do you feel the algorithm design questions help streamline the design process for you?

PART 2

Understanding Programs

There are two ways to write error-free programs; only the third one works.

GOAL

In the previous part of this book, I introduced algorithmic thinking as the way we organize our logic to solve a problem.

In this part, I aim to show how we present that logic formally by discussing the two main elements of a program: data and code. While the first part of the book covered how programmers think, this part shows how they talk.

PART 2

Understanding Programs

Data

Program's Information

OVERVIEW

In the last three chapters, I laid out the foundation we need for programming. With the basic understanding of how computers work and how to think algorithmically to solve a problem, the next step is to learn how to present our algorithms in formal ways that computers can read and execute. That is through programming languages.

The choice of a language is not always made by the programmer since there are organizational requirements and other factors. But in general, as a programmer, you need to know the common categories of programming languages and their advantages and drawbacks. I start this chapter with a quick review of the common types of languages we work with. I focus on three important languages: C/C++, Javascript, and Python. They are very good examples of the languages you may be using later and show common trends and variations in programming language design.

As I mentioned in the previous part of this book, programs are made of two basic parts that are conceptually distinct but may be mixed together in the program: data and code. In this chapter, I focus on the representation of data in our selected programming languages. A program is mainly a set of instructions to process data. So, it is important to start the design of a program by proper selection and representation of that data. The key concept here is what we call a **variable**. In the previous chapter, while discussing algorithms, I introduced the Golden Rule of Programming #1 and the need to clearly identify data items. A variable is a data item of any type that is identified with a name and has a value that can change over the course of the program.

Understanding the use of variables and the methods of presenting and operating on data prepare us for the next chapter, which is about code and basic ways of processing data.

3.1 HIGH-LEVEL PROGRAMMING LANGUAGES

In the previous chapter, I discussed the concept of logical creativity. We saw how algorithms could help us define a solution logically and clearly. Algorithmic thinking is the foundation of programming. Once we master the ability to think algorithmically, we have taken the most important step for writing programs. But as we saw at the beginning of this book, as much as the thinking process is important, we still need certain rules and conventions to express our thoughts in a way that a computer understands. Programming languages are collections of such rules and conventions.

Let's revisit the program we started with, considering what we have learned so far. **The line numbers for all sample codes in this book are just for reference purposes. The actual code should not include the line numbers.**[1]

Sample Code #1

```
1. <p>Hello, World!</p>

2. <p id="demo">?</p>

3. <input id="data">

4. <button onclick="demo.innerHTML=data.value*2;">Double</button>

5. <script>

6.    document.write("End of page");

7. </script>
```

---

[1] To have a digital copy of the source code files, which you can use directly, go to the Companion Website. See the information at the start of the book before the Introduction.

This program is written in HTML and Javascript, both primarily designed for describing web pages. The Javascript code can be embedded inside an HTML file, as shown above, or be a separate file with the extension .js. For example, we can have the following text in the HTML file:

```
1. <script src="code.js">
2. </script>
```

Files containing source code (human-readable code as opposed to machine code) are usually called **source files**. Table 3.1 shows the source file extension for the programming languages used in this book.

There are a few important things about Sample Code #1 and other programs in HTML and/or Javascript that we should pay particular attention to. They help us identify different categories of programming languages.

3.1.1 High Level vs. Low Level

HTML and Javascript programs are not machine-readable or written for a specific CPU. We refer to HTML, Javascript, and other languages that have this feature as **high level**. In contrast, machine code and Assembly languages that we discussed in Chapter 1 are **low level**.

- Machine code is binary and can be read directly by CPU.

- Assembly program is human-readable text but written for a particular CPU.

3.1.2 Declarative vs. Imperative

HTML programs don't describe the operations that the computer has to perform. Instead, they describe the desired output. We refer to such languages as **Declarative**. Languages such as Javascript, on the other hand, describe the operations (the process of getting to that desired output). We refer to them as **Imperative**. In Sample Code #1, we see how a Declarative language is used to describe the known part of the output, and an Imperative language is used to describe the operations that create a dynamic output: the paragraph text changes when we click on button depending on what we have entered in the edit box.

TABLE 3.1 File Extensions for Common Source Files

Language	File Extension	Comments
HTML	.html	Can include Javascript and other languages
Javascript	.js	
C	.c	
C++	.cpp	Can be used for C programs too
Python	.py	
C#	.cs	
Java	.java	

3.1.3 Compiled vs. Interpreted

Neither HTML nor Javascript code runs as a standalone application. Another program reads every line of them, translates it to machine code, and pass that to CPU to run. This is in contrast to the browser that runs them or Microsoft Word that I used to write this manuscript. These programs are translated into Machine Code as a whole and then run as a standalone application. We refer to HTML, Javascript, and similar languages as **interpreter-based** or **interpreted**. The interpreter is the program that reads and executes the code.

Python and Basic are other examples of popular interpreter-based languages. The interpreter can be an application that uses the interpreted code to enhance its operation (such as the web browser) or an application whose sole purpose is to run the interpreted code. In the former case, we sometimes refer to the interpreted code as **script** and writing such code as **scripting**.[2]

There are other languages that require the whole program to be translated into Machine Code and saved as a new file, which will then run as a standalone program. This act of translation to Machine code is done for a particular "platform," which is a combination of CPU and the Operating System (OS). For example, the machine code generated for Windows on an Intel CPU will not work on Linux on the same CPU or Windows on a different CPU type (although many CPUs are compatible). The translation to machine code is called **compiling**, and these languages are called **compiler-based** or **compiled**. C and C++ (pronounced "C Plus Plus") are popular examples of this group of languages. Compiled programs are generally faster because translation happens only once, and then, the machine code will be executed every time we run the program. Compiling is also a more efficient way of translating than interpreting due to various optimizations that are beyond the scope of this book.

3.1.4 Virtual Machines

Programming languages such as Java and C# belong to a third category called **Virtual Machine (VM)-based**. Programs in these languages are compiled but not for a real CPU. The target CPU is a Virtual Machine. Every real computer that needs to run these programs requires a program that implements the VM and then runs the programs within the VM.

☞ **Key Points**
- High-level languages are human-readable and not CPU-dependent.
- Declarative languages describe the output, but Imperative languages describe the operation.
- Interpreted languages are read and executed by a standalone program, but compiled languages are translated into standalone machine code.
- Virtual Machines run programs that are compiled but not for any specific target platform.

---

[2] For example, Visual Basic can be used as a scripting language for Microsoft Office, same as Javascript for most browsers, or Python for Maya (a modeling and animation program).

SIDEBAR: WEB AND VIRTUAL MACHINES

In the early years of the Internet, the web pages were fairly static HTML-based content. web designers very soon realized that the sites required user interaction. For example, you search for a subject and need to see a dynamically created page with your customized search results. Any online form, similarly, required a customized set of information to be sent to the web server (form content) and a customized response to be received by the client (the browser). A common approach for creating custom content and allowing user interaction was server-side executables through standards such as Common Gateway Interface (CGI) that allowed the form content to be processed by a program running on the server computer. This program (written in C/C++ or other languages) would then send back a response in the form of a new web page.

Server-side executables are very powerful but mean that all interactions with the user have to happen through the web server. To allow browsers to have local operations, different methods of client-side executables were developed. For example, Microsoft implemented ActiveX objects, written in C/C++ and any other supporting language. These objects were special executable code that would be embedded in the web page, downloaded by the browser, and then run on the client computer. The first problem with this approach was security, as a C/C++ code could pretty much do anything on the client computer. The second problem was compatibility; a code compiled for Windows could only run with browsers running on Windows. **Virtual Machine (VM)** is a solution that addresses both these problems.

Java programming language was created in the early 1990s and released in 1996 by Sun Microsystems. Java programs are compiled for Java Virtual Machine (JVM) and can run on any computer that has JVM installed, a feature that is commonly referred to as Write Once, Run Anywhere (WORA). Java also included various security measures by limiting what the programs can do. Removing many "low-level" features such as direct memory/disk access and adding various permission mechanisms, Java program is a lot safer than C/C++, although that comes at the cost of reduced performance.

Later, Microsoft released **.NET framework** as its own VM and created **C#** as the native language for it. An open-source version framework compatible with .NET, called Mono, was also developed that help spread this VM-based approach. Many programs such as Unity Game Engine use this VM, and as such, the programs made with them can be executed on any platform.

Throughout this book, you will see many examples of programs in Javascript, which is a very common language, not just for the web but also for other applications. Almost all these examples are integrated into a web page, and as such, you may need to know a little bit about HTML. The basic information you may need to have about HTML can be found in many online and printed resources such as those mentioned in "End-of-Chapter Notes," but we will rarely need anything more than the paragraph, button, and edit box that we have used so far. Otherwise, the focus of this book is on Imperative languages, which are what most people consider when talking about programming. In addition to Javascript, the examples in this book are from Python and C/C++. These languages are chosen as they are commonly used and are good representatives of interpreted and compiled languages. Also, C and C++ are the basis of many other languages such as VM-based Java and C#.

3.1.5 Development Environments and Tools

To write programs using different languages, we need a certain number of tools. These, in general, provide the ability to

- Write the text with common abilities such as edit, find, and replace, and more advanced ones such as highlight, suggest, and auto-correct.

- Manage all the source and data files related to the program.

- Build the executable code and show possible errors in the program.

- Run the program and show the results, memory usage, and other information that can help improve the code.

- Collaborate with a team of developers.

- Control different versions of the code.

While this is a long list and can be even extended, the minimum requirements for writing programs can be much less. For example, we saw that to create HTML and Javascript programs, we only need a text editor. Running the programs needed an interpreter, which by default would be a web browser. There are text editors specifically designed with programmers in mind. They can simplify programming with features such as highlighting a different part of the code with different colors (called "syntax highlighting") and formatting the text according to language rules and practices. NotePad++ (https://notepad-plus-plus. org/) is a free open-source text editor for Windows, and TextMate (https://macromates. com/) is one for Mac. There are also those such as UltraEdit (http://www.ultraedit.com/) that support multiple operating systems.

On the other hand, more advanced tools exist that bring many more functionalities together, particularly building and running the code. Such tools are referred to as the Integrated Development Environment (IDE). While different graphical and command-line tools exist for editing and compiling programs, it is now common to use IDEs for this purpose due to the complexity of the process, especially in languages like C and C++. Microsoft Visual Studio (https://visualstudio.microsoft.com/) is one of the most popular IDEs that is available as a paid professional version and also a free community version. Visual Studio supports developing programs in multiple languages, including HTML, Javascript, C, C++, and Python, which I use primarily in this book. Other popular IDEs include XCode (https:// developer.apple.com/xcode/) for Mac and Eclipse (https://www.eclipse.org/), both of them free to use. Adobe DreamWeaver (https://www.adobe.com/products/dreamweaver.html) is a commonly used commercial IDE for web development. You may use most of these tools for the majority of examples in this book, but all examples that include information about the development tool assume the use of Visual Studio. I will discuss many of the common features of IDEs throughout the book (Exhibit 3.1).

☞ **Key Points**
- Integrated Development Environment (IDE) is a tool that allows writing, editing, compiling, running, and debugging programs.
- Visual Studio is the main IDE used in the examples of this book, although most examples can be used with other IDEs as well.
- Some languages, such as Javascript, can be easily used with only a text editor and a web browser. Others like C/C++ can also be written with a text editor and compiled using a command-line program, but using an IDE is strongly recommended.
- Discussing the command-line method for compiling C/C++ programs is beyond the scope of this book.

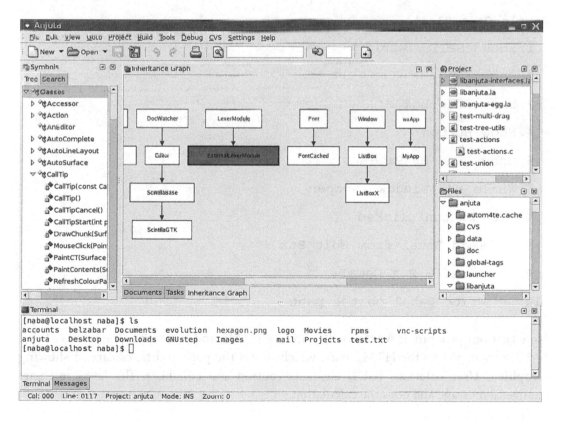

EXHIBIT 3.1   A typical IDE.

```
                    ┌─────────────────────────┐
                    │      Displayed Data      │
                    └─────────────────────────┘


    ┌──────────────────────────┐   ┌──────────────────────────┐
    │   Input Data (Edit Box)   │   │          Button          │
    └──────────────────────────┘   └──────────────────────────┘
```

EXHIBIT 3.2   Visualization of Sample Code #1.

3.2 DATA AND VARIABLES

The primary job of computers is information processing. As such, we defined programs as a set of operations (code) performed on information (data). In the Introduction, we wrote a simple program that was revisited in the previous section (Sample Code #1). The operations in this program included showing some messages and also performing calculations. Its data, on the other hand, included the messages, the number entered by the user, and the number calculated by the program to be shown on the page. Exhibit 3.2 shows the visualization of this program representing main data elements, and the simplified algorithm is as follows:

Sample Code #1-a[3]

```
1. Show "Hello, World!"

2. Show "0"

3. Show Edit Box

4. Show Button

5. Show "End of Page"

6. While the window is open

7.    If button clicked

8.       Read Data1 from Edit Box

9.       Data2 = 2 * Data1

10.      Write Data2 on the page
```

Note that our program is written with a mix of HTML and Javascript. The lines 1–4 and 6–7 are managed by the HTML part, which draws the page and takes care of showing **Graphical User Interface (GUI)** elements (button and edit box). The Javascript part changes a message and does the calculation.

---

[3] Throughout the book, if a sample is algorithm for another sample, I identify them with the same number and a suffix -a for the algorithm. Similarly, if multiple versions of a code are given in different languages, I identify them with -j, -p, and -c for Javascript, Python, and C/C++.

In Chapter 1, I continued the view of programs as a combination of data (information) and code (operations on the data) within the context of games, and in Chapter 2, I expanded this idea and introduced the data-centered design approach (seeing the program as "what we have" and "what we do"). I also introduced what I called the **Golden Rule of Programming #1:**

> For every piece of information, define a variable that has a name, holds that information, and can be used (and change) later.

With that approach in mind, let's identify which variables our program is using ("what we have"). The variables, in this case, are those HTML elements that have a name (id):

- The variable demo is a paragraph, and demo.innerHTML is its text. We assigned a variable to it because it changes. The other paragraph that has the text "Hello, World!", on the other hand, has no associated variable because it never changes, and we never refer to it. Note that even if we need to read the value of something later, we should assign a variable (a name) to it.

- The variable data is an edit box, and data.value is its text. In the program, we need to read the value, so we assign a variable to it.

- All Javascript programs that are embedded inside HTML have a variable called document that refers to the web page. Note that if a variable has multiple parts, to access one of them, we use a . symbol, such as demo.innerHTML, data.value, and document.write (used in the sample code given in Introduction). This is common among most programming languages.

☞ **Key Point:** Golden Rule of Programming #1: For any important piece of information, define a variable.

Now imagine we want to calculate a total price plus tip in a restaurant or store. For simplicity, assume we only have two types of items and can order multiple of them, and that there is no tax. The first type costs \$5 per item, and the second \$2. We modify the Sample Code #1 to have demo displaying the result with two edit boxes (data1 and data2) for the number items of each type and two buttons. The first button calculates the total price, and the second adds an optional tip.

Sample Code #2

```
1. <p>Hello, World!</p>

2. <p id="demo">0</p>

3. <input id="data1">

4. <input id="data2">
```

Hello, World!

total price is 8.4

| 1 | | 1 | | Price | Add 20% Tip |

EXHIBIT 3.3   Output for Sample Code #2.

```
5. <button onclick="demo.innerHTML = 'total price is ' + (data1.
   value*5 + data2.value*2);">Price</button>

6. <button onclick="demo.innerHTML = 'total price is ' + (data1.
   value*5 + data2.value*2) * 1.2;">Add 20% Tip</button>
```

Don't be alarmed by the longer text for the buttons. The only different thing is the onclick code:

```
"demo.innerHTML = 'total price is ' + (data1.value*5 + data2.
   value*2);"
```

It displays a message such as "the total price is 15." Note that the whole code for onclick has to be inside a " " pair (it is an *attribute* for the button *element*) and it includes a text itself ("the total price is"). To avoid confusion and error, the internal quote is single '. Exhibit 3.3 shows the output for this code.

You notice that our program uses similar calculations with data1 and data2 for both buttons. The first button calculates the price and then adds it to the text ("the total price is") and writes to the page. When we click on the second button, the calculated price is no longer available because we didn't store it anywhere. So, we need to recalculate the price.[4] This is an example of the usefulness of Golden Rule #1. Price is a piece of information we calculate once (when clicking on Price button) and need later (when clicking on Add Tip button), so we should assign a variable to it.

Sample Code #3

```
1. <p>Hello, World!</p>

2. <p id="demo">0</p>

3. <script>

4. var price=0;

5. </script>

6. <input id="data1">

7. <input id="data2">
```

---

[4] We could extract the price information from the paragraph text, but the calculation would be even more complicated.

8. `<button onclick="price = data1.value*5 + data2.value*2; demo. innerHTML = 'total price is ' + price;">Price</button>`

9. `<button onclick="demo.innerHTML = 'total price is ' + price*1.2;">Add 20% Tip</button>`

Line 4 of the above code shows how to add a variable to a Javascript code. It starts with the word var, followed by a name and an optional value. The word var is pre-defined in the language, as opposed to names such as demo that we define in our code. The pre-defined words are called *keyword*, and when using a language, it is a good idea to be familiar with the common keywords. One of the advantages of using modern IDEs compared to a regular text editor is that they use a technique called *syntax highlighting* that shows different parts of the code with different colors, as illustrated in Exhibit 3.4.

In Chapter 2, when I first introduced the concept of variables, I listed **three variable-related questions** (or **Three HOW Questions**) you should be asking when defining a variable:

- How is the variable initialized?

- How is the variable changed?

- How is the variable used?

```
BlankCordovaApp1
BlankCordovaApp1 JavaScript Content Files                          <global>
  1      // For an introduction to the Blank template, see the following documentation:
  2      // http://go.microsoft.com/fwlink/?LinkID-397704
  3      // To debug code on page load in cordova-simulate or on Android devices/emulators: launch yc
  4      // and then run "window.location.reload()" in the JavaScript Console.
  5    (function () {
  6        "use strict";
  7
  8        document.addEventListener( 'deviceready', onDeviceReady.bind( this ), false );
  9
 10        function onDeviceReady() {
 11            // Handle the Cordova pause and resume events
 12            document.addEventListener( 'pause', onPause.bind( this ), false );
 13            document.addEventListener( 'resume', onResume.bind( this ), false );
 14
 15            // TODO: Cordova has been loaded. Perform any initialization that requires Cordova h
 16            var parentElement = document.getElementById('deviceready');
 17            var listeningElement = parentElement.querySelector('.listening');
 18            var receivedElement = parentElement.querySelector('.received');
 19            listeningElement.setAttribute('style', 'display:none;');
 20            receivedElement.setAttribute('style', 'display:block;');
 21        };
 22
 23        function onPause() {
 24            // TODO: This application has been suspended. Save application state here.
 25        };
 26
 27        function onResume() {
 28            // TODO: This application has been reactivated. Restore application state here.
 29        };
 30    } )();
```

EXHIBIT 3.4   Syntax highlighting. Javascript code created and shown in Visual Studio.

Giving an initial value to a variable is optional in almost all programming languages. The general rule is that if the first time the variable is used in your program, it is getting value, then you don't need to give it an initial value. In Sample Code #3, the first time `price` is used is on line 8 when it gets a value (`data1.value*5 + data2.value*2`). So, the initial value of 0 is not required. But in line 9, we do have a calculation that uses the value of the price. While line 9 comes after line 8, when running the program, it is possible to click on the Tip button first. You will get an error from your browser (it will display the value as NaN, Not a Number). In some other programs, using a variable that has no value can cause more serious problems, so it is a good practice to always give an initial value to (initialize) your variables.

Another thing to pay attention to is where the script element with the definition of the new variable is located. It used to be at the end of our code, but in Sample Code #3, I moved it to the top (lines 3–5). If you leave it at the end, you will notice that the program won't work properly. The reason is lines 8 and 9 are using the variable `price`. If it is defined after those lines, the program has no variable with that name when the `onclick` code is running. This brings us to the **Golden Rule of Programming #2**:

> Always define items before you use them.

To understand the reason for this rule, you need to remember that programs are executed sequentially by default. In Chapter 2, we learned about constructs such as selection and iteration that control the flow of the program, so it doesn't always run linearly and in one direction. But regardless of that, always remember to apply Golden Rule #2. We will see later that some languages allow a certain level of flexibility in this rule. For example, Python automatically defines a new variable the first time the name is used, making this rule seems unnecessary. But it is important to remember that the rule still applies but behind the scene, and the main concept will remain valid.

☞ **Key Points**
- In Javascript, we define a new variable using the keyword var.
- Giving initial value to a variable is called initialization.

☞ **Key Point:** Golden Rule of Programming #2: Define before you use!

☑ **Practice Task:** Modify Sample Code #3 and make a simple calculator. You will need two input boxes for operands and four buttons for operations.

Here is another example of using variables in Javascript, based on the algorithm in Sample Code #13 of Chapter 2:

Sample Code #4-a (algorithm for the following Javascript code)

1. Get DATA1

2. MAX = DATA1

3. Get DATA2

4. If DATA2 > DATA1

5. MAX = DATA2

Sample Code #4

```
1. <p>Hello, World!</p>
2. <p id="demo">0</p>
3. <script>
4. var d1 = 90;
5. var max = d1;
6. var d2 = 70;
7. if(d2 > d1)
8. max = d2;
9. demo.innerHTML = "max of " + d1 + " and " + d2 + " is " +
   max;
10. </script>
```

This code is not particularly helpful now because we know the values of data items, but soon we will revise it to work with data that the user enters; that is, lines 4 and 6 will be replaced by methods to get data from the user such as an edit box.

In Javascript and most other programming languages, the keyword if is used to create conditional statements, similar to what we did in the previous chapter. In Javascript, the keyword is followed by a condition within ( ) and then a block of code (one or more lines) that will be executed if the condition is true. This is the programming construct called *selection* or *decision-making* that we reviewed in the previous chapter and discuss in more detail in the next chapter.

☑ **Practice Task:** Modify the Sample Code #4 to enter the grades of 5 students in a class and find the top performer.

Reflective Questions
- What problems happen when we enter many data, such as the above task? In Chapter 2, we discussed iteration. Can you see how helpful it will be to use that instead of writing the same code again and again? We will discuss the iteration and selection in the next chapter.

3.3 PROGRAMMING IN PYTHON

Python is another interpreter-based language. The first version of Python was released in the early 90s. The main idea behind the design of Python language was to make it more human-readable by making it closer to natural writing. As such, Python's syntax is a little different from Javascript, but the basic concepts stay the same.

Before you can write programs in Python, you need to install its interpreter as most computers don't come with Python installed. The good news is that Python is freely available to download and fairly easy to install. We can go to its official website (https://www.python.org) and download the version we need based on the operating system of our computer. See The Companion Website for some details on how to set up Python on your computer.

Python can be integrated with IDEs such as Visual Studio, but for now, let's see how we can use it without any other tool. Use your text editor, and create a new file with the following code. Remember that the line numbers for all sample codes in this book are just for reference purposes. The actual code should not include the line numbers.

Sample Code #5

```
1. print("Hello, World!")
2. d1 = 6
3. max = d1
4. d2 = 10
5. if d2 > d1:
6.     max = d2
7. print(max)
```

Not that Python also uses the keyword if for creating conditional statements. The syntax is slightly different from Javascript, as we have no ( ) but a : at the end of the condition.

Save this text as Hello.py, and then open the command prompt or terminal program on your computer. Go to the folder where you saved the file, and type "py Hello.py." You should see a result similar to Exhibit 3.5.[5]

---

[5] While Python is an interpreted language that runs through its run-time environment (the "py" program), Python programs can be converted to standalone executables. Tools such as Py2Exe can do such conversion.

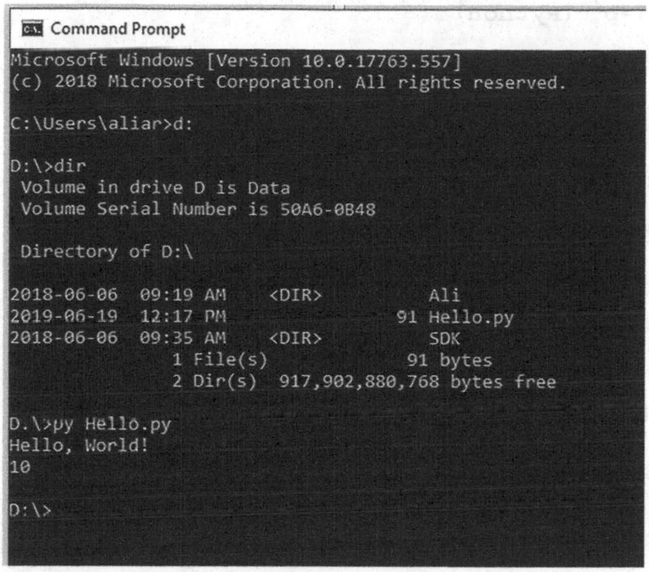

EXHIBIT 3.5 Python output for Sample Code #5.

While the concepts and main logic of the program are the same in Python and Javascript, you can see the following differences in the syntax:

- Python instructions do not end with the ; symbol.

- Python if the statement does not require a ( ) for the condition but has a : at the end

- The variables are created by simply using their name. If you assign a value to a name that does not exist, Python interpreter simply assumes it is a new variable. So, the keyword var is not used.

  - If you use a variable, as in print(max), that is not defined earlier. Python will give you an error. A new variable is created only when you assign a value.

Another difference that may not be quite obvious is the use of indention at the start of different lines. In Javascript, this is optional and only to make your code more readable. But in Python, using indention creates a block of code that is related to the preceding line, which ends with a : symbol.

> ☞ **Key Point:** Python has similar concepts to Javascript, but the syntax is a little different. In particular, there is no ; at the end of each line, no var to define a new variable, and indents instead of { } to define a block of code.

For example, the following Python and Javascript code compare A and B and print messages based on that comparison.

Sample Code #6-p[6] (Python)

```
1. if A == B:
2.    print("A and B are equal")
3. if A > B:
4.    print("A is greater than B")
5. print("end of comparison")
6. print(" Bye ")
```

Sample Code #6-j (Javascript)

```
1. if(A == B)
2. document.write("A and B are equal");
3. if (A > B)
4. {
5. document.write("A is greater than B");
6. document.write("end of comparison");
7. }
8. document.write(" Bye ")
```

Note that

- In Python, regardless of how many lines of code we have in the conditional block of code, we use indention to identify them. Anything that is indented is part of the conditional block.

- In Javascript, a conditional block of code is identified with { } if it is more than one line. If we only have one line of conditional code, we can write without { } or with them.

- Indention and spaces have no meaning in Javascript. They are only for the human reader.

- In both cases, the "Bye" message is printed regardless of the comparison, so it is written outside the block of code for the conditions.

---

[6] Throughout the book, if a sample is algorithm for another sample, I identify them with the same number and a suffix -a for the algorithm. Similarly, if multiple versions of a code are given in different languages, I identify them with -j, -p, and -c for Javascript, Python, and C/C++.

- In both languages,

  - If an operation (like print or write) needs data, it will be given inside ( ).

  - All names and keywords are case-sensitive.

  - We use == to compare two numbers or variables.

Table 3.2 has a list of commonly used operators that exist similarly in both Javascript and Python. The following examples demonstrate the use of some of these operators for detecting if a number is odd or even. I have included the HTML part for Javascript and used the demo paragraph instead of document.write, which is not a common method and can delete the whole page if done after it is fully loaded.

Sample Code #7-j (with required HTML)

```
1. <p id = demo>?</p>
2. <script>
3. var x = 67;
4. if(x%2 == 1)
5.    demo.innerHTML = "odd";
6. else
7.    demo.innerHTML = "even";
8. </script>
```

TABLE 3.2 Common Operators

- +, −, *, and / are four basic arithmetic operations
- = is the assignment operator
- +=, −=, *=, /= are short versions combining = with the four arithmetic operators when the left side is the same as one of the operands on the right side. For example,
  - "X = X + 1" is the same as "X += 1".
  - "X = X / 5" is the same as "X /= 5".
- ==, !=, >, <, >=, <= are comparison operators. They don't change the value of operands and result in true or false.
- % is the modulus or remainder operator. For example,
  - 15% 4 is the remainder of dividing 15 by 4 which is 3.
- In Javascript (and C/C++), we have ++ and −− as the short versions of +=1 and −=1 (increase and decrease by 1). For example,
  - "X++" means "X += 1" or "X = X + 1".
  - **Python does not have the ++ and −− operators.**

Sample Code #7-p

```
1. x = 67
2. if x%2 == 1:
3.   print("odd")
4. else:
5.   print("even")
```

Operators are keywords in the language, which means they are words that the interpreter (or compiler) knows and can turn into machine code. Operators are provided by the language to perform very basic operations that programs commonly need to perform. But if you look at the Javascript and Python examples, you will notice that they both include operations that are not done through operators but commands such as `write` or `print`. These are modules of code pre-written and provided with the interpreter or compiler. We refer to these modules of code as **functions**.

We can also define our own functions. In both cases, the advantage is that we don't have to repeat that code. Instead, we reuse that module by simply referring to its name. Here is another example of two useful functions in Python, which will allow the user to enter a value, turns it into a number, and then sets variable x to that numeric value.

```
x = int(input())
```
[7]

We will discuss this and other functions very soon, but for now, you can use a line like this to get numbers from users in a Python program similar to using an HTML edit box in Javascript.

Functions are essential in organizing the code. While we can use functions in Javascript and Python, in languages such as C and C++, we have to use them. Another different feature in C and C++ compared to Python and Javascript is the use of data types. We will learn about C/C++ programming and data types in the remaining sections of this chapter. Functions will be discussed in Chapters 4 and 5.

☞ **Key Point:** Operators are keywords understandable by compiler/interpreter. Functions are modules defined by the programmer.

☞ **Key Point:** Functions are modules of code that we can define in our program or are pre-defined. We can use these modules in our program, so we don't have to repeat that code.

---

[7] Notice that functions are always followed by ( ). We will see soon what can go inside those parentheses.

☑ **Practice Task:** Use the `int(input())` functions and Sample Code #14 in Chapter 2 to write a Python program that allows the user to enter 5 numbers and show the maximum and minimum values.

✋ **Reflective Questions:** Do you understand the difference and similarity of operators provided by the language (such as those in Table 2) and functions? Why can't a language have only one of these? Operators provide a small base for programming. They cannot be extended as they use pre-defined language "keywords." But functions are there to create new operations. Can you find analogies for this in other areas? Think about modularization that I introduced in the first chapter.

3.4 PROGRAMMING IN C AND C++

C is a very popular compiled programming language and the base for many of the modern languages such as C++, Javascript, C# (pronounced "C Sharp"), and Java (not the same as Javascript). In fact, C++ can be considered an extension to C, and most C compilers are really C++ compilers. For that reason, many programmers simply create C++ programs (files with extension .cpp, instead of .c, which is for C programs) even if they are not using the added features of the C++ language.[8] Throughout the book, I use C/C++ when the program works in both languages. I use C or C++ when I am talking about a feature specific to one of these languages. I also assume you are using Visual Studio on Windows for C/C++ programming.

To write your first C/C++ program, start Visual Studio. See the Companion Website for a review of how to install and use this IDE. Visual Studio provides a hierarchical structure

- The solution is a combination of related projects. The information for a solution, such as the list of projects in it, is stored in a file with extension .sln.

- Project is a combination of source files that are required to create an executable program. Project information is stored in a file with extensions such as .vcxproj (for C and C++) and .pyproj (for Python).

- Source files are individual modules of code

Exhibit 3.6 shows the interface for Visual Studio. The main window shows the code, but other parts of the interface include

- Solution Explorer shows the hierarchical file structure (from solution to project to source and other files). If you don't see it, go to the View menu and select it. You can drag the Solution Explorer and place it at any side of the main window or even have it floating.

---

[8] Using .cpp for a C program does have some minor effects on the compilation process, but most beginners can safely ignore them.

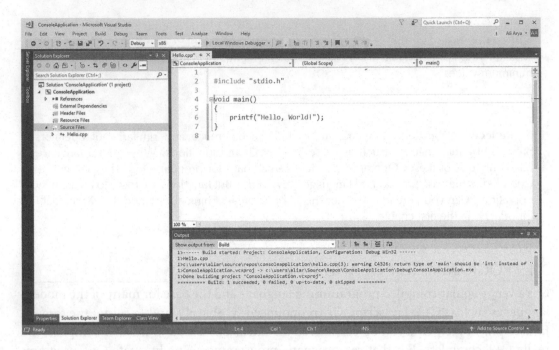

EXHIBIT 3.6   Visual Studio User Interface with project view, source file, and output windows.

- Output window

- Errors

The first step in using Visual Studio is to create a new **Project**. Keep in mind that a single source file cannot be compiled or executed by itself. We will discuss creating projects and writing simple programs in the next section, but first, let's have a better understanding of the process of building an executable program from our source code.

To create a new program, follow these steps (see the Companion Website for details and images):

- Go to the File menu and select New > Project.

- Choose Visual C++ > Windows Desktop > Windows Desktop Wizard.

- Choose a name and location.

- Check the option for "Create Directory for Solution" if you want your solution file to be in a different folder from your project files.

- Click on OK.

- In Project Properties, select Console Application, check Empty Project, and then click Finish.

- Visual Studio creates different types of projects. Console Application is a simple text-based program in C/C++ that is portable to other operating systems, IDE's, and compilers. Standard C/C++ has no support for graphics operations, such as drawing and showing images. These operations are supported through extensions that are not portable. We will discuss them later.

- In Solution Explorer, right-click on your project or Source Files section and choose Add > New Item > C++ Source, and choose a name such as Hello.cpp.

> ☞ **Key Point:** Visual Studio requires a Project and does not work with a single source file.

Your Visual Studio solution should look like Exhibit 3.5. Go to Windows File Explorer, and see the files created. If you can't find them, right-click on Project and select Open Folder in File Explorer. Now add the following text to the source file:

Sample Code #8 (C/C++) [9]

```
1. #include "stdio.h"
2. void main()
3. {
4. printf("Hello, World!");
5. }
```

This program can be used as a base code for many C/C++ programs we write. It is not much different from the examples in Javascript and Python.

- C and C++ also have a ; symbol at the end of each instruction.

- C and C++ are also case-sensitive.

- Strings of text are shown with a " " pair.

- The code can be modularized into functions. A function is identified primarily by the ( ) pair after its name, which holds the input information for it (data needed for the operation). For example, the above code defines a function called main.

---

[9] If the code is given in multiple languages, I will identify them with the same number and the suffix -c, -p, etc. In this case, the code is only for C/C++. If the code is specific to C or C++, I will refer to that language. Otherwise, I use C/C++ which means the code can is compatible with both.

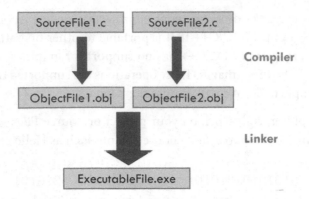

EXHIBIT 3.7   Build process for C programs.

- Blocks of code, such as the code for functions or conditional blocks of code for if and while statements, are grouped with a { } pair (similar to Javascript).

- Function and variable names are a combination of letters, numbers, and the _ (underscore) symbol.

Every C/C++ program is required to have a function called main, so your first step in writing any C/C++ program should be to add an empty main(). While in Javascript and Python, there are functions that your program can immediately use (such as write and print), C/C++ treats all functions as external to your program, so the compiler needs to be "told" where these functions are defined. The compiler only understands the keywords and basic operators (Table 2).

More complicated operations provided as functions are grouped into "libraries," and when using a library function, the programmer has to reference the library. For example, the printf() function is defined in a library called stdio (Standard Input and Output) and allows outputting data to screen.[10] We will see more about the libraries and the include command (line 1 of Sample Code #10) in the next chapter.

> ☞ **Key Point:** C/C++ syntax is very similar to Javascript, but all C/C++ programs need a function called main.

3.4.1 Compiling and Running C/C++ Programs

Exhibit 3.7 shows the full process of building an executable file from C source code files (illustrated for Windows operating system where executable files have extension .exe). This is a typical process for compiled languages and involves two steps:

---

[10] The f at the end of printf stands for "formatted" as the function allows formatting the output using new line, tab, and other commands.

- Compiling translates each source file to machine code. The resulting file is usually referred to as *Object* file.

- Linking puts together all the object files and creates a single executable file (with extension .exe).

  - Programs frequently involve multiple source files to allow better management of the code. Even if there is only one source file, C programs (and most other languages) require linking the object file to many system resources. We will discuss these in detail later.

If your C/C++ compiler and linker detect errors in the program (which even for simple programs can happen due to typos), they will identify them as **Compile-Time Error** that you can fix. Unfortunately, there are errors that they can't detect (such as assigning a wrong value to your data or performing an unwanted operation). These may cause issues only when you run the program and are called **Run-Time Errors**.

You can run a C/C++ program by locating the .exe file and double-clicking on it or select the Start command in the Debug menu. You have two options: with debugging (F5 shortcut key) and without it (Control-F5). We will talk about debugging later in this book. For now, just keep in mind that the second option will hold the program window open at the end so you can see the results.

Run your sample program with both options, and see the resulting message on the screen. When you use the without-debugging option, press any key on the keyboard to close the program window and go back to Visual Studio.

3.4.2 Comments

Let's see how we can use the simple print operation and do something fun. The following code draws a simple shape on the screen with the * symbol. The \n at the end of a text string tells the printf function to go to the next line.

Sample Code #9

```
1. #include "stdio.h"

2. //This is the main function

3. void main()

4. {

5.    //the following lines draw a rectangle

6.    printf("*************************************\n");

7.    printf("* *\n");

8.    printf("* *\n");
```

```
 9.   printf("* *\n");

10.   printf("* *\n");

11.   printf("* *\n");

12.   printf("* *\n");

13.   printf("*********************************\n");

14. }
```

You notice that lines 2 and 5 start with //. When our code starts to get longer and more complicated, it will be hard to understand what is happening in it. A common programming practice is to "describe" the code by adding *comments*. A comment is part of the text that will be ignored by the interpreter or compiler, and won't be executed.

In most programming languages, text can be marked as a comment in two ways: single line and multi-line.

In single-line comments, adding a particular symbol in any line marks the remaining text on that line as a comment. The comment can start at the beginning of the line or anywhere else.

- C/C++ and Javascript use // for single-line comment.
  - `X = 5; //this is a comment`
- Python uses # for single-line comments.
  - `X = 5 #This is a comment`
- HTML has no separate single-line commenting method. The multi-line method is used for both cases.

In multi-line comments, adding a particular symbol in any line marks the beginning of a comments section. This section can end on any line with a terminating symbol.

- C/C++ and Javascript use /* and */ for starting and ending the comment section.
  - `int X,Y; /* this is the start of comment section`
  - `X = 5;`
  - `Y = 6;`
  - `*/`
- Python uses """ for starting and ending multi-line comments.
  - `""" this is the start of comment section`
  - `X = 5;`

- • `Y = 6;`

- • `"""`

- HTML uses <!—and --> for starting and ending comments.

There are multiple advantages to using comments:

- The programmer can describe the text for future use when she may forget what was done and why.

- The programmer can describe the code for others who may read it in order to make changes.

- The programmer may want to temporarily remove part of the code without deleting it. *Commenting out* is a common practice for testing programs when we want to see the effect of a part of the code.

The examples throughout the book have minimum comments to make them shorter and easier to read. If you go to the Companion Website, you will find the files with proper comments which is the recommended practice.

☞ **Key Point:** Comments are the most basic way of documenting a program. They also serve the purpose of temporarily disabling a part of the code.

Now expand Sample Code #9, and try to draw different shapes. Add proper comments to describe your code. Comments are the easiest way of documenting programs, and documentation is essential in successful software projects (and is much more than just adding comments).

Using text characters to draw shapes is sometimes called *ASCII Art*. The abbreviation ASCII is for American Standard Code for Information Interchange, a well-established standard for assigning numeric codes to textual characters. Keep in mind that computers store everything as binary numbers.

☑ **Practice Task:** Use the text output function in C/C++, and draw different shapes.

✍ **Reflective Question:** When working on these examples and tasks, did you think of operators and/or functions that could help you do things easier? Make a "wish list," and then see if they are actually available.

> ✋ **Reflective Questions:** How do you feel about seeing examples of multiple languages together? Is it helping you understand the differences in syntax while learning the common concepts?

SIDEBAR: COMPILER OPTIMIZATION

Imagine that you are giving directions to a friend who is driving and that there are five main steps:

1. Going straight for 500 m on a highway.
2. Take the exit.
3. Driving for another 200 m.
4. Turn left.
5. Drive 700 m.

If you have experienced driving or navigating, you probably know that the best way to give these directions is to let the driver know what is coming up so they can be prepared. If the driver doesn't know that after driving for 500 m, they have to take the exit, we may miss it because of being in the wrong lane or just not acting quickly enough. On the other hand, when the driver has a general idea of all five steps, we will optimize our driving by taking a suitable lane and having appropriate speed.

Interpreted programs act like step-by-step directions. The interpreter reads the code line by line, translate it to machine code, and then pass it to CPU to run. This happens when a browser reads Sample Code #1. When it reads line 2, the browser sets the value of the paragraph "demo" to "?". But when it reaches the line 6, it changes the value. The browser does not optimize your code for performance speed, efficient use of memory, or other factors, as it never looks at the program as a whole.

The ability of the compiler to read the whole code and translate allows it to perform various optimizations, some dependent on the CPU and some independent of it. A very common example is removing repeated code to optimize for speed. Consider the following C code, which looks very much like Javascript:

```
X = A + B + C
Y = A + B + D
```

This can be converted to

```
Z = A + B
X = Z + C
Y = Z + D
```

While the new code has more lines, it has fewer operations and as such takes less time. Here is another example:

```
while(X > B-2)
{
    A = 5
    X = X - 1
}
```

You notice that the operations A = 5 and B - 2 are inside the loop and repeated in every iteration, but their values don't really change. So, we can modify the code by taking them outside and reduce the time spent in the loop:

```
A = 5
C = B - 2
While(X > C)
{
    X = X - 1
}
```

3.5 DATA TYPES

In addition to name and value, variables also have a type. For example, in the code we started this chapter with, demo was a paragraph, data1 and data2 are edit boxes, and price is a number. Knowing the type of data is important for the CPU and the program as they need to allocate memory for the data and use that memory correctly based on how the data is stored. A number and an image require a different amount of memory, and even with the same size, an image and a paragraph of text are stored and used differently.

In Javascript or Python, the programmer is not required to specifically define the type of a variable. When you create variables and give them a value, the program automatically recognizes the type of data based on the first value that was assigned to it. Any consequent use of that variable assumed the same type. This is convenient for programmers but may not be very flexible or efficient. Languages such as C and C++, on the other hand, require the programmer to identify the type of each variable.

Sample Code #10

```
1. #include "stdio.h"
2. void main()
3. {
4.    int x = 9;
```

```
5.    int y = 8;

6.    int z;

7.    z = x + y;

8. }
```

The lines 4–6 should remind you of creating variables in Javascript. This is done the same way as in C/C++ except that instead of var, we are more specific and use the data type for the variable. While all data is stored as binary numbers by the computers, most CPUs make a distinction between integer (no decimal point) and non-integer (with decimal point) numbers. 2, 10, and -89 are integer numbers, while 2.3 and 56.09 are not.

Storing integer numbers is fairly straight-forward: after converting to binary, every digit gets a bit in memory. Non-integer numbers, on the other hand, need to identify where the decimal point is. This can be done in two methods:

- Fixed-point, where there are a certain number of bits, is allocated for the left and right sides of the decimal point.

- Floating-point, where the number of bits may vary to allow flexibility in storing numbers.

Almost all computers these days use the floating-point method. As such, in C/C++, the standard data types

- int for integer. Depending on the total size, this can be

  - char: usually 1 byte

  - short (a.k.a short int): usually 2 bytes

  - int (a.k.a. long int): usually 4 bytes.

- float for non-integer. Depending on the total size, this can be

  - float: usually 4 bytes

  - double: usually 8 bytes.

- bool for Boolean. Named after English mathematician, George Boole, this type is commonly used to hold true/false values. It requires only one bit, but the actual implementation is usually an integer. A value of zero means false, and anything else is true.

Note that the sizes listed above are dependent on the compiler and platform (CPU and Operating System). The values I have listed are the most common ones. C/C++ standard function sizeof() can show you the exact number for any variable or type, if you don't want to assume these values. The size of char type is determined by the system setting

CHAR_BIT, which is in most personal computers eight bits (one byte). The sizeof(char) is assumed to return 1, and all sizes are measured in terms of the size of char. So, to truly measure the size of any variable or type, you should use a code like the following:

```
int n_int = sizeof(int) * CHAR_BIT / 8; //size in bytes

float data;

int n_data = sizeof(data) * CHAR_BIT / 8; //size in bytes
```

Unless you are programming special devices, it is generally safe to assume CHAR_BIT is 8. It is also common (but less sure) that char, short, int, float, and double types are 1, 2, 4, 4, and 8 bytes, respectively. Float types are always signed (positive or negative). Integer types can be unsigned, which changes the range of supported values. For example, the 1-byte char type can have 256 possible values (0–255 binary, or all ones). The default is to consider one bit as the sign bit, and we will have 0–127 (for 7 bits), positive or negative. The type unsigned char, on the other hand, is 0 to 255, positive only.

> ☞ **Key Point:** In C/C++, all variables and functions need to have a data type. Basic types include integer (without a decimal point) and float (with a decimal point), and their variations based on size.

Sample Code #11

```
1. #include "stdio.h"
2. void main()
3. {
4. char t = 0;
5. int x = 9;
6. float y = 8.6;
7. double z = 10.5;
8.
9. t = 'A';
10.   printf("%d \n", x);
11.   printf("%f \n", y);
12.   printf("Character code for %c is %d \n", t, t);
13. }
```

The lines 4–7 of the Sample Code #11 should look very familiar. They are the same as Javascript code for defining new variables except for specifying the data type. Line 9 shows a new operator. The single quote converts a character to its numeric code. The value of variable c changes from 0 to the code for A (65 if using ASCII standard). In C/C++ programs, don't confuse single quote (for a single character) with double quote (for a string of text).

The lines 10–12, on the other hand, show how to use the `printf` function for showing the values of variables. Note that

- A `%` sign within the printed string means the value of a variable will be printed.

- `%d`, `%f`, and `%c` represent the value in integer, non-integer, and character formats, respectively.

- If printing the value of a variable, the string is followed by the name of the variable separated with a "," symbol.

- You can have as many variables as you want and can combine them with constant text, as shown in line 12.

The character codes for computers (regardless of the programming languages) are defined using different standards. ASCII is one of the oldest ones that uses 7 bits per code, so it can have values from 0 to 127, as shown in Table 3.3. With a 1-byte character code, ASCII systems allowed codes from 128 to 255 to be used for secondary languages. Due to the limited number of codes, older computers using this standard could not use multiple languages at the same time. The more recent UNICODE standard allows up to 1,114,112 codes (in 21 bits, or up to 4 bytes) to be used simultaneously. UNICODE has been implemented in computers using different methods such as UTF-8 (characters have a variable length from 1 byte to 4), UTF-16, and UTF-32 (4 bytes for all characters). The latest standard for C/C++ supports using UNICODE and a range of new types to manage characters, such as `wchar_t`, `char16_t`, and `char32_t`. A detailed discussion of characters sets and their implementation is beyond the scope of this book. All these methods are backward-compatible with ASCII, so you can safely use the 1-byte ASCII codes in most cases as long as you are not using any language other than English.

TABLE 3.3 ASCII Codes

Dec	Char	Dec	Char	Dec	Char	Dec	Char
0	NUL (null)	32	SPACE	64	@	96	`
1	SOH (start of heading)	33	!	65	A	97	a
2	STX (start of text)	34	"	66	B	98	b
3	ETX (end of text)	35	#	67	C	99	c
4	EOT (end of transmission)	36	$	68	D	100	d
5	ENQ (enquiry)	37	%	69	E	101	e
6	ACK (acknowledge)	38	&	70	F	102	f
7	BEL (bell)	39	'	71	G	103	g
8	BS (backspace)	40	(	72	H	104	h
9	TAB (horizontal tab)	41	)	73	I	105	i
10	LF (NL line feed, new line)	42	*	74	J	106	j
11	VT (vertical tab)	43	+	75	K	107	k
12	FF (NP form feed, new page)	44	,	76	L	108	l
13	CR (carriage return)	45	-	77	M	109	m
14	SO (shift out)	46	.	78	N	110	n
15	SI (shift in)	47	/	79	O	111	o
16	DLE (data link escape)	48	0	80	P	112	p
17	DC1 (device control 1)	49	1	81	Q	113	q
18	DC2 (device control 2)	50	2	82	R	114	r
19	DC3 (device control 3)	51	3	83	S	115	s
20	DC4 (device control 4)	52	4	84	T	116	t
21	NAK (negative acknowledge)	53	5	85	U	117	u
22	SYN (synchronous idle)	54	6	86	V	118	v
23	ETB (end of trans. block)	55	7	87	W	119	w
24	CAN (cancel)	56	8	88	X	120	x
25	EM (end of medium)	57	9	89	Y	121	y
26	SUB (substitute)	58	:	90	Z	122	z
27	ESC (escape)	59	;	91	[	123	{
28	FS (file separator)	60	<	92	\	124		
29	GS (group separator)	61	=	93	]	125	}
30	RS (record separator)	62	>	94	^	126	~
31	US (unit separator)	63	?	95	_	127	DEL

3.5.1 Type Casting

Let's explore some operations with various data types (Exhibit 3.8).

Sample Code #12

```
1. #include "stdio.h"

2. void main()

3. {

4.    int x = 9;
```

```
ConsoleApplication                    ▼   (Global Scope)                      ▼
   1     #include "stdio.h"
   2     void main()
   3     {                              C:\WINDOWS\system32\cmd.exe
   4         int x = 9;                8
   5         int y = 8.6;              4
   6         printf("%d \n", y);       4.000000
   7         y = x / 2;                4.500000
   8         printf("%d \n", y);       Press any key to continue . . .
   9         float z = x / 2;
  10         printf("%f \n", z);
  11         z = x / 2.0;
  12         printf("%f \n", z);
  13     }
  14
```

EXHIBIT 3.8   Output for Sample Code #12.

```
 5.    int y = 8.6;

 6.    printf("%d \n", y);

 7.    y = x / 2;

 8.    printf("%d \n", y);

 9.    float z = x / 2;

10.      printf("%f \n", z);

11.      z = x / 2.0;

12.      printf("%f \n", z);

13. }
```

There are a few things in the Sample Code #12 that are worth paying special attention to.

Assigning a float value to an int variable will result in the loss of the non-integer part, as the int variable cannot hold that part (lines 5–6 and 7–8). For example, 9 divided by 2, stored in an int variable, is 4, not 4.5. This is a very important point. The act of assigning a value of a certain type to a variable of a different type is called *typecasting*. It can be used in our programs intentionally, as we will see later but can cause errors if not used properly.

3.5.2 Integer and Float Operations

Dividing an int variable by another int variable (or an int value) will also result in an int number. This is why the result of line 9 is again 4.0 and not 4.5, even though it is stored in a float variable. To understand this, remember that the assignment operator (=) calculates the right side first and then writes it to the left side. This means we have 9 divided by 2 as int (4) and then write to a float variable to have 4.0.

To make sure an operation is done in float, you have to include at least one float operand. See line 11 as an example.

☑ **Practice Task:** Assuming that there is no % operator to calculate the remainder of division, write a C/C++ code that uses typecasting to calculate the remainder.
Hint: If you divide 7 by 2 in integer mode, the result is 3, and in float, it is 3.5. The difference (0.5) multiplied by 2 is the remainder.

✋ **Reflective Questions:** Why do we need to have the data types specified? What do you think the advantages and disadvantages are? How and when do you think you will be using this in your programs?

Let' use what we learned to write a code for detecting if a number is of or even without the use of % (remainder) operator.

Sample Code #13

```
1. #include "stdio.h"
2. void main()
3. {
4.    int x;
5.    scanf _ s("%d", &x);
6.    int y = x / 2;
7.    if (y * 2!= x)
8.       printf("odd ");
9.    else
10.      printf("even ");
11. }
```

This example starts by defining a variable. Not that we are not giving this variable an initial value because the next line is an input operation that provides a value for the variable. Inputting data in C can be done in different ways. One of them is to use the scanf() function defined in stdio library. This function is similar to input() that we saw earlier in Python but allows different types of data to be entered. Due to security concerns for getting data into the program, the original scanf function is replaced with a "safe" version called scanf _ s. Note that when using this function, you need to identify the type of data (for example, %d for int), and the name of the variable receiving the data (preceded by a & symbol). This function can be used with a more complicated input string and more than one variable, but I recommend you stick to the simple version demonstrated here.

Line 6 of the above code defines a new variable and gives it an initial value. You can guess that the non-integer part of the value will be lost. So, if the input data is 7, we will have 7 divided by 2, which is 3.5 but stored in y as 3. Then line 7 compares y*2 and x. If there was a data loss (which happens for odd numbers), then these two won't be equal.

> ☑ **Practice Task:** Use the Sample Code #13 and the calculator programs you created in Python and Javascript, and develop a simple calculator in C/C++. Ask the user to enter two operands, and then, ask for an operation. Assume numbers 1 to 4 mean four arithmetic operations, for example, `if(operation == 1)`, and then, add.

HIGHLIGHTS

- High-level languages are human-readable and not CPU-dependent.

  - Declarative languages describe the output, but Imperative languages describe the operation.

  - Interpreted languages are read and executed by a standalone program, but compiled languages are translated into standalone machine code.

  - Virtual Machine (VM) runs programs that are compiled but not for any specific target platform.

- Integrated Development Environment (IDE) is a tool that allows writing, editing, compiling, running, and debugging programs, for example, Visual Studio.

- Golden Rule of Programming #1: For any important piece of information, define a variable.

- In Javascript, we define a new variable using the keyword var.

- Golden Rule of Programming #2: Define it before you use it!

- Python has similar concepts to Javascript, but the syntax is a little different. In particular, there is no ; at the end of each line, no `var` to define a new variable, and indents instead of { } to define a block of code.

- Operators are keywords understandable by compiler/interpreter. Functions are modules defined by the programmer.

  - Functions are modules of code that we can define in our program or are pre-defined. We can use these modules in our program, so we don't have to repeat that code.

- Visual Studio requires a Project and does not work with a single source file.

- C/C++ syntax is very similar to Javascript, but all C/C++ programs need a function called main.

- Comments are the most basic way of documenting a program. They also serve the purpose of temporarily disabling a part of the code.

- In C/C++, all variables and functions need to have a data type. Basic types include integer (without a decimal point) and float (with a decimal point), and their variations based on size.

END-OF-CHAPTER NOTES

A. Things I Should Mention

- Java and C# and the most popular Virtual Machine-based languages out there. We will have some examples of Java and C# later, but they are not covered as much as C/C++, Python, and Javascript in this book. Their syntax is very similar to C++ (and I will mention major differences). They do have some new features, but most concepts we discuss in C/C++ apply to Java and C#.

- There are many other high-level programming languages, many of them particularly suitable for specific purposes, such as PHP and Ruby for web development, Swift for iOS/macOS, and SQL for databases.

B. Self-Test Questions

- What are the main categories of high-level programming languages?

- What is a variable, and how do we define it?

- What are the main similarities and differences in the syntax for C/C++, Python, and Javascript? Which two look more similar?

- What is a data type, and which languages explicitly define it for variables?

C. Things You Should Do

- Check out these web resources:

  - https://www.w3schools.com – This is a great place to learn about many programming languages. It has the cool feature of letting you write the code and run it online to see the result.

  - https://www.tutorialspoint.com – This is another good place for tutorials on many programming languages.

  - http://www.cplusplus.com

  - http://www.cprogramming.com

- Try an online development environment, such as

  - https://repl.it/languages – This site has a simple IDE for many languages.

  - https://www.onlinegdb.com – This site is only for C/C++.

- Take a look at these books:

  - *Programming Language Explorations* by Ray Toal, Rachel Rivera, Alexander Schneider, and Eileen Choe

  - *Introduction to Programming Languages* by Arvind Kumar Bansal

D. Reflect on the Experience of Reading This Chapter

- What did you expect from this chapter before reading it?

- What was it about, and what did you learn?

- What tasks did you perform, and what difficulties did you face?

- How did you feel about the material and tasks presented in this chapter?

- How can you improve your learning experience?

- How do you see this topic in relation to the goal of learning to develop programs?

- How do you feel about learning multiple languages at the same time?

- Do you think comparing multiple languages can help you understand basic concepts and the reason for differences better? Take variables and data type as an example.

CHAPTER 4

Code

Program's Operation

> **Topics**
> - Sequential Execution and Control Flow
> - Functions, Selection, and Iteration

> **At the end of this chapter, you should be able to:**
> - Write simple linear programs
> - Use selections to create branches and multiple options in programs
> - Use iterations to repeat parts of the code
> - Understand common constructs for selection and iteration
> - Understand naming conventions for variables and functions

OVERVIEW

In Chapter 3, we learned about defining and using variables as the main way of managing data in our programs. In this and the next chapter, we focus on managing the code.

Programs are a series of instructions given to computers to execute. A key point in this definition is the combination of "series" and "execute." By default, programs run step by step, or instruction by instruction, from start to end. This is a fundamental concept in programming, commonly referred to as **sequential execution**. But as fundamental as this concept is, most programs deviate from it because they require some form of non-linear execution at some point.

In Chapter 2, we reviewed many examples of algorithms that needed to repeat something, skip an operation, or choose among multiple options. All these are examples of what we call **control flow**. Considering that programs are controlling the behavior of the computer, every instruction passes this control to the next, and control flow is a series of statements that together determine the program flow or the order of execution. In the absence of any specific control flow statements, this order is linear and sequential. A control flow statement changes this order by altering passing the control of the computer to another part of the code instead of the next instruction.

In this chapter, we review common methods of control flow. We start with the simple, old, and infamous GoTo **statement**, and see why it was not a good way of managing the flow of a program. Then, we discuss selection and iteration as two of the most basic control flow mechanisms in modern languages. Together with modularization, they are the basis of what we call **Structured Programming (SP)**.

As we saw in the Introduction, modularization is a concept that goes beyond programming and involves breaking down an entity into smaller components. These components (or modules) are simple but can be used to create complex more objects. Modularization of code is done primarily through functions which we will discuss in the next chapter.

4.1 SEQUENTIAL EXECUTION AND PROGRAM CONTROL

4.1.1 GoTo Statement

As discussed in Chapter 2, the GoTo statement is the most basic and simplest way of performing control flow. GoTo is the high-level equivalent of a very common machine code or Assembly instruction called *Jump* or *Branch*. It is an unconditional change of control to another location in the program. An example of using GoTo in C/C++ is shown in Sample Code #1. The C/C++ keyword goto is followed by a name that does not represent a variable or function but is a line name or *label*. The label has to be defined in the program and identified with a : after the label, as in line 1 of the sample code. The goto statement simply stops the sequential execution and "jumps" to that line.

Sample Code #1:

```
1. get _ data: printf("please enter a number: ");

2. scanf _ s("%d", &number);

3. goto get _ data;

4. printf("after goto ");
```

While easy to use, the GoTo statements can be very confusing when reading the program as it is not clear which points "go to" a label, or where the label is for a given GoTo operation. Back in 1968, Edsger Dijkstra wrote an article arguing that this operation was a common source of error. While GoTo still exists in C/C++ as a legacy, most modern languages do not include it anymore, and C/C++ programmers are strongly discouraged from using it. Instead, more structured control flow methods are practiced, particularly

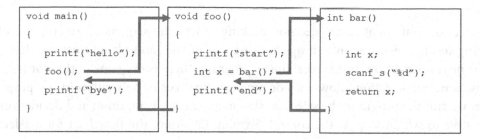

EXHIBIT 4.1   Function call and return.

functions, selections, and iterations. These are the basis of what is called **Structured Programming (SP)**.

SP is one of the most commonly used programming paradigms, which are categories and styles of programming. **Programming Paradigms** define high-level approaches to software design and development that control the general organization and style of programs. We will discuss SP in this and the next part of the book, while the following parts are dedicated to **Object-Oriented Programming (OOP),** which introduces a new module of code called **Object** and is another popular alternative. C, Fortran, BASIC, and Pascal are examples of languages that aim at facilitating SP (sometimes called SP languages), while C++, Javascript, Python, C#, Java, and Visual Basic are examples of the OOP paradigm.[1]

SP relies on three major components to organize a program:

- **Selection** that allows choosing between multiple options based on a condition.

- **Iteration** that allows repeating parts of the code as long as a condition is true.

- **Modularization** of the program primarily using functions but also other methods that we will discuss later.

4.1.2 Structured Programming and Control Flow

4.1.2.1 Function

Functions temporarily transfer the control of the program to another point. The program then continues the sequential execution at that point until the function ends. Then, the control is given back to the original part of the code, and the program continues from the next instruction. This is illustrated in Exhibit 4.1. The act of moving to a function is referred to as a **function call** and coming back from the function as "return," which may or may not include a result. We will see some simple examples of functions in this chapter and will discuss them in detail later.

---

[1] When talking about the SP languages I listed, you may hear the term "Procedural" which is not exactly the same as Structured. Procedural refers to the notion of top-down sequential execution with function (or procedure) calls. SP shares this concept but also emphasizes on the use of program control constructs such as iteration and selection. Languages like C can be considered both Procedural and Structured.

4.1.2.2 Selection

The selection statement (a.k.a. decision-making) alters the sequential execution, not by jumping to another location, but by choosing one of the multiple alternatives that exist for the next instruction. In a typical linear program, there is only one option for the next instruction, but selection allows two or more alternatives to be written in the program. When we run the program, the selection statement checks a condition and decides on the alternative to which to pass the control. Exhibit 4.2 shows the flowchart for a selection operation.

4.1.2.3 Iteration

Iterations are also conditional similar to selection, but allow repeating a part of the code. This means that at the end of the repeating part, the code goes back instead of moving to the next instruction. Iteration creates a structure called a loop in the program. The condition for an iteration can be similar to a selection except that the conditional code will be executed as long as the condition is true, for as many times as needed. The condition for an iteration can also be a certain number of times that loop has to be executed. We refer to these two iterations as indefinite and definite loops, respectively. Exhibit 4.3 shows the structure of an iteration.

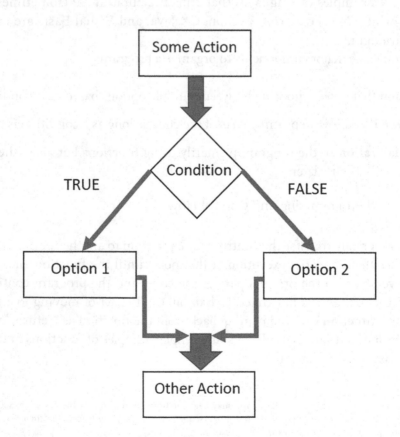

EXHIBIT 4.2   Selection.

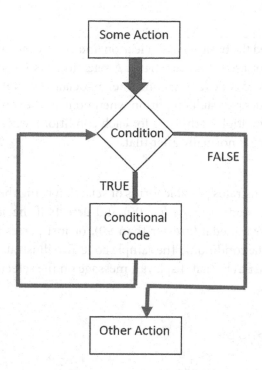

EXHIBIT 4.3   Iteration.

The use of functions, selections, and iterations in SP brings us to the **Golden Rule of Programming #3**:

*A program is a linear sequence of instructions by default. Change that linear sequence by: *Functions (to temporarily move to another place in code)*

*Selections (to allow conditional execution)

*Iterations (to allow repeated execution)

4.1.2.4 Blocks of Code and the Lexical Scope

All the structured control flows described above involve certain groups of instructions (or a "block of code") that are performed without following the sequential execution. These blocks of code are identified in different ways based on the syntax of a particular language. For example, in Javascript and C/C++ we use { } and in Python indents. The blocks of code not only identify which instructions are related to the function, selection, or iteration but also defines what is called a **lexical scope**, an area of code where variables are defined. We will talk about functions in Chapter 5 and discuss selection and iteration in the following sections.

☞ **Key Point:** Golden Rule of Programming #3: Programs run sequentially and linearly, by default. Various control flow mechanisms can alter this, including functions, selection, and iteration.

4.2 SELECTION

In Chapter 2, I explained the basic idea of selection and conditional statements. Exhibit 4.2 shows the basic design of a selection construct. A selection has at least one condition (c1 in Exhibit 4.2) and a block of code (at least one line) associated with it. Other conditions are optional and that includes a default condition when none of the other conditions are true. The selection provides multiple paths (one for each condition), which all merge at the end, and the program continues normally after that.

4.2.1 If/Else

Sample Code #2-j demonstrates possible forms of selection using the keywords **if** and **else**. The program receives a student grade (0–100) and detects if the student has failed (less than 50), should receive a medal (greater than 90), or just passes the course. Line 13 is executed regardless of the conditions. The sample code also illustrates the use of a common Javascript function, `alert()`, that displays a message on the screen.

Sample Code #2-j

```
1. <p id="demo">?</p>
2.
3. <script>
4. function Grade()
5. {
6.     var grade = data.value;
7.     if(grade >= 90)
8.             demo.innerHTML = "Congrats! You get a medal.";
9.     else if(grade < 50)
10.            demo.innerHTML = "Sorry. You failed.";
11.    else
12.            demo.innerHTML = "Good. You passed."
13.    alert("Bye");
14. }
15.
16. </script>
17. <input id="data">
18. <button onclick="Grade();">Grade</button>
```

We will discuss functions in the next chapter. For now, we just need to know that they are defined by the following parts:

- A keyword specifying the type
  - In C/C++, we specify the data type. All examples in this chapter use void, and we discuss all the options in the next chapter.
  - In Python and Javascript, functions and variables don't have explicitly defined types. Defining functions is by using the keyword function in Javascript and def in Python.
- Name
- Input data inside a pair of ( )
- Block of code
  - In C/C++ and Javascript, the code is inside { }.
  - In Python, the code is identified with indents.

Notice that lines 9–12 are not required by the language. An **if** statement does not need to have any **else** part, but it can. Sample Code #3-j demonstrates this with an added complexity of having more than one line for the conditional block of code. You see that in such cases, we need to use { } to identify the conditional code. Without them, only the first line will be associated with the condition, and line 10 will happen regardless of the condition.

☞ **Key Point:** In C/C++ and Javascript, the conditional code has to be inside { } if it is more than one line. It is a good and safe practice to always use { } even for one line.

Sample Code #3-j

```
1. <p id="demo">?</p>

2.

3. <script>

4. function Grade()

5. {

6.        var grade = data.value;

7.        if(grade >= 90)

8.        {

9.                demo.innerHTML = "Congrats! You get a medal.";

10.                    alert("Congrats!");
```

```
11.      }
12.      alert("Bye");
13. }
14.
15. </script>
16. <input id="data">
17. <button onclick="Grade();">Grade</button>
```

The C/C++ code for implementing selection is almost identical to Javascript, although input and output methods are very different as we have seen in previous examples.

Sample Code 3-c

```
 1. #include "stdio.h"
 2. void Grade()
 3. {
 4.     int grade;
 5.     printf("please enter a number: ");
 6.     scanf_s("%d", &grade);
 7.
 8.     if (grade >= 90)
 9.             printf("Congrats! You get a medal.");
10.     else if (grade < 50)
11.             printf("Sorry. You failed.");
12.     else
13.             printf("Good. You passed.");
14. }
```

The Python code is conceptually the same, but the syntax is different as we have seen earlier:

- The conditions don't use ( ) but have the : symbol at the end.

- Indents are used to identify the block of code.

- While in Javascript and C/C++ else and if can be used together, in Python, there is a new single keyword, elif, for that purpose.

The Python library function input combines lines 5 and 6 in the C/C++ code as it has input plus a prompt, but it only receives a text. The function int converts the text to an integer number. Similarly, the function float converts a text to the floating point non-integer type. These two functions are examples of operations that produce a result data, as opposed to those such as print that simply perform an action. The result can be used to give value to a variable, as in lines 1 and 2. We will discuss functions in more detail in the next chapter.

Sample Code 3-p

```
1. x = input("enter a grade: ")

2. grade = int(x)

3. if grade >= 90:

4.          print("Congrats! You get a medal.")

5. elif grade < 50:

6.      print("Sorry. You failed.")

7. clse:

8.      print("Good. You passed.")
```

☑ **Practice Task:** Investigate what happens if we use if instead of else if.
☑ **Practice Task:** In the field of Artificial Intelligence (AI), a Conversational Agent (CA) is a program that can have conversations with the user. This is usually done through a series of rules, such as if the user says X, then the answer is Y. Simulate a simple CA using if/else construct. Limit the user choices to a series of sentences, print those options ("1- Hello, 2-How are you? etc."), and ask the user to choose one by entering the correct number. Then, print an appropriate response.
Note: In more advanced CA programs, the AI will process the freely entered input text and prepares the right answer.

4.2.2 Switch/Case

Command Processing is a very common task in computer programs. It involves receiving a command from the user and performing an appropriate action in response. The resulting program structure is usually referred to as **Command Processor**. The Conversational Agent example that we had as a task in the previous section was such a Command Processor. As a simplified case, imagine that the command is given using a number (menu item), and the action is simply to print a message. Sample Code #4-c demonstrates such a program.

Sample Code 4-c

```
1. #include "stdio.h"

2. void main()

3. {

4. int command;

5. printf("Enter a command number:\n");

6. printf("1-Hello, 2-GoodBye, 3-How are you?\n");

7. printf(": ");

8. scanf _ s("%d",&command);

9. if(command == 1)

10.         printf("Hello\n");

11.    else if(command == 2)

12.         printf("Goodbye\n");

13.    else if(command == 3)

14.         printf("How are you?\n");

15.    else

16.         printf("Invalid command\n");

17. }
```

In addition to the use of if/else, the Sample Code #3-c illustrates a couple of important software design considerations:

- When asking the user to do something, always provide proper information. Good **User Interface (UI)**, including displaying a menu and prompting the user properly to enter data, as shown in lines 5–7, is essential in reducing the user's error and frustration.

- Allow for wrong user input and handle it properly, as done in lines 15–16. In general, **Error Handling** is a very important part of programming that will save you much time fixing issues in the code.

SIDEBAR: THE EVOLUTION OF USER INTERFACE (UI)

UI is the part of computer hardware and software that interacts with the user and allows data input and output.

The keyboard is the traditional (and oldest) input systems in computers. It dates back to the older computers, such as mainframes that were not capable of doing any graphics and displayed or printed text as the standard output. Advanced in computer hardware and software allowed multimedia output such as graphical displays and also other input devices such as light pen and later mouse. This resulted in the era of Graphical User Interface (GUI) pioneered by the Xerox Palo Alto Research Center (PARC) in the early 1970s.

Initial personal computers had a text-based UI similar to that of mainframe terminals that showed a command prompt where the user could enter commands. GUIs were later popularized by Apple computers and Windows operating system which still kept the terminal interface and allowed programming for both graphical and text-based (a.k.a. console) applications.

Newer UI methods include 3D and immersive displays, haptic input and output, motion and gesture tracking, touch input, speech recognition, and many other emerging technologies.

Despite the changes in technology, some guiding principles remain valid for UI design, such as clarity, simplicity, feedback, consistency, tolerance, and reuse. Some great books to read on the subject of UI design are:

- *Don't Make Me Think* by Steve Krug
- *Design of Everyday Things* by Don Norman
- *Human-Computer Interaction* by Alan Dix et al.

Looking at the above code, you notice that all conditions (and we may have more of them if we have more commands) are similar and include comparing the same variable with a series of possible values. This is a special case of conditional statements and, in most languages, has a special construct called **switch/case**. Sample Code #5-c demonstrates the use of switch/case in C/C++:

Sample Code 5-c

```
1. #include "stdio.h"
2. void main()
3. {
4.    int command;
5.    printf("Enter a command number:\n");
6.    printf("1-Hello, 2-GoodBye, 3-How are you?\n");
7.    printf(": ");
```

```
8.    scanf _ s("%d",&command);
9.    switch(command)
10.         {
11.             case 1 :
12.                 printf("Hello\n");
13.                 break;
14.             case 2 :
15.                 printf("Goodbye\n");
16.                 break;
17.             case 3 :
18.                 printf("How are you?\n");
19.                 break;
20.             default:
21.                 printf("Invalid command\n");
22.         }
23. }
```

The switch/case construct starts by defining the variable inside the ( ) and after the switch keyword. The set of cases follows each with a value, a : symbol, and then the body of code for that case. The program checks the variable and all case values and moves to the code for matching value, or default if there is no match. Notice the break at the end of each case. Without that, the program will move to the next case. The break keyword ends the current selection or iteration regardless of the conditions. You can use it at any time to jump out of the current block of code, skipping the remaining code in that block.

The switch/case is used in Javascript exactly the same way as C/C++ and with the same syntax. Python, on the other hand, does not have a built-in switch/case, but we can have the feature in Python programs using a **Dictionary,** which we will see later.

Keep in mind that all conditional statements can be written using if/else, and you don't need to use switch/case at all. But those selections that have the following requirements can be written with switch/case:

- Use the same variable

- Compare that variable to see if it is equal to a series of possible values.

> ☞ **Key Point:** The most common selection constructs are if/else (most general with any type of condition) and switch/case (a special situation where conditions are all comparing the same variable with different possible values).

4.3 ITERATION

Repeating a part of the code, usually the same operation but with different data or slight variations, is very common in programming. This is primarily because the ability to define a big and complicated task as a set of small and simple ones is an efficient approach to programming. Adding 1000 numbers sounds quite complicated. But if you think about it as adding two numbers and then repeat that 1000 times, the task is quite simple. Also, computers are pretty fast at performing certain tasks such as arithmetic calculations. So repeating the addition takes a computer a lot less time than us.

4.3.1 While Loops

Almost all programming languages provide support for iterations as another means of deviating from the sequential execution. Just like selection, iteration involves a condition and a block of conditional code. The main difference is that the code will run **as long as the condition is true**, not just **once if it is true**.

In Chapter 2, when talking about a simple task of calculating the average, we saw how using loops can help. Without them, we need thousands of lines of code and variable names just to get a new number and add. Consider the algorithm below. Lines 1–7 show the process of adding numbers without a loop. You can see that if we have 1000 numbers, the code is simply not manageable as it becomes too long. Alternatively, lines 9–14 show the use of a loop that can be extended to as many numbers as we want without a single line of code added.

SAMPLE CODE #6-a

```
1. SUM = 0

2. Get A

3. SUM = SUM + A

4. Get A

5. SUM = SUM + A

6. Get A

7. SUM = SUM + A

8.

9. SUM = 0

10.      COUNT = 0
```

```
11.   While COUNT < 50
12.        Get A
13.        SUM = SUM + A
14.        COUNT++
15. AVERAGE = SUM / 50
```

The Sample Codes #6-c and #6-p show the same operation is C/C++ and Python. Javascript code is almost identical to C/C++, with the difference only in defining functions and variables, as we have seen before. Here are some quick notes about the C/C++ and Python code:

- Always add comments to explain the code.
- Remember that += and ++ operators are used as shorthand version:
  - A += B is the same as A = A + B
  - A++ is the same as A = A + 1.
- In C/C++, numeric data can be integer or float. Since all our data types are int, line 24 will result in an integer division.

SAMPLE CODE #6-c

```
1. #include "stdio.h"
2. void main()
3. {
4.        //variable to receive the numbers
5.        int a;
6.        //variable to hold sum
7.        int sum = 0;
8.        //variable to count numbers
9.        int count = 0;
10.           //loop
11.           while (count < 50)
12.           {
13.                   //get a new number
14.                   printf("Enter a number: ");
```

```
15.                    scanf_ s("%d", &a);
16.
17.                    //add to sum
18.                    sum += a;
19.
20.                    //increment count
21.                    count ++;
22.            }
23.        //calculate average
24.        int average = sum / 50;
25.        printf("average = %d\n", average);
26. }
```

For both C/C++ and Python codes, notice how while has a syntax similar to if. In the case of Python (below), pay attention to how while does not use ( ) and has a : instead. Also, notice how we combined the int and input functions into one line, compared to lines 1 and 2 in Sample Code #2-p. The result of input is passed to int. We will talk about the functions in the next chapter.

SAMPLE CODE #6-p

```
1. count = 0
2. sum = 0
3. while count < 50 :
4. a = int(input("enter a number"))
5. sum += a
6. count += 1
7. average = sum / 50
8. print(average)
```

☑ **Practice Task:** Modify the above sample codes, and write a program that calculates factorial. You wrote an algorithm for it in Section 2.2.

Here is another example of using `while` loops where no counter is needed. In this case, we continue the iteration for an indefinite number of times until a certain number is entered by the user. The example involves using selection and iteration together. It detects if the input number is odd or even, and repeats until the user enters zero.

Sample Code #7-c

```
1. #include "stdio.h"
2. void main()
3. {
4.     int number = 1;
5.     while(number!= 0)
6.     {
7.          printf("Enter a number (zero to end: \n");
8.          scanf _ s("%d",&number);
9.          switch(number%2)
10.             {
11.             case 0 :
12.                    printf("Even\n");
13.                    break;
14.             case 1 :
15.                    printf("Odd\n");
16.                    break;
17.             default:
18.                    printf("Invalid number\n");
19.             }
20.     }
21. }
```

☑ **Practice Task:** In the above code, when the user enters zero, the current iteration continues. Change the code, so we no longer do anything if zero is entered.

When using iterations, it is important to pay attention to which part of the code needs to be inside the loop. In this case, line 4 (defining the variable that holds the number) does not need to be repeated, so it is outside the loop. We use the same variable every time and change its value by getting new user input. The old value is no longer needed.

> ☞ **Key Point:** A while loop is similar to an if statement except that the code runs as long as the condition is true, not just once.

4.3.1.1 Forever Loops
We can use a loop without any condition. Leaving the ( ) empty:

```
while()
```

Or using the keyword `true`:

```
while(true)
```

This will result in a loop that never ends because the condition is always true. We usually refer to this as a forever loop. Such a loop, while technically possible and sometimes helpful, should be avoided or used very carefully as they may result in a program that is stuck at the loop. If using a forever loop, you should generally have a way of getting out. For example,

```
if (number == 0)

break;
```

Using such loop in Sample Code #7-c has the advantage that when we enter 0, the program gets out of the loop and doesn't detect odd/even or print a message.

Note that a C/C++ program running as a Console (text-based) application can be terminated by pressing Control-C

> ☞ **Key Points**
> • Using the while construct, we can have definite loops (using a count variable and a condition) or indefinite loops (using only the condition).

> ☑ **Practice Task:** Write a C/C+ program that stays in a forever loop, receives a number N from the user, and prints it and its square value (N*N) with a message such as "the square value of 5 is 25."
> Implement a way to end the loop.

4.3.2 For Loops

Sample Codes #5 illustrates a common case when we know how many times a loop is repeated. Considering the common use of this case, most programming languages have another iteration method that simplifies definite loops. This new method is usually implemented using the keyword for, as shown in Sample Code #8-c.

Sample Code #8-c

```
1. #include "stdio.h"
2. void main()
3. {
4.    for(int count=0; count< 10; count++)
5.            printf("*****\n");
6. }
```

A for loop consists of the following parts:

1. The for statement:

   1.1. Keyword for

   1.2. Action at the start of the whole loop, for example, defining a counter and initializing it to zero

   1.3. Loop condition, for example, counter less than a certain number

   1.4. Action at the end of each iteration, for example, incrementing the counter

2. Conditional block of code

The Javascript syntax is exactly the same as C/C++ (except for data type that is not needed and replaced with var). The <br> tag in HTML is equivalent of \n in C/C++. Since the output of the code is written into an HTML paragraph, we use that to create a new line.

Sample Code #8-j

```
1. <p id="demo">?</p>
2. <script>
3.    var text = "";
4.    for(var i=0; i < 10; i++)
```

```
5.    {
6.            text += "*****<br>;
7.    }
8.    demo.innerHTML = text;
9. </script>
```

For loops in Python work with a set of data such as a string of text or a list of numbers. We will discuss them later when we talk about the modularization of data.

You notice that the difference between for and while loops is the addition of the Parts 1.2 and 1.4. There is an action that happens at the start of the whole loop (in the above code, defining a variable) and another action that happens at the end of each iteration (in the above code, incrementing that variable). Note that these actions can be anything that means for loops don't have to be definite. Instead of defining a counter, we can do any other initial and iterative actions that are needed in our program. For example, the Sample Code #9-c uses a for loop to detect the odd and even numbers:

Sample Code #9-c

```
1. #include "stdio.h"
2. void main()
3. {
4.    int number = 1;
5.    for(printf("Enter a number (zero to end: \n"); number!= 0;)
6.    {
7.            scanf _ s("%d",&number);
8.            switch(number%2)
9.            {
10.                    case 0 :
11.                            printf("Even\n");
12.                            break;
13.                    case 1 :
14.                            printf("Odd\n");
15.                            break;
```

```
16.              default:
17.                  printf("Invalid number\n");
18.          }
19.      }
20. }
```

You see that in this code, we have no counting variable at all. The initial action of the for loop is printf(), and there is no ending action. It is common practice, though, to use for loops with a counter as in Sample Code #8-c and #8-j, and I suggest you use them that way.

☞ **Key Point:** In addition to a condition, for loops include an action to be performed at the start and an action to be performed at the end of each iteration.

☑ **Practice Task:** Modify the Sample Code #7 to ask user for a number and then run the loop for that many iterations.
☑ **Practice Task:** Write a program that asks the user for the number of courses, then stays in a loop, and receives the grade for each course. Then, modify the code to calculate the average.
   Modify the code, so instead of asking the number of courses, we use a forever loop and end it if the user enters –1.

✋ **Reflective Questions**
• What do you think about the differences and similarities of selection and iteration constructs in different languages? Is the approach of learning these languages together helping you understand the common concepts?

4.4 COMBINING SELECTION AND ITERATION

Most programs need to use both selection and iteration, in many cases combined together. In this section, we will see a few example programs that demonstrate such a need. These examples also provide a context for us to put together all we have learned, from the algorithmic thinking and problem-solving to language syntax and control flow constructs.

4.4.1 Guessing Game

For our first example, let's make our very first game. Guessing games are very simple in logic and implementation. We will follow our data-centered design approach to first visualize and design our algorithms. Then, we write the code in multiple languages.

What We Have (Our Major Data Items)

- The target number. When playing against the computer, this has to be randomly generated. For now, we assume a single-player game.

- A range for the target number (for example, less than 100). We can hard-code this and assume it is fixed, or have it entered by the player.

- The player guesses. This number changes throughout the game, depending on the game logic.

What We Do (Major Operations on Data)

1. Initialize the target randomly

2. Ask for guess

3. If not correct, then

   a. Give a hint (higher or lower)

4. Repeat until the guess is correct

If you are not clear about the algorithm, always ask yourself the three questions about how to initialize (line 1), when to use (line 3), and when to change variables (line 2). Exhibit 4.4 shows the visualization of this game.

☑ **Practice Task:** Write the pseudocode for the guessing game.

The C/C++ code for this game is shown in Sample Code #10-c.

Sample Code #10-c

1. #include "stdio.h"   //for standard input/output

2. #include "stdlib.h"   //for rand()

3.

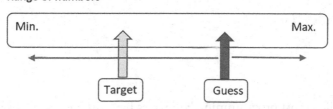

EXHIBIT 4.4   Guessing game.

```
4. void main()
5. {
6.     //parts of the code that runs only once
7.
8.     //target number
9.     int target = rand() % 100;
10.
11.        //player guess
12.        // no need to initialize
13.        // as the player will enter value)
14.        int guess;
15.
16.        while (true)      //forever loop
17.        {
18.            printf("enter a number between 0 and 100: ");
19.            scanf _ s("%d", &guess);
20.            if (target > guess)
21.                printf("go higher\n");
22.            else if (target < guess)
23.                printf("go lower\n");
24.            else
25.            {
26.                printf("correct!");
27.                break;
28.            }
29.        }
30. }
```

As mentioned earlier, most programming languages have different sets of functions, in addition to language keywords and operators, to facilitate common tasks. These are called **libraries**, provided either as part of the standard distribution package for the language or through third-party developers. The `stdio.h` (Standard Input/Output) is an example of such libraries. It is provided by any C/C++ compiler and supports functions such as `printf()`. Another example is `stdlib.h` (Standard Library). It provides some other functions, including `rand()` that generates a random integer number (line 22). The remainder operator applied to the result of `rand()` creates another random number guaranteed to be less than 100.

Notice that if the guessed value is not less than or greater than the target, it has to be equal, and that is when we print the winning message. We also need to end the loop. In order to do that,

- The conditional code is inside { } to be more than one line.

- The keyword `break` is used to end the loop.

☑ **Practice Tasks**
- Write the Python and Javascript versions of the Guessing Game.
- Using forever loops is not a good programming practice. Can you think of a variable that can be used as a condition for the loop in line 29? See the next example for an idea.

✌ **Reflective Question.** Do you see any strengths or weaknesses in different languages we are discussing?

4.4.2 Simple Calculator

In Section 4.1, we talked about a type of program called command processor, which involves receiving a command and performing a process in response. While a calculator can be a complicated program with all typical features, here we consider a simplified case with four commands corresponding to four standard arithmetic operations, each requiring two operands.

Exhibit 4.5 shows the visualization of this simple calculator. It has three major data items and four possible operations. The program stays in a loop and updates the data items and then performs the required action using a command processor-style selection construct. Sample Code #11-p shows the code in Python.

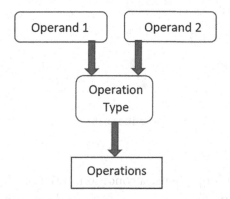

EXHIBIT 4.5   Simple calculator.

Sample Code #11-p

```
 1. quit = False
 2. while quit == False :
 3.     #get the operation
 4.     print("0-exit 1-add 2-subtract 3-multiply 4-divide")
 5.     op = int(input("enter 1-4 to choose an operation: "))
 6.
 7.     #get the first number
 8.     n1 = float(input("enter a number: "))
 9.
10.     #get the second number
11.     n2 = float(input("enter a number: "))
12.
13.     #command processor
14.     if op == 1:
15.         print(n1+n2)
16.     elif op == 2:
17.         print(n1-n2)
18.     elif op == 3:
19.         print(n1*n2)
20.     elif op == 4:
21.         print(n1/n2)
22.     elif op == 0:
23.         quit = True
24.     else:
25.         print("invalid operation")
```

Here are some important notes:

- Instead of the forever loop, we use a condition. If the condition can be easily defined based on your existing data items, you can use them. But if you can't immediately think of something or there are multiple things that can result in the loop ending, then you can think of ending the loop as a piece of information itself and define a variable for it, as per our Golden Rule #1. The variable quit in this code tells us that we need to end the loop. It is initialized with False.

- Note that in Python, we have False and True, but in C/C++ and Javascript, we have false and true).

- We give the user an option to end the game (command zero). If that option is selected, we set quit to True, so the loop ends on the next iteration.

☑ **Practice Task:** The loop ends if we choose command zero, but it still finishes the current iteration and asks for n1 and n2, even though there is no operation. Modify the code, so the loop ends immediately if we choose to exit.

✋ **Reflective Questions:** What have you learned about proper UI? What information and functionality should be provided to users, in addition to the main task? How about checking if the user has made an error, such as invalid input? We will discuss UI and error handling more in the next chapters.

4.5 NAMING CONVENTIONS

In this and previous chapters, we saw how to define new variables and functions. A common part of defining new items in all programming languages is choosing a name. While there are some rules, such as being case sensitive and limited to letters, numbers, and few symbols like underscore, the programmers are usually free to use any name. Different practices have grown out of the experience of programmers over that last few decades that can be summarized into two main approaches:

- Using short and simple names is the older approach and was based on the fact that programmers did not have access to editors with features such as auto-complete, syntax highlighting, and graphical interfaces. Having shorter and simpler names meant shorter and easier to read code.

- Using meaningful and possibly longer names is a more recent approach motivated by the availability of modern IDE's and also more emphasis on the readability of the code, collaborative work, and software quality standards.

While there are those who argue in favor of both these methods and various combinations, the following guidelines can help have a more readable, understandable, and memorable code:

- Be consistent with your naming method, whatever it is.

- For variables that are used frequently in the program and have important roles (key data items), use meaningful names. This can apply to all variables, but especially the frequently-used and important ones.

- For less important and rarely used variables, you may use short and common names. For example, it is common among programmers to use names i and j for the counting variables of a for loop.

  - for (int i=0; i < 10; i++)

  - Use common names such as count (for a variable that counts something), temp (for a variable that holds a data temporarily), and num (for a number or the number of some items).

- Separate different parts of a name using underscore or upper case letters, for example, user _ data, userData, or UserData

- Differentiate between function and variable naming. For example, variables can always start with lower case letters and functions with upper case, as in userData vs. UserData().

There are also naming conventions (semi-standard styles accepted by a large group of programmers) that you may adopt. A common one is the **Hungarian Notation**. There are variations for this naming convention, but the general idea is that the variable names start with a lower case part showing their data type plus one or more title-cased parts (the first letter is upper case) that for meaningful names. Examples are iUserData for an integer user data and sName for a string of text showing a name.

Your company or development community may have its own standard or convention, and you may develop your own, but either way, the key is to be as consistent as possible, so your code is more understandable.

☞ **Key Point:** There are different naming conventions, but the most important guideline is to be consistent and use meaningful names when possible.

✋ **Reflective Question:** What is your preferred naming method? Short or long? Why?

4.5.1 #define in C/C++

C/C++ compilers support a feature called **preprocessor directives**. These are commands that the compiler receives and affect the way the program is compiled. These directives are identified with # at the start. We have already seen one example that allows the use of libraries:

```
#include "stdio.h"
```

Preprocessor directives are not compiled into machine code. They simply tell the compiler to do something. In the above example, the compiler will add a file named stdio.h to the program. We will talk more about this feature in the next chapter.

Another example of preprocessor directives is #define, which defines a label. For example, the following line defines the label HEIGHT as 15:

```
#define HEIGHT 15
```

Note that the line does not end with ; as it is not a C/C++ instruction. It does not create a new variable named HEIGHT. It simply tells the compiler that within the current source file, anywhere that the text HEIGHT is used, it should be considered the same as 15. This is identical to use your text editor's Replace All feature and replace all HEIGHT with 15.

When using #define, it is common (not required) to use all-caps text, and use underscores to separate parts, such as HEIGHT and DATABASE _ SIZE.

Sample Code #12-c demonstrates the use of a defined label (SIZE). You can notice the following advantages of using this label:

- Since the value is used multiple times, if we decide to change it, we only need to change the #define line.

- The code is more readable and understandable because it refers to meaningful names and labels instead of random-looking numbers.

Sample Code #12-c

```
1. //This program is not complete.
2. //only the related parts are shown
3. #define SIZE 10
4.
5. //somewhere in the code
6. //drawing a line made of * characters
7. for(int i=0; i<SIZE; i++)
8. printf("*");
9.
10. //somewhere else in the code
11. //drawing another line of the same length
12. for(int i=0; i<SIZE; i++)
13. printf("=");
```

C and C++ also have #ifdef (if defined) and #endif directives that allow conditional inclusion of code in the compiling. In the following code, if the label SIZE has been defined, it will be used to create an array. Otherwise, an array of size 10 will be created. Note that the final executable does not have any conditional statement. This is a command for the compiler.

```
#ifdef SIZE
int data[SIZE];
#else
int data[10];
#endif
```

Here is another example using #ifndef (if not defined) that gives a value to SIZE if it is not defined already:

```
#ifndef SIZE

#define SIZE 10;

#endif
```

Note that #ifdef and #ifndef have to have a matching #endif. #else and #elif (else if) are optional.

HIGHLIGHTS

- Golden Rule of Programming #3: Programs run sequentially and linearly, by default. Various control flow mechanisms can alter this, including functions, selection, and iteration.

- In C/C++ and Javascript, the conditional code has to be inside { } if it is more than one line. It is a good and safe practice to always use { } even for one line.

- Programs have sequential execution by default; that is, they run linearly forward step-by-step.

- Selections, iterations, and functions allow us to deviate from the sequential execution.

- The most common selection constructs are if/else (most general with any type of condition) and switch/case (a special situation where conditions are all comparing the same variable with different possible values).

- A while loop is similar to an if statement except that the code runs as long as the condition is true, not just once.

- Using the while construct, we can have definite loops (using a count variable and a condition) or indefinite loops (using only the condition).

- In addition to a condition, for loops include an action to be performed at the start and an action to be performed at the end of each iteration.

- Any iteration can be written using either `for` or `while` loops. But some are easier to do with one of these methods.

- There are different naming conventions, but the most important guideline is to be consistent and use meaningful names when possible.

END-OF-CHAPTER NOTES

A. Things I Should Mention

- **SP** is one of the most popular paradigms that was established in the late 1960s. **OOP**, another popular programming paradigm, was introduced in the late 1950s. Despite being more recent, SP is generally considered a simpler and more universal programming paradigm that most languages support. For that reason, SP is discussed before OOP in this book, as in many others.

- I don't have any preferred naming convention but, in general, tend to start function names with upper case letters and variables names with lower case ones. I also tend to avoid using an underscore to separate different parts of names and use title case instead. But that's just me.

B. Self-Test Questions

- What is Structured Programming, and what are its main elements?

- How is the sequential execution controlled in cases of functions, selection, and iteration?

- What is a block of code, and how do we identify it in C/C++, Javascript, and Python?

- What is the difference between `if/else` and `switch/case` constructs?

- What is the difference between for and while loops?

- What are the similarities of `if` and `while` constructs?

- What are definite and indefinite loops? Are they the same as for and while loops?

C. Things You Should Do

- Check out these web resources:
  - https://www.tutorialspoint.com
  - https://www.onlinegdb.com/online_c_compiler
- Take a look at these books:
  - *Game Programming for Artists* by Jarryd Huntley and Hanna Brady
  - *C: From Theory to Practice* by George Tselikis and Nikolaos Tselikas.

D. Reflect on the Experience of Reading This Chapter

- What did you expect from this chapter before reading it?

- What was it about, and what did you learn?

- What tasks did you perform, and what difficulties did you face?

- How did you feel about the material and tasks presented in this chapter?

- How can you improve your learning experience?

- How do you see this topic in relation to the goal of learning to develop programs?

- Do you see the relationship between different selection and iteration methods?

- Why do you think a `for` loop is necessary, as opposed to using only `while` loops?

- What naming convention makes more sense to you?

Functions

OVERVIEW

So far, in this part of the book, we have seen how to define data and perform basic operations on it. We reviewed selection and iteration as two of the basic control flow mechanisms in Structured Programming. In this chapter, we will talk about functions as the primary means of **modularization**.

Almost any program uses functions, and that is why it is essential to understand how to define and use them, and more importantly, how to share data between them. Remember that a program is really about processing data. So, if you divide a program into multiple modules (functions), then you need to have a way to pass your data to them and receive data from them.

Modularization is a key concept in programming. Not only it is one of the basic features of Structured Programming, it is the key for defining and understanding other programming paradigms such as Object-Oriented Programming and Functional Programming. Modularization is also fundamental in defining various **Software Architectures** (higher-level design of complex software systems).

In this book, I use modularization as the thread that connects many concepts in programming. We will see how these concepts, such as arrays, structures, and objects, can be seen as different ways of achieving modularization in programs. While these are the topic of the following parts of the book, in this chapter, I focus on functions.

5.1 DEFINING AND USING FUNCTIONS

If you pay attention to Sample Code #2-j and #3-j in the previous chapter, you notice that we defined a function called `grade()` to be executed (called) when the user clicks on a button. We defined functions as modules of code, and these two examples demonstrate one of the most important advantages of using such modules. It is hard to write all that code in the `onclick` attribute of the button. Functions allow us to simply refer to a body of code by a simple name any time we need it, instead of writing all that code.

In everyday life, when we say "boil the water," "turn on the car," or "go to school," we don't specify all the steps needed to perform those tasks. We use **names** that we know refer to a **set of actions** such as (1) fill up the pot, (2) put the pot on the stove, (3) turn the stove on, and (4) set the stove to high heat. It is a lot easier to just say "boil the water" when we know what it involves. Using the name also gives us flexibility to perform a modified set of actions without changing the general name. We can still say "boil the water" even if change the process to include camp fire.

In the case of Sample Code #3-c, our C/C++ program had a function called `grade()` as well. But you can imagine that we probably want to perform the "grade" operation multiple times in that program. Again, repeating that code multiple times is not a very efficient use of our time and resources. Referring to that code by a simple name is a lot more convenient.

Convenience is not the only advantage of using functions. As we saw in the case of `print` and `write`, we can have pre-defined modules of code that we can use in our program. These modules of code allow us to perform tasks that we might not know how to do or tasks that have to be done in a certain way, so programmers are not allowed to do them on their own. Operating systems frequently provide functions for accessing system resources such as screen, keyboard, or network, as there are very strict rules for using these resources.

Last but not least, once we define our functions we can reuse them easily in other programs and make complicated applications using existing modules. This is similar to how we use parts from simple LEGO® pieces, and complex objects from parts. Those pieces and parts are modules in our LEGO construction, and functions are key to modularization of the code. They allow us to have programs made of smaller parts, and so more manageable and reusable. Together with selection and iteration, they are the fundamental control flow methods in programs, and the proper use of them is essential in one of the most popular programming paradigms called **Structured Programming**.

To learn more about functions, we start from an example similar to what we did in the previous chapter. Sample Code #1 defines a function in Javascript and uses a call to it as the code for a button's `onclick` event.

In Javascript, we define a function by

- Using the keyword `function`

- Adding the name of the function plus a pair of ( ) for any input it may need

- A pair of { } to hold the code for the function (**body** of the function).

Sample Code #1-j

```
1. <p id = demo>?</p>

2. <input id="data">

3. <button onclick="OddEven();">Odd or Even</button>

4. <script>

5. function OddEven()

6. {

7.     var x = data.value;

8.     if(x%2 == 1)

9.         demo.innerHTML = "odd";

10.    else

11.        demo.innerHTML = "even";

12. }

13. 1</script>
```

Note that in Javascript, the function can be defined before or after the line of code that has the function call (line 3 above) if the call is actually done after the page is fully loaded. This does not contradict the Golden Rule #2 (define before use) because Javascript does not look for the function OddEven at line 3 when it is named. It does that when we load the page (all the way through line 13) and then click on the button. Some other languages like C and C++ are more restrictive, though.

Also, note that we can have many functions defined in the program, but they won't be executed until we execute a line of code that calls those functions. So, without line 3, and an actual click on the button, the function OddEven() will not run. When the browser loads the HTML file, it reads the content. When it reaches the script section, it will execute lines that are not part of any function. You can use this to perform actions that have to happen at the start.

Python programs work in a similar way. We saw that any C/C++ code starts with a function called main(). Python code, on the other hand, starts with the first line that is not part of any function. In Python, we define a function by

- Using the keyword def (short for "define")

- Adding the name of the function plus a pair of ( ) for any input it may need

- Adding a : symbol

- Using indention to add the code (body of the function)

  - Line 7 below is not indented, so it won't count as part of the function OddEven(). This line is the first line that is not part of any function and as such will be executed first. It calls the function.

Sample Code #1-p

```
1. def OddEven():
2. x = 67
3. if x%2 == 1:
4.     print("odd")
5. else:
6.     print("even")
7. OddEven()
```

> ☞ **Key Point:** Functions can be easily identified by the pair of ( ) after a name (not a keyword such as if and while). When defining a new function, the ( ) is followed by a body of code inside { }. When calling an existing one, it can be followed by ; to end the statement or have other code in the same statement.

It is important to remember that a function, no matter where it is defined, is not executed until it is called by a line in the program. In Sample Code #1-j, when we load the HTML file, the browser reads the definition of the function but will not run the code until we click on the button. In Sample Code #1-p, the function is defined at the top of the program, but it won't run until the line 7 where it is called. Both Javascript and Python programs start from the top. Every line that is creating a variable will be executed but function definitions (including their local variables) will not.

☞ **Key Points**
- Javascript and Python programs start from the top. They execute the lines that are not part of any function.
- Functions run when they are called, not when they are defined.

Here is another example of using functions in Javascript that simplifies the `onclick` action of the buttons from the sample code for Sample Code #3 in the previous chapter (calculating total price plus tip). Note how line 4 defines a variable that is used in the program. This line will be executed as soon as the HTML page is loaded because it is not part of any function.

Sample Code #2-j

```
1. <p>Hello, World!</p>

2. <p id="demo">0</p>

3. <script>

4. var price;

5. function SetPrice()

6. {

7.      price = data1.value*5 + data2.value*2;

8.      demo.innerHTML = 'total price is ' + price;

9. }

10. function AddTip()

11. {

12.      demo.innerHTML = 'total price is ' + price*1.2;

13. }

14. </script>

15. <input id="data1">

16. <input id="data2">
```

17. `<button onclick="SetPrice()">Price</button>`

18. `<button onclick="AddTip()">Add 20% Tip</button>`

☑ **Practice Task:** A polynomial is a mathematical function such as $A*X^4+B*X^3+C*X^2+D*X+E$. In a polynomial, X is the variable, and numbers such as A–E are the parameters. The degree of a polynomial is the highest power of the variable (4 in this example). Write a Javascript function called Polynomial that can calculate polynomials up to a degree of 4 by reading parameters and the value of X from a series of edit boxes. An HTML button shall call the function.

- *Hint:* `Math.pow()` function can calculate power or you can just multiply the number by itself.

🖐 **Reflective Questions**
- Do you see the value of functions as the primary way of modularizing the code?
- How do you think we can write a reusable function?

5.1.1 Returning from a Function

Just like the whole program, a function also starts from the beginning and follows sequentially. By default, the functions end when they reach the last line, but we can force a function to end earlier by using the keyword **return**, available in C/C++, Javascript, and Python (and many other languages). This forces the function to terminate and cause the program to return to where it was called. For example, the following Python code returns if the user input is not acceptable:

Sample Code #3-p

```
1. def Divide():
2.      x = input("enter a grade: ")
3.      num1 = int(x)
4.      x = input("enter a grade: ")
5.      num2 = int(x)
6.      if num2 == 0:
7.          return
8.      num3 = num1/num2
9.      print(num3)
10. Divide()
```

Python (and C/C++) also have a standard function called `exit()`, which causes the current function, and the whole program, to terminate.

The second number (num2) in the above sample code cannot be zero as it results in a divide-by-zero error. As such, we will return from the program if such input is entered. This is a good example of input and error checking. It is a good practice that can save you much effort trying to find out why a program crashes or doesn't work properly. I strongly recommend that you never assume a user has entered a proper input if it can cause errors in the program.

☑ **Practice Task:** How else can we write the code #3-p? Is the return needed?

✋ **Reflective Question:** The Sample Code #3-p has one function, and the only other line of code (line 10) simply calls that function. What do you think is the value of defining that function instead of just writing the code as the main program without any function? Think about modularization and reuse.

While a function can define many variables and perform many calculations, in most programming languages, a function is allowed to choose one data item as a **result**. The keyword `return` can be used to specify this result. Sample Code #4-p demonstrates this by modifying #3-c to return a result.

`Sample Code #4-p`

```
1. def Divide():
2.     x = input("enter a grade: ")
3.     num1 = int(x)
4.     x = input("enter a grade: ")
5.     num2 = int(x)
6.     if num2 == 0
7.         return -1
8.     num3 = num1/num2
9.     return num3
10. result = Divide()
11. print(result)
```

In this case, the function does not print any output. It returns the result to be printed by the code that called the function. This result can also be further used by the code. Note that line 10 sets a variable to a function. This is possible only if the function returns a result, as it happened in line 2. Otherwise, the variable will not have a proper value. Also, note the practice of returning an **error code** when something happens in the function. In line 7, we return -1 if we detect an invalid input. We could return any other number, as long as in the program, we consistently use that number as an error code. The program can have a set of error codes for different purposes.

We can use the keyword `return` similarly in Javascript code. We will see an example of that in future sample codes.

☞ **Key Point:** In most programming languages, the keyword `return` can be used to end a function and provide a result.

☑ **Practice Task:** Remove the return from Divide function and then see what will be printed. Use `return` in Javascript.

✍ **Reflective Questions:** What happens if we need to return multiple data items? What other programming methods and syntax are needed for that? Would global variables help? What other language features would be on your wish list for this functionality?

Using functions is not required in Javascript and Python, although most programs are too complicated to organize without functions. Some other languages, on the other hand, require the programmers to use functions. C and C++ (and Java, C#, and many others) are among those.

5.2 FUNCTIONS IN C/C++

Defining and using a function in C/C++ is very much like Javascript. The main difference is that functions, similar to variables, need to have a data type. You may wonder what it means to have a data type for an operation. It makes sense to talk about data type for a variable. The function type in C/C++ is simply the type of data it is producing.

A function can work with many variables and use and create different data. In C/C++ and many other languages, only one of these can be treated as the formal "result" of the function. A C/C++ function identifies this using the keyword `return`. If a function has a "result," then its type is the same as that result (a.k.a. the **return value**). If a function does not have a result, then its type is identified as **void**.

Let's see an example of using functions and their types and return values.

Sample Code #5-c

```
1. #include "stdio.h"
2. void Hello()
```

```
3. {
4.      printf("Hello, World!\n");
5. }
6. int GetData()
7. {
8.      int number;
9.      printf("please enter a number: ");
10.     scanf _ s("%d", &number);
11.     return number;
12. }
13. void main()
14. {
15.     Hello(),
16.     int x;
17.     x = GetData();
18.     printf("%d\n", x);
19. }
```

The above code defines three functions:

5.2.1 void main()

This is the required function in all C/C++ programs. The program starts by calling the main() function and ends when this function reached its end. Any other function is executed only if it is called within main or another running function.

The main() function can be defined as void (no return value) or int (returning an int value). Since the end of main() means the end of the program, the return value from this function has no use in the program itself. This value is used by the operating system to know the reason for program termination. A return value of 0 usually means a normal termination. Other values identify various error codes. These values are program dependent and will be reported to the user. For example, a database programmer may choose to return error code 1 to identify insufficient disk space to save data. The reported value will let the programmer know what the internal problem was when the program ends unexpectedly. This code may not mean much to an end user, though.

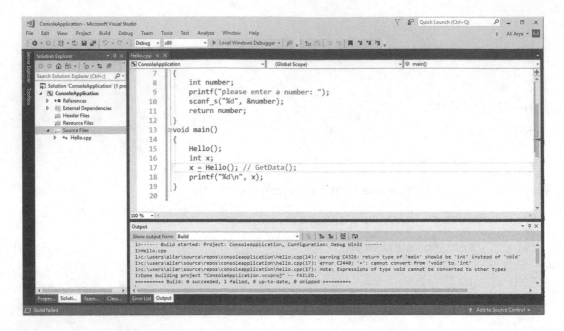

EXHIBIT 5.1   Compiler error.

5.2.2 void Hello()

This function is called by main(). It performs an action and then ends without returning any result.

5.2.3 int GetData()

This function shows a message to the user and receives an int value. This value is stored in a variable called number and then returned as the result. The main() function has a variable called x, and it is set to the return value of GetData().

In line 17, recall that the=(assignment) operator calculates the value on the right side and sets it to the variable on the left side. When calculating the value on the right side, the function is simply replaced with its return value. You can imagine that a void function cannot be used in an assignment operation as it has no value. Try replacing GetData() with Hello() in line 17, and see the error message from the compiler as shown in Exhibit 5.1.

As we can see, the keyword return is used in C/C++ in the same way it is used in Javascript and Python. The only difference is C/C++ functions have a defined type. This means that if we add a return value, then we have to modify the function type. A void function can use return but without a value (only to terminate the function). A non-void function, on the other hand, has to have a return line with a value of the right type.

☞ **Key Point:** Similar to variables, C/C++ functions need to have data type both for the return value and for parameters.

☑ **Practice Task:** Try removing the return line or changing the return value to different types. Then, notice the compiler errors.

✋ **Reflective Question:** Why do you think C/C++ need a `main()` function? If that function is not called from anywhere in the code, what is the point of return value for `main()`? Recall that `main()` can be void or int.

The `exit()` function in C/C++ (defined in `stdlib` library) also requires a parameter, for example, `exit(0)`. This is similar to the return value from `main()`. These values are reported to the operating system or Integrated Development Environment (IDE) that has started the program. They are methods for reporting why and how the program terminated.

5.3 LOCAL AND GLOBAL VARIABLES

Imagine we are writing a simple program that asks the user to enter first name and last name and then creates a single string of text as the full name. If we try to visualize this program, we will have something as illustrated in Exhibit 5.2, with three major data items: first name, last name, and full name.

The algorithm will be fairly simple as

```
Sample Code #6-a
```

```
1. Get FIRSTNAME
2. Get LASTNAME
3. FULLNAME = FIRSTNAME + LASTNAME
4. Print FULLNAME
```

The Python code is shown below and is divided into three sections: Input, Process, and Output. Remember that in Python, comments are defined using # as opposed to // in C/C++ and Javascript.

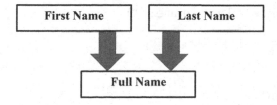

EXHIBIT 5.2   Full name program visualization.

Sample Code #6-p

```
 1. #input
 2. print("enter first name: ")
 3. firstName = input()
 4. print("enter last name: ")
 5. lastName = input()
 6. #process
 7. fullName = firstName + " " + lastName
 8. #output
 9. text = "your full name is " + fullName
10. print(text)
```

In Chapter 4, when discussing switch/case, we talked about a program that receives a set of commands and performs appropriate actions. This is a common structural pattern in programs, and we called it a **Command Processor**. The example above is another common pattern in programs that we may call **Input-Process-Output (IPO)**. The IPO pattern is a linear and extremely useful program organization. In this simple example, it may not be clear why separating these three parts is helpful. But once the program grows in size and complexity, each of these three parts can become longer and more complicated.

One of the primary purposes of functions is to organize the code into modules, as we will see more in the next chapter. To start with this idea, let's split the above code into three functions and then call them. The Sample Code #6-p has two parts: the functions and the main code that calls the functions similar to the main() function in a C/C++ program.

Sample Code #6-p-v2

```
 1. #functions
 2. def Input():
 3.     print("enter first name: ")
 4.     firstName = input()
 5.     print("enter last name: ")
 6.     lastName = input()
 7.
 8.  def Process():
 9.     fullName = firstName + " " + lastName
```

```
10.
11. def Output():
12.     text = "your full name is " + fullName
13.     print(text)
14.
15. #main code
16. Input()
17. Process()
18. Output()
```

If we run the above code, we notice that the program will have an error that only shows up when running the program (a **run-time error**), as shown in Exhibit 5.3.

If you remember, when we were talking about defining blocks of code in the previous chapter, I mentioned that they are also the basis of what we call a lexical scope, i.e., a part of the code where names have meaning. A variable (or a function) is generally accessible and meaningful only within the block of code where it is defined. In particular, variables in a source file are categorized into two groups:

- Local variables are those defined inside a function and are only accessible within that function.

- Global variables are those that are defined outside all functions and are available to all functions.

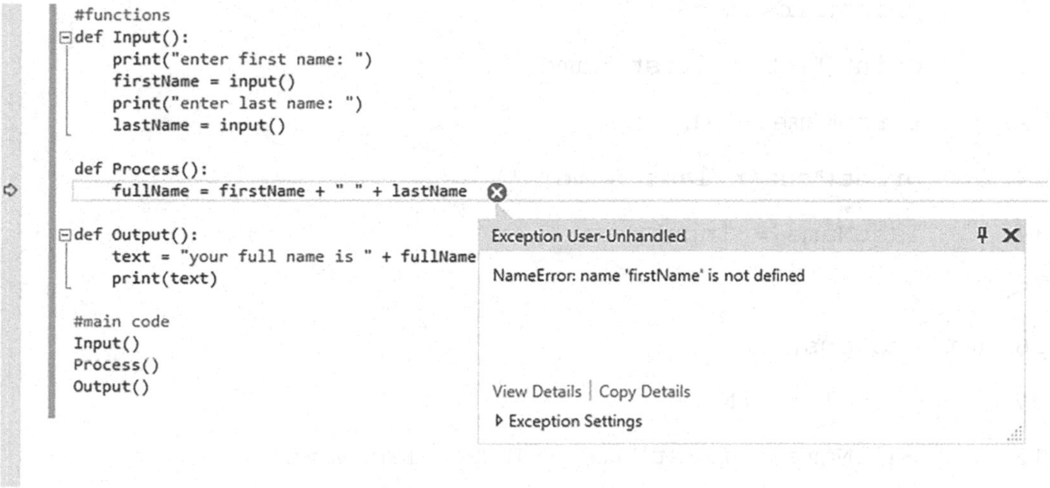

```
#functions
def Input():
    print("enter first name: ")
    firstName = input()
    print("enter last name: ")
    lastName = input()

def Process():
    fullName = firstName + " " + lastName  ⊗

def Output():
    text = "your full name is " + fullName
    print(text)

#main code
Input()
Process()
Output()
```

Exception User-Unhandled   ⏻ ✕

NameError: name 'firstName' is not defined

View Details | Copy Details
▷ Exception Settings

EXHIBIT 5.3   Undefined variables.

Think of local vs. global variables as individual names; any family can have a John or Jane (local variables), but when you use the names of celebrities like Picasso or Beyoncé, they are globally known. Even though there may be other people with those names, we assume those celebrities by default.

The Sample Code #6-p does not work because Python creates all variables as local and, as such, Process() has no variable named firstName and lastName, and Output() has no fullName.

Functions can have local variables with the same name without any conflict as each uses its own variable. They can also have local variables with the same name as a global variable. In such cases, the program uses the local variable. If you have data that needs to be used by multiple functions, the easiest way to share the variables among all functions is to define them globally. Here is how the name program works with global variables.

Sample Code #7-p

```
1. #global variables
2. firstName = ""
3. lastName = ""
4. fullName = ""
5.
6. #functions
7. def Input():
8.      global firstName
9.      global lastName
10.     print(firstName)
11.     print("enter first name: ")
12.     firstName = input()
13.     print("enter last name: ")
14.     lastName = input()
15.
16. def Process():
17.     global fullName
18.     fullName = firstName + " " + lastName
19.
```

```
20. def Output():
21.     text = "your full name is " + fullName
22.     print(text)
23. #main code
24. Input()
25. Process()
26. Output()
```

Remember that in Python, there is no explicit way to define a new variable. The first time that you assign a value to a variable that does not exist, Python creates a new variable for it. As such, an instruction like the one in line 12 will create a new local variable unless Python knows that you intend to use the global variable. For that reason, functions Input and Process that are assigning values to global variables need to declare those variables as global (lines 8, 9, and 17). On the other hand, the function Output only reads the value of a global variable, so there is no confusion about creating a local one, and so no need for declaring `fullName` as global.

> ☞ **Key Points**
> - Global variables are those defined outside all functions. They are accessible anywhere in the source file. Local variables are defined inside a function and are only valid within that function.
> - Two functions can have local variables with the same name.

Declaring a variable as global inside a function is not necessary for Javascript and C/C++ as they define their local variables explicitly using keywords, so it is clear when we want to use a global variable or create a new local one. Sample Code #7-c shows how to use global variables and the IPO pattern in a C/C++ program that processes student data. Using text data in C/C++ requires some features that we will discuss in the next chapter, so this example shows the use of global numeric variables.

Sample Code #7-c

```
1. #include "stdio.h"
2. //global variables
3. int id;
4. int grade1;
```

```
 5. int grade2;

 6. float gpa;

 7. //functions

 8. void Input()

 9. {

10.         printf("enter student ID: ");

11.         scanf _ s("%d", &id);

12.         printf("enter the first grade: ");

13.         scanf _ s("%d", &grade1);

14.         printf("enter the second grade: ");

15.         scanf _ s("%d", &grade2);

16. }

17. void Process()

18. {

19.         gpa = (grade1 + grade2)/2.0;

20. }

21. void Output()

22. {

23.         printf("ID=%d: G1=%d, G2=%d, GPA=%f \n",

24.         id, grade1, grade2, gpa);

25. }

26. //main code

27. void main()

28. {

29.         Input();

30.         Process();

31.         Output();

32. }
```

> ✋ **Reflective Questions:** If the use of global variables is not encouraged, and sharing local variables is not allowed, then do you think there should be a way to share a variable among a selected group of functions? This is the basic idea behind "objects," which I will discuss in Chapter 9. Think about how such a feature may be defined and used.

5.4 FUNCTION PARAMETERS

Developing good software is not just about writing a code that "works," as in "does what it needs to do." It is also about how well it works (performance), how easy it is to use (usability), how easy it is to change and upgrade (maintainability), how easy it is to read and understand (readability), and how much it can be used in other programs (reusability), among other factors. If you look at the Sample Codes #7, you notice two important issues:

• If we inspect the "main" part, and without reading all the code, it is hard to know what type of information we are processing. Each of the functions is using some global variables (and not all of them), and those variables can be defined anywhere in the code. This makes the code hard to read and understand, and so hard to maintain.

• The functions rely on external data, so each of them is not a self-sufficient module. We cannot simply copy a function to another program and hope to use it. This makes the code less reusable.

Global variables are easy solutions for exchanging and sharing data between multiple functions, but they are not the best way to do so. Functions are modules of code that perform actions. They can have input and output data. In previous discussions, we saw how a function could return a result as the output. This result can be a simple number or a complicated type, such as an image or even a database, as we will see later. The functions can also receive data as input, as have seen in cases like print(). The input data for a function are usually called **function parameters** or **arguments**.[1]

Parameters are information required to perform an operation. If you tell someone to move, they probably need to know where, or in which direction and how far. These are the parameters of the action "move." If a function requires parameters, we need to include them on two different occasions:

• When the function is defined, we need to specify what type of data it needs.

• When the functions called, we need to provide appropriate data. This is called **passing parameters**.

---

[1] Later, I will talk about how parameters can be used to give input to a function and receive output from it by providing a container to hold the output.

Parameters are local variables of a function but have to be given values when the function is called. Sample Code #8-c shows how to define and use a function with parameters. It is based on the Sample Code #7-c but without the use of global variables.

Sample Code #8-c

```
1. #include "stdio.h"
2.
3. //functions
4. int Input()
5. {
6.     int num;
7.     printf("enter a number: ");
8.     scanf_s("%d", &num);
9.     return num;
10. }
11. int Process(int num1, int num2)
12. {
13.     int num3 = num1 + num2;
14.     return num3;
15. }
16. void Output(int num1, int num2, int num3)
17. {
18.     printf("%d + %d = %d \n", num1, num2, num3);
19. }
20.
21. //main code
22. void main()
23. {
24.     int num1 = Input();
25.     int num2 = Input();
```

```
26.

27.     int num3 = Process(num1, num2);

28.

29.     Output(num1, num2, num3);

30. }
```

> ☞ **Key Point:** A parameter (or argument) is a local variable in a function that has to be initial-ized by the code that calls the function. This act of initialization is called "passing parameters."

In the Sample Code #8-c, after line 13 inside Process(), add code to change the value of num1 and num2 to zero. In lines 26 and 28, inside main(), print the value of these variables and see if they change? Notice that even though we have changed them in Process() (line 27), their value remains unchanged in main(). This is because num1 and num2 in Process() are not the same as num1 and num2 in main(). The variables in Process() are initialized with the same value as the ones in main() at the time Process() is called. After that, they are completely separate.

Instead of printing these values, most IDE's allow programmers to pause the program and inspect the value of different variables at any time. This is part of the functionality provided by a tool called **Debugger**. A detailed description of debugging techniques is out of the scope of this book. See the sidebar for some quick information.

SIDEBAR: DEBUGGING

Debugging is the act of testing and inspecting a computer program in order to iden-tify and correct problems. A debugger is a program that helps with this process. Most modern IDEs such as Visual Studio include a series of debugger features such as

- Breakpoint: specifying any line in the code so that the program pauses upon reaching that line.
- Watch: different forms of inspecting the value of variables including all variables in a line of code, pre-defined set of variables that are being tracked, or any manually selected variable.
- Step-by-step execution: running the program line by line after a breakpoint.
- Stepping in and out of functions: inspecting the code inside a function that is being called in the current line or finishing the current function and moving to the next line after the function call

Depending on the availability of debugging information, debuggers can show the source code of the libraries used by the program as well as the program itself. Most compilers create two versions of the executable code: Debug (with debugging information such as source code) and Release (without that information). The Debug version is generally larger in size and slower in performance. It is only suitable for programmers during the testing. Once that is done, a Release version will be compiled and made available to clients.

> ✋ **Reflective Questions**
> - Do you see the difference between a parameter and a normal local variable?
> - Why do you think we need parameters?
> - What is your preferred way of sharing data between functions?
> - Could you imagine a different method of sharing data between functions? You may investigate and see if it exists. For example, in Chapter 10, we will talk about passing parameters by reference and by value. It is a way of passing parameters without local data.

```
void main()
```

The main() function can also be defined with parameters.

```
int Input()
```

This function does not require any input data, so it has no parameters. When defining or calling the function, there is nothing inside the ( ). If the function is defined with no parameters, calling it with some data (passing parameters to it) will result in a compiler error.

On the other hand, the function shares its result by returning a value. When calling the function, we may set it equal to a variable of the same type to use the output of the function (lines 24 and 25). This is not required, though, and it sometimes happens that the function performs an action and also creates a result, but we are not interested in the result. We will see examples like this later.

```
int Process(int num1, int num2)
```

This function receives two parameters. These parameters are local variables, which means they cannot be used outside the function. Other functions may have variables with the same name, but they will be separate data. So num1 in Process() and num1 in main() are two separate variables. If Process() changes the value of num1, the num1 in main will not change.

When the function is called (line 27), the caller (the main function) has to provide two values to be used to initialize num1 and num2 in Process(). The caller can use a variable, a constant data, an operation, or even another function that returns a value.

```
void Output(int num1, int num2, int num3)
```
This function receives three parameters and returns none. Again, the caller can use any value for these parameters, so any of the following lines will be correct:

```
Output(2, 6, 8);

Output(2, 6, 2+6);

int x=9;

int y=10;

Output(x, y, x+y);

Output(Input(), Input(), 0);
```

Note that the last line above is syntactically correct, but since the function does not calculate num1+num2 and we have not stored the result of two Input() function calls, we have no way of providing a correct num3 to output().

5.4.1 Simple Board Game Simulator

Board games involve various information and activities. Here, we consider a very simplified version with only the most basic part: two pieces that can move along a path independently with the movement controlled by throwing dice. Exhibit 5.4 visualizes our program.

What We Have (Our Major Data Items)

- The location of two players. These are two integer numbers that start from zero and increment by random numbers (dice) at each round.

- The dice value. This is the random number generated by throwing the dice.

- Some sort of visual representation. Since we have not discussed computer graphics, this will be done simply through a line with increasing size, as shown in Exhibit 5.5.

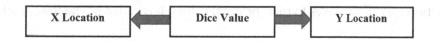

EXHIBIT 5.4   Simple board game simulator.

```
-------------------------------------------------------X won!

---------------------------------------Y

[                              ]  Run One Round   Reset
```

EXHIBIT 5.5   Sample output of the board game.

What We Do (Major Operations on Data)

1. Initialize the locations to zero

2. Throw dice

3. Move player 1

4. If reached the end, then

5. Say "you win"

6. Repeat 3-5 for Player 2

7. Repeat 2-6 if no player has won

The Sample Code #9-j shows the code in Javascript. The front end is similar to our previous Javascript examples. Two lines represent the progress of two players, and characters X and Y show the current locations. In addition to location and dice variables, we also define a new variable that tells us the game is over (someone has won). Although we didn't include that in "what we have" above, after writing the algorithm, it is clear that we need this information to decide if we continue throwing dice or not (line 7 above). As per Golden Rule #1, we should define a variable for this information. It will be initialized to false (a keyword in most languages generally meaning zero).

The program is made of three functions.

`Dice()`

This function simulates throwing dice and creates a random number. `Math.random()` is equivalent of `rand()` in C/C++, but it creates a float data between 0 and 1. Multiplying by 12 makes it between 0 and 12, and then, the function floor() rounds it up so we will have 1–12.

`Reset()`

This function sets all variables to their original values. It is called by the Reset button.

`Game()`

This function is the main function that is called by the Run button and calls `Dice()` itself.

Sample Code #9-j

1. `<!-- front end: X and Y represent two players -->`

2. `<p id="player1">`

3. `X`

```
 4. </p>
 5.
 6. <p id="player2">
 7. Y
 8. </p>
 9.
10. <input id="userData">
11. <button onclick="Game()">Run One Round</button>
12. <button onclick="Reset()">Reset</button>
13.
14. <!-- Javascript code -->
15. <script>
16. var dice,
17. var p1=0;
18. var p2=0;
19. var gameOver = false;
20. function Dice()
21. {
22. dice = 12*Math.random();
23. dice = Math.floor(dice);
24. dice++;
25. }
26. function Reset()
27. {
28. p1 = p2 = 0;
29. player1.innerHTML = "X";
30. player2.innerHTML = "Y";
31. gameOver = false;
32. }
```

```
33. function Game()
34. {
35.     if(gameOver == true)
36.                 return;
37.
38.     Dice();
39.     p1 += dice;
40.     var line1 = "";
41.     for(i=0; i < p1; i++)
42.     {
43.                 line1 += "-";
44.     }
45.     line1 += "X";
46.     if(p1 >= 50)
47.     {
48.                 line1 += " won!";
49.                 gameOver = true;
50.     }
51.     player1.innerHTML = line1;
52.
53.                 if(gameOver == true)
54.                 return;
55.
56.
57.     Dice();
58.     p2 += dice;
59.             var line2 = "";
60.     for(i=0; i < p2; i++)
```

```
61.       {
62.                       line2 += "-";
63.       }
64.       line2 += "Y";
65.       if(p2 >= 50)
66.       {
67.                       linc2 += " won!";
68.                       gameOver = true;
69.       }
70.       player2.innerHTML = line2;
71. }
72. </script>
```

5.4.2 BMI Calculator

Body-Mass Index (BMI) is calculated using the following formula:

$$BMI = \frac{weight\_In\_Kilo\_gramo}{height\_In\_Meters \times height\_In\_Meters}$$

Let's write a program that asks the user to enter `weightInKilogram` and `heightInMeters`, calculates and prints `BMI`, and then identifies and prints the BMI category:

- Underweight, less than 18.5

- Normal weight, between 18.5 and 24.9

- Overweight, between 25 and 29.9

- Obese, greater or equal 30.

The data items are the input weight and height plus the calculated BMI. The operations are mainly (1) getting the input, (2) calculating BMI, and (3) giving a health report. It makes sense to have a main module that calls these three operations either as part of the main code or a separate function. Sample Code #10-c shows a possible implementation of this program in C/C++.

Note that w and h are local variables in both functions, and having the same names doesn't make them the same variable or sharing the same value automatically. As parameters, w and h in BMI are initialized by whatever value is passed to the function when it is called. So, any of the following lines would work:

```
BMI(75, 1.75)     //or any other number

BMI(w, h) //or any other variable
```

Sample Code #10-c

```
1. #include "stdio.h"
2.
3. //functions
4. float BMI(float w, float h)
5. {
6. return w / (h*h);
7. }
8.
9. //main code
10. void main()
11. {
12.        float w = 1;
13.        float h = 1;
14.
15.        //input data
16.        printf("enter weight (kg): ");
17.        scanf_s("%f", &w);
18.        printf("enter height (m): ");
19.        scanf_s("%f", &h);
20.
21.        //calculate BMI
22.        float bmi = BMI(w,h);
```

```
23.          printf("Your BMI is %f\n", bmi);

24.

25.          //health report

26.          if (bmi < 18.5)

27.                  printf("underweight\n");

28.          else if (bmi < 24.9)

29.                  printf("normal\n");

30.          else if (bmi < 29.9)

31.                  printf("overweight\n");

32.          else

33.                  printf("obese\n");

34. }
```

☑ **Practice Tasks**
- Add a new function that receives a float BMI value and prints the health report.
- Add the necessary code to check if the input height is zero (user error) and prevent the BMI calculation if that happens.

5.4.3 Prime Number Listing

While the previous example demonstrated how to break a program into smaller (and possibly reusable) functions, in this example, we will see another case of passing parameters to functions and receiving results from them.

A prime number is greater than 1 and cannot be calculated by multiplying smaller integer numbers. We would like to write a program that asks the user for a number and then shows all the prime numbers that exist in the range from 1 to the user number. For example, if the user enters 20, the result will be 2, 3, 5, 7, 11, 13, 17, and 19.

What We Have (Our Major Data Items)

- The number that the user enters. We call this N.

- A number that changes from 1 to N.

What We Do (Major Operations on Data)

- The main program asks the user to enter N and then loops over all the numbers from 2 to N to see if they are prime. This requires detecting if a single number is prime.

- A Prime function can detect if any single number is prime or not. This requires to loop over all numbers from 2 to the input number. If anyone is divisible, then the input number is not prime.

- A Divisible function that receives two numbers and see if the division has a remainder.

The algorithm and Python code are shown below. Notice that in the Prime function, if we get to the end of the loop without returning False (no divisible numbers found), then the input number is prime, and we return True.

Sample Code #11-a

```
1. Divisible (A module to detect if A is divisible by B)
2.        Divide A by B
3.        If any remainder
4.            Return false
5.        Else
6.            Return true
7. Prime (A module to detect if a number X is prime)
8.        For i=2 to X-1
9.            If Divisible(X,i)
10.               Return false
11.       Return true
12. Main program
13.       Get N
14.       For i= 2 to N
15.           If Prime(i)
16.               Print i
```

Sample Code #11-p

```
1. #functions
2. def Divisible(a, b):
3.      if a%b == 0:
4.          return True
5.      else:
```

```
6.          return False

7.

8. def Prime(a):

9.      i = 2

10.     while i < a :

11.          if Divisible(a,i) == True :

12.               return False

13.          i += 1

14.     return True

15.

16. #main code

17. N = int(input("enter a number: "))

18. i=3

19. while i < N:

20.     if Prime(i) == True:

21.          print(i)

22.     i += 1
```

☑ **Practice Tasks**
- Write a function that calculates A to the power of B without using any library functions. Then, write a full program that uses the function and asks the user to enter two numbers and show the power result.
  - Use a loop to implement power. A to the power of B means A multiplied by itself B times.
  - Add a main loop to your program to keep asking the user to enter new numbers. Make sure there is an option to end the program.
  - Use any language you prefer.
- The C/C++ function rand() creates an integer random number. We saw how to use % to make a number between 0 and N. Write a function to return a random number between A and B (parameters to the function).
  - *Hint*: If we add 5 to a random number between 0 and 10, we will have a random number between 5 and 15.
  - Add a main loop to allow the user to enter a range and get the random number by calling this function. End the loop in an appropriate way.

HIGHLIGHTS

- A function is a module of code with a given name.

- Running a function is done by "calling" its name in any other part of the code.

- Functions can be easily identified by the pair of ( ) after a name (not a keyword such as if and while). When defining a new function, the ( ) is followed by a body of code inside { }. When calling an existing one, it can be followed by ; to end the statement or have other code in the same statement.

- Javascript and Python programs start from the top with the first code that is not a function definition. C/C++ programs start with the function main().

- Functions run when they are called, not when they are defined.

- In most programming languages, the keyword return can be used to end a function and provide a result.

- Similar to variables, C/C++ functions need to have data types both for the return value and for parameters.

- Global variables are those defined outside all functions. They are accessible anywhere in the source file. Local variables are defined inside a function and are only valid within that function. Two functions can have local variables with the same name.

- A parameter (or argument) is a local variable in a function that has to be initialized by the code that calls the function. This act of initialization is called "passing parameters."

END-OF-CHAPTER NOTES

A. Things I Should Mention

- One of my early programming mentors used to say that a function should not be longer than your screen size. If you don't see all of it, you should probably break it into smaller ones. It is not an exact rule and more of a rather loose guideline. But it does give us a general idea that functions should be easily readable, so we know what they are doing. In the next part of the book, we will talk more about the design of good functions and conflicting opinions that programmers have about it.

- There are ways to have multiple results from a function. More common ones involve using structures (Chapter 7), objects (Chapter 9), and pointers (Chapter 10).

- There is a programming paradigm called **Functional Programming** that is based on the notion of not using variables. It is beyond the scope of this book.

B. Self-Test Questions

- What do the terms "call" and "return" mean in the context of functions?

- What is the type of a C/C++ function?

- How do functions provide a "result"?

- What are the global and local variables?

- Can two functions have local variables with the same name?

- What is a function parameter?

C. Things You Should Do

- Try breaking down the programs you use into a separate function. Identify reusable ones. For example, what can be shared reused within Office applications?

- Check out these web resources:

  - https://www.tutorialspoint.com/cprogramming/c_functions.htm

  - https://www.python.org/

  - https://docs.python.org/3/tutorial/controlflow.html#defining-functions

- Take a look at these books:

  - *C Programming Language* by Brian Kernighan and Dennis Ritchie

  - *Functional Programming in JavaScript* by Luis Atencio

  - *Practical Programming: An Introduction to Computer Science Using Python 3* by Paul Gries, et al.

D. Reflect on the Experience of Reading This Chapter

- What did you expect from this chapter before reading it?

- What was it about, and what did you learn?

- What tasks did you perform, and what difficulties did you face?

- How did you feel about the material and tasks presented in this chapter?

- How can you improve your learning experience?

- How do you see this topic in relation to the goal of learning to develop programs?

- A program can be a single long function or be broken into many small ones. What do you think is a good approach for deciding which part of the code should go to a separate function?

- Does the difference between global variables, local variables, and parameters make sense to you? Can you imagine using a different method to share data between functions?

PART 3

Structured Programming

There are 10 types of people in this world.

Those who understand binary and those who don't.

GOAL

In the previous parts of this book, I introduced algorithmic thinking, the common structure of computer programs, and the important features of three popular programming languages.

In this part, I aim to discuss structured programming as one of the most fundamental programming paradigms. The goal is to revisit modularization, selection, and iteration as the means of structuring a program, and see how to create modules of data and code to solve some problems commonly seen in programming.

CHAPTER 6

Types, Files, and Libraries

> **Topics**
> - User-defined data types
> - Libraries and programs with multiple source files
> - Graphics programming

> **At the end of this chapter, you should be able to:**
> - Understand and define simple user-defined types
> - Understand and write C/C++ header files
> - Divide programs into multiple source files
> - Use function libraries
> - Write simple graphics programs

OVERVIEW

This part of the book is about problem-solving using **Structured Programming (SP)**. In previous chapters of this book, I discussed how to think algorithmically and how to represent an algorithm through code and data. The examples in the last chapters may not represent the complexity of real-world problems, as the goal was to explain the main programming concepts. Now is the time to put what we have learned into action and see how programmers can solve real software problems commonly seen in typical applications. Particularly, in this part of the book, I will show how to achieve modularization of data and code as a key to developing manageable and reusable software. But first, there are a couple of other foundation topics to cover.

More complicated programs, which I will be discussing from now on, generally involve more complicated data items and operation. For example, a game can involve manipulating

Two-Dimensional (2D) and **Three-Dimensional (3D)** objects, visual effects such as explosion and smoke, network connections and data, and various **Human–Computer Interaction (HCI)** technologies. These program elements are generally developed by multiple people, or multiple groups or organizations, and then integrated together. All these require a few important features, such as

- The ability to define new data types such as database, image, and game object
- The ability to organize the program into multiple files for more efficient management,
- The ability to use predefined (or externally defined) functions and libraries
- The ability to perform graphics operation in addition to simple text output.

I discuss these topics in this chapter, starting with a quick look at data types.

6.1 DATA TYPES REVISITED

6.1.1 Predefined and User-Defined Types

Computers store data as numbers, and all programming languages, explicitly or implicitly, define the type and size of the data they are processing. In some languages like C/C++, this is done explicitly by the programmer when a variable is defined:

```
int x;      //integer data with 4 bytes

float y;   //non-integer data with 4 bytes

char z;    //integer data with 1 byte
```

In some other languages such as Javascript, we simply use a keyword like var to define a variable (for example, Javascript). But the data type and size are decided "behind the scene" by the program and based on the value assigned to the data. When you use a line such as var x = 7; in Javascript, the program creates a new numeric data based on the value 7. Initializing x with "hello" would create it as a string of text, instead. There are also other languages such as Python, where the programmer doesn't even need to use a keyword. If the name does not exist already, the code simply defines a new one.

```
X = 0
```

Each programming language has a series of **standard** or **predefined data types**. In C/C++, these are basically integer and floating-point in different lengths. String or text is another common predefined data type in many languages such as Javascript, Python, C#, and Java. C does not have a specific string type, although it can deal with strings of text.[1]

---

[1] The char data type, as we saw before, is simply a 1-byte integer.

C++, on the other hand, adds a string type. I will talk about text processing in more detail in the next chapter.

> ☞ **Key Point:** Data types define the format and size of data. They can be defined explicitly by the programmer or behind the scene by the language.

In addition to these standard types, programs usually need to deal with other, more complicated data types. Some of these cannot be predicted by the language designers, or even if they can think of the need, there may be too many types to support in the language itself. For example, a program may deal with many students and require a new unit of data, called Student, that combines everything that is needed for a student (such as name, ID, and grades). Another program may work with images and needs a collection of data (such as width, height, and color) for that purpose. As such, most programming languages allow the programmer to create a new type by combining and managing existing ones. This is called a **User-Defined Type (UDT)**. User, in this case, refers to the programmer, the user of the language. While there are a variety of ways to define new types, two common methods are **enumerations** and **structures** (or **classes**[2]).

> ✍ **Reflective Questions:** Can you think of a programming language that doesn't need UDTs? What would be the restriction? Why should a UDT be a combination of existing ones?

6.1.1.1 Enumeration

An enumeration is a set of symbolic values associated with constant numeric values. More simply said, it is a type that has a limited number of possible numeric values with specific names. In a sense, enumerations are like naming our fingers; instead of first, second, to fifth finger, we say thumb, index, middle, ring, and pinky. Names make it easier to work with items.

In C/C++, enumerations are defined using the keyword enum. The following line defines an enumeration called Meal with three values and then defines a variable called x of type Meal with an initial value of Breakfast:

```
enum Meal {Breakfast, Lunch, Dinner};

Meal x = Breakfast;
```

Note that the first line is not defining a variable but a **type**. Defining a new variable starts with a type, so it is the second line that defines a variable. The keyword enum defines a new type.

---

[2] Structures and classes are not the same but have many similarities. The term "structure" is used primarily in C/C++ and represents a subset of classes. We will see the differences soon.

Assigning any value that is not from Meal enumeration to x will result in a compiler error. This helps programmers not make mistakes in giving the wrong value to restricted variables. Using enumerations also makes the program more readable compared to using plain integer numbers.

Here are a few lines from Sample Code #4-c in Chapter 4:

```
if(command == 1)

    printf("Hello\n");

else if(command == 2)

    printf("Goodbye\n");

else if(command == 3)

    printf("How are you?\n");
```

You notice that the variable command had a limited set of "allowed" values (1–3), but the code as done in the above lines is not very readable because we don't see clearly what those numbers are, and it is easy to make mistakes. Compared to that, Sample Code #1-c is more readable and less likely to have errors.

Sample Code #1-c

```
1. enum Meal { Breakfast=1, Lunch, Dinner };
2. void main()
3. {
4.     Meal userMeal;
5.     bool quit = false;
6.     while (quit == false)
7.     {
8.         printf("Choose meal type
9.          (1-breakfast, 2-lunch, 3-dinner, 0-exit): ");
10.        scanf _ s("%d", &userMeal);
11.        switch (userMeal) //or use if/else
12.        {
13.        case Breakfast:
14.                printf("Breakfast selected. Processing...");
```

```
15.              break;
16.         case Lunch:
17.              printf("Lunch selected. Processing...");
18.              break;
19.         case Dinner:
20.              printf("Dinner selected. Processing...");
21.              break;
22.         case 0:
23.              quit = true;
24.                  break;
25.         default:
26.           printf("invalid");
27.           break;
28.         }
29.      }
30. }
```

Behind the scene, what really is happening is that the program is defining any `Meal` variable such as x as an integer. Three values of `Meal` are simply 0, 1, and 2, which now have textual labels (names). If you use `printf` to show the value of x for the above code, you will see that it is an integer number. This means that a C/C++ programmer can still ignore the enumeration and use `userMeal` as an integer variable, as in line 10. This is an example of the flexibility that C/C++ offers, which, of course, comes with a higher risk of mistakes. You will see later that other languages take away some of the flexibilities that C/C++ has to reduce the risk of programming mistakes and the programmer overhead for taking care of many issues.

The numeric equivalents of enum values are set by default to 0, 1, 2, etc. But that can change. For example, the following definition will start from 8 instead of zero. Lunch will be 9, and then, `Dinner` is set to Lunch+10, which is 19.

```
enum Meal { Breakfast=8, Lunch, Dinner=Lunch+10 };
```

> ☞ **Key Point:** Enumerations assign limited numeric values to textual labels for increased readability of the code.

In Javascript, enumerations are defined and used in a similar way but with some differences in syntax.

```
var Meal={Breakfast:0,Lunch:1, Dinner:2};
var x = Meal.Lunch;
```

You notice that

- There is no specific keyword, and the new enumerated type is simply defined as a variable.

- The numeric values have to be specified. There is no default value.

- To assign a value to another variable, we have to specify the enumeration followed by a period (the "." symbol) and then the name of the value.

  - Using a period is common for accessing "parts" of a complex item. We saw that previously when calling `document.write` in Javascript and will see more about it in the following chapters.

Enumerations in Python are defined using **classes** which I will discuss later.

6.1.1.2 Structure

When we refer to an object as a "chair," we don't list all the parts that make up a chair, such as legs, a place to sit, and something to lean back to. If we work in a school, we refer to students without every time mentioning what sections exist in a student file. We just have a general understanding that each student has information such as ID, name, grades, and a Grade Point Average (GPA). This is the concept of modularization that I talked about multiple times earlier. Small parts make more complicated ones, and once that is done, we don't have to constantly deal with the "components" that are combined together.

Enumerations are a simple example of user-defined data types, but they are not really a new type but more of naming and restriction on the integer type. **Structures** offer a new type that is a combination of existing ones into new **modules**. Most programs require different data items that are somehow related and used together. For example, a program may have an ID, a series of grades, and an average (GPA) for a student. If we are dealing with two students, our data items will look like this:

```
//data for student #1
int ID _ s1;
int G1 _ s1;
int G2 _ s1;
float GPA _ s1;
//data for student #2
int ID _ s2;
int G1 _ s2;
int G2 _ s2;
float GPA _ s2;
```

This code causes issues in many cases:

- Every time that we need to have a new student, we will have to repeat those four lines.
- If, at some point in time in future we have to modify the data needed for a student (add, remove, or change), then we will have to modify all places in our code that the above lines were used.
- If we need to pass the student data to a function, we will have to pass four parameters.
- If we want a function to return a new set of student data, we cannot simply return one variable.

In all such cases, it makes sense to have a new type of data that we can call "student," defined once in our code. This is similar to our everyday use of names for more complex objects that replace a full description. In such casual cases, we rely on a general understanding that may not be very precise. For example, we may not always be on the same page as to what a "chair" is or what information is associated with a "student." For computers, as we know, things have to be well defined.

The Sample Code #2-c shows how to use the keyword struct (short for Structure) in C/C++ to create a UDT that is a combination of other data types.

Sample Code #2-c

1. struct Student
2. {

```
3.          int    ID;

4.          int    G1;

5.          int    G2;

6.          float GPA;

7. };

8.

9. void School()

10. {

11.         Student s1;

12.         s1.ID = 1000;

13.         s1.G1 = 90;

14.         s1.G2 = 80;

15.         s1.GPA = (s1.G1 + s1.G2) / 2.0;

16.

17.         Student s2;

18.         s2.ID = 1001;

19.         s2.G1 = 80;

20.         s2.G2 = 95;

21.         s2.GPA = (s2.G1 + s2.G2) / 2.0;

22. }
```

A structure is a module of data with a series of members that

- Are related
- Have different names
- Can have different types.

Once the structure is defined, it can be used just like any other data type, as shown in line 11. While Student is a type, s1 and s2 are variables of that type, each with their own set of members. These variables that are created based on a UDT are called **instances** of that type. The members of a structure are accessed using the name of the structure variable, followed by a period (the "." symbol), and then the name of the member, as shown in lines 12 to 15. These members can be used just like any other variable of that type. For example, s1.ID and s2.ID are two separate int variables.

> ☞ **Key Point:** In C/C++, structures are data types that combine other types into a single module. Variables can be defined based on structures, just like standard data types. Such variables are called instances of that type.

> ☑ **Practice Task:** Define and use a C/C++ structure that holds information for a bank account. To simplify, assume each account has an integer ID, a float balance, and an integer overdraft limit.

> ✋ **Reflective Question:** How do enumerations and structures relate to the concept of modularization?

Python and Javascript don't use the term **structure**. They use the term **class** for a UDT and the term **object** for a variable defined based on a class. Class and object are more common terms in programming languages for a combination of different members. Historically, C programming language did not use these terms and had structures as a combination of data elements. Later, C++ added both terms and defined class as a combination of data and functions (what we will discuss in Chapter 9).[3] Note that class and structure are types while an object is the variable made of those types. Javascript and Python include the keyword class for creating UDTs. Javascript also allows creating objects without any special keyword. Sample Code #2-j shows how to do that using a function that returns an object.

Sample Code #2-j

```
1. function Student()
2. {
3.      this.ID = 0;
4.      this.G1 = 0;
5.      this.G2 = 0;
6.      this.GPA = 0;
7. }
8.
```

---

[3] In C++, structures are in fact a type of class with easier access to members as we will see in Chapter 9.

```
 9. function School()
10. {
11.      var s1 = new Student();
12.      s1.ID = 1000;
13.      s1.G1 = 90;
14.      s1.G2 = 80;
15.      s1.GPA = (s1.G1 + s1.G2) / 2.0;
16. }
```

🖐 **Reflective Question:** A Javascript function that constructs an object is similar to a C structure. Can you see the similarity in concept while noticing the difference in syntax?

Javascript objects are defined using a function that has variables (similar to the members of the structure). This function is the **constructor** of the object. The keyword this shows that these are not local variables but members of the object that is being created (lines 3–6). Anywhere else in the program, we can define a variable of this new type using the keyword new and the constructor function (line 11).

Note that the keyword struct in C and the corresponding function in Javascript define a new type of (template for) objects. The actual items of that type (such as s1 in the above example) are created when we define a variable using the type (line 11). In Python, the keyword class is used to define that type or template, as shown in Sample Code #2-p. In C++, we can use both class and struct with some minor differences that I will discuss in Chapter 9.

Sample Code #2-p

```
1. class Student:
2.      ID =  0
3.      G1 =  0
4.      G2 =  0
5.      GPA = 0
6. def School():
7.      s1 = Student()
8.      s2 = Student()
9.      s1.ID = 1000
```

```
10.        s1.G1 = 90

11.        s1.G2 = 80

12.        s1.GPA = (s1.G1 + s1.G2) / 2.0;
```

A Python class is similar to a C/C++ structure, but to create an instance (a variable), we use it as a function, Student(), as shown in lines 7 and 8. This is similar to Javascript, but the keyword new is not necessary in Python. You notice that, despite different syntax, the basic concept in all these languages is the same:

- We define a new type as a combination of simpler elements.

- We define variables of that type anywhere else in the code.

- Each variable has its own members, which can be accessed separately, just like any data.

- Members are accessed using the name of the main variable and the member name, such as s1.ID.

Structures and classes can have constructor functions to initialize them, similar to Javascript. We will see this feature in Chapter 9 when we discuss Object-Oriented Programming (OOP).

> **Key Points**
> - In Python, UDTs are created using the keyword *class*.
> - In Javascript, UDTs are created using the keyword *class* or a constructor function.
> - In C/C++, UDTs are created using the keywords *struct* (both C and C++) and *class* (only C ++).
> - In C, structures include only data elements.
> - In C++, structures and classes can include both data and functions.
> - Objects are variables defined from a class type.

> **Reflective Questions:** Can a UDT such as C structure include functions as well as data? What would that mean? We will see this later in Chapter 9 when we discuss OOP.

6.1.2 Memory Address and Pointer Type

When talking about data, we saw that using variables is the common way of defining data in programs. Variables have name, value, and type, which can be defined explicitly in the program or implicitly based on the value. On the other hand, we know that all the information used by a computer is stored as binary values in memory. With a large memory that

Address	Content (Value)
0	1100 0010
1	0011 1000
...	...
1023	0000 0000

EXHIBIT 6.1   Computer memory.

can have billions of bytes of data, computers need a very efficient way to identify a piece of information to read from or write to memory. This is where the concept of **memory address** comes in.

Consider the example of moving around in a city. You probably have a few commonly visited places that you identify with specific names: home, work, school, mom and dad's house, and so on. But these names are not useful for the postal service as they are hard to follow and can be repeated by different people for different locations. Street addresses and postal codes are used to uniquely identify a location and where exactly it is located. Another example is when you use a storage facility. You may put your belonging in a box and add a label on it for your identification purposes, such as Kitchen Stuff or My Books, but with thousands of boxes stored in a big room, it is impossible for the storage staff to find your box using that label. Your name and label will be translated to a unique number identifying that box and where it is, for example, shelf number 1005.

A memory address is a linearly increasing number that identifies locations in memory, as illustrated in Exhibit 6.1. Software modularization can create a hierarchy of data elements from complicated items such as a document to simple ones such as an integer number. Computer memories, on the other hand, don't have a hierarchical structure at the hardware level and are made of bytes in a linear order. This succession of data locations is simply numbered from zero to the size of memory hardware. For example, a 1 K-Byte[4] memory will end at the address 1023, as 1K has 2^{10} or 1024 bytes, which is from 0 to 1023, as illustrated in Exhibit 6.1.

SIDEBAR: COMPUTER ARCHITECTURE AND MEMORY ACCESS

The terms **Computer Architecture** and **Computer Organization** refer to the general structure of a computer, including various components and how they are connected. On the main circuit board for a digital computer, the data transfer between CPU and other parts such as memory is done through three sets of electronic connections, called **Data Bus, Address Bus,** and **Control Bus**. Each bus is a set

---

[4] Kilo-byte (1024 bytes), mega-byte (1024 * 1024 bytes), giga-byte (1024 * 1024 + 1024 bytes).

EXHIBIT 6.2   Electronic connections on a CPU circuit board.

of connections, each carrying a single bit. For example, a 64-bit CPU has a data bus with 64 connections (wires or lines on the circuit board). Exhibit 6.2 shows an example of such connections.

Data bus holds the actual data, and so the number of lines it has depends on the size of data the CPU can work with. For an 8-bit CPU, the data bus is usually 8 bits, although the designer can choose to use a smaller number so that the data has to be transferred in multiple phases.

The address bus holds the address of the memory location that is accessed. The number of lines it has depends on the memory space the CPU can have. For an 8-bit CPU with maximum 64K bytes of memory, the address bus is 16 bits. Remember that

- 2 to the power of 6 is 64, so with 6 bits, we can have a number from 0 to 63.
- 2 to the power of 8 is 256, so with 8 bits, we can have a number from 0 to 255.
- 2 to the power of 10 is 1024 (or 1K).
- 2 to the power of 16 is 64K.

The control bus holds the lines that select different memory and input/output devices and the type of operation. Keep in mind that many devices will connect to the same buses, so at any given time, only one of them can be selected to prevent two devices from trying to "write to the bus" at the same time.

A simplified model of memory access for an imaginary CPU is shown below. The CPU is 4 bit, can have only 8 memory locations (3 bits for address bus), and connects to only one device (a memory unit). Control bus has two lines: one for the type of the operation (1 means Read and 0 means Write) and one for selecting the device (1 means Enable and 0 mean Disable) (Exhibit 6.3).

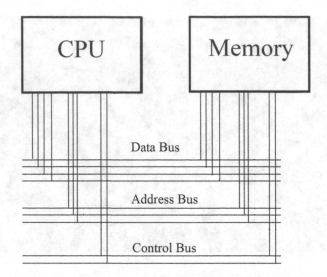

EXHIBIT 6.3 Memory access through data, address, and control buses.

The Central Processing Unit (CPU) of a computer does not access data using names. It only uses addresses, which is a more efficient way to access data. But memory addresses are not practical or convenient for the programmer to use. To achieve optimal use of memory, the Operating System (OS) allocates programs to different areas in memory every time they run, based on which programs are running at that time. So, programmers cannot know where a data item is going to be. That is why variables are given names to use in a program. Behind the scene, all these names will be translated to actual memory addresses every time we run a program.

☞ **Key Point:** A memory address is a number that specifies a particular location in computer memory, as opposed to a variable name that is more readable for humans but less useful for the CPU.

Most programming languages isolate the programmer from the actual addresses of their variables. This is to simplify programming and reduce the chances of accessing incorrect addresses. But the ability to access data directly through an address has some advantages. It can be used in the following cases:

- Accessing data in different ways. For example, having the address of structure s1 in Sample Code #2-c, we can simply read the integer value of ID from that address. This is a rather advanced use of addresses that can create more efficient code but is not recommended for new programmers as it is easy to make mistakes when dealing with addresses, and such mistakes can cause data corruption and program failure.

- Creating dynamic data items, as I will discuss in Chapters 7 and 8.

- Exchanging data with functions with more flexibility and efficiency, as we see in the next section.

To allow using memory addresses, C/C++ supports a built-in data type called a **pointer**. Stored as an integer number, a pointer is the memory address of a data item and is represented with the * symbol after a standard data type, as shown below. The first line defines an int variable and the second line an integer pointer.

```
int a;
int* pa = &a;
```

The & operator, which we saw when using scanf(), takes a data item and returns its address. So, the above line initializes the new pointer with the "address of" the variable a. Once the pointer is created, we can use it to access the data it is pointing to. The * operator takes us to the data pointed to by an address, so it is the opposite of the & operator, as demonstrated below and in Exhibit 6.4.

```
a = 0;
*pa = 0; //equivalent
```

The above lines access the same data, identified with the name a and pointed to by the address in pa. Exhibit 6.4 shows how two variables a and pa are related.

Pointers can be made and used similarly for other types:

```
float b;

float* pb = &b;

Student s;

Student* ps = &s;

ps->ID = 100;

int* px = (int*)ps; //convert Student* to int*

*px = 0;  //write to int variable at that address
```

Description	Name	Address	Value	
An int variable	a	1100 0000	0000 0001	The value of pa
A pointer	pa	1100 0100	1100 0000	is the address of a.

EXHIBIT 6.4   Pointers.

Note that in C/C++, if you have a pointer to a structure, you need to use -> instead of . to access its members. This is to make the code more readable, so when you look at the above code, it is clear that ps is a pointer. The last two lines of the code show how the use of pointers adds flexibility in accessing data. The variable ps is a memory address pointing to a Student. The variable px is an integer pointer. But since they are both memory addresses, we can use the same address in ps to initialize px. Now the program assumes this address is pointing to an integer data and can write to that data. The structure Student starts with an Integer Data (ID), so the line *px=0; will write 0 to ID. Using this notation can be simpler than s.ID or ps->ID, and also more versatile as px can be changed at any time to point to another data.

> ☞ **Key Point:** C/C++ pointers are memory addresses that also specify the data type at that location. So, a char pointer and a float pointer are not the same, even if they have the same value.

> ☑ **Practice Task:** Write a simple C program with a few variables of different types. Use the & operator and print the address of these variables. Initialize a pointer with those addresses, and then, print the value and address of that pointer.

> ✋ **Reflective Questions:** Pointers can be confusing. They are addresses but still data themselves. Just like any other data, they have name, type, and value (and even their own address). Can you visualize how a pointer works? Can you see how Exhibit 6.4 shows the relationship between type, value, and address?

Keep in mind that pointers, while very powerful tools to make more efficient programs, are very tricky to use. For example, if a pointer is not properly initialized, using it can crash the program by trying to read or write a memory location that is not available.[5] As such, it is important to never set a direct value for a pointer and instead use functions and operators that return a valid address. It is also a good practice to set a non-initialized pointer to NULL (0), so we know it cannot be used yet.

```
void* pv = NULL; //NULL means 0
```

The above code also demonstrates the use of void as a pointer type. A **void pointer** is a memory address, but we don't know what data type is stored at that location. It can be used to access data as a series of bytes.

---

[5] Operating system provides each program with a memory space where the program is allowed to do read and write. This is to protect simultaneously running programs from each other, so, for example, a faulty under-debug program doesn't corrupt the data belonging to our word processor. A running program and its allocated memory space define a system element called a **process**.

☞ **Key Point:** A void pointer is a pointer to an unknown data type. It can be used as a generic memory address.

Using pointers to access data in different types requires some experience, and I don't recommend you do that until you are more comfortable with pointers. A more common use of pointers is in exchanging data with functions through what we call **passing parameters by reference**, as opposed to the default method, which is **passing parameters by value**. Most other programming languages (including Javascript and Python) do not support pointers as a data type, but they have methods to achieve the concept of **references** discussed in the next section.

6.1.3 Passing Parameters to Functions

Imagine sharing a computer file with a friend. If your file is stored locally on your computer, you will have to make a copy and send it to her. She will then store that file on her computer. At this point, you both have a file on your computers with the same name and content. Once your friend starts changing her copy (for example, editing the text if the file was a document), the content will no longer be the same as yours as her changes will not reflect in your copy (the original file).

Passing data to a function is similar to sharing a file with your friend. Variables defined in a function are local to that function. Parameters of a function are local variables but get their initial value when the function is called. Sample Code #3-c shows this in a C program.

Sample Code #3-c

```
1. void Report(int data)
2. {
3.     printf("data is %d\n",data);
4.     data = 0; //reset the data after use
5. }
6. void main()
7. {
8.     int data;
9.     scanf_s("%d",&data);
10.    Report(data);
11.    int y = data; //use the data after report. No change
12. }
```

In the above sample code, the variables data in Report() and main(), despite having the same name, are different variables. The program would work exactly the same if these were named differently. When the function Report() is called, the data in Report() gets its value from the data in main(). So, we will have two copies of that data. Now, if the function changes the value of that variable (line 4), and once we go back to the first function (line 11), the value of original data is still the same as it was originally.

> ☑ **Practice Task:** Try printing the value of data in Sample Code #3-c before and after the call to Report()

This way of providing data to a function is called **Passing Parameters by Value**, because we don't send the actual data but a copy of it (the original value). Now imagine that you want to share a file with your friend for the purpose of collaboration. In this case, you want the changes made by each person to be reflected in the copy that the other person has, or even better, you want to have only one copy, and you both access it. In file-sharing, we have two options:

- Using file-sharing services such as Dropbox and Google Drive: These services will keep the original file on a server and make local copies but automatically sync them. This method is good, but there are still two copies, and you may end up with problems due to network connections and version confusion.

- Using cloud services such as Google Docs: These services no longer need to keep a local copy, and all the files can be stored on a server.[6] You will receive a link and can access the original file directly through that link.

Providing data to a function (or sharing data between two functions) in a way similar to using a cloud service is called **Passing Parameters by Reference**. In this method, we don't create copies and instead share a link to the original data, or a **Reference**. Many programming languages have the option of passing parameters by reference, and they use different notations and methods to implement them. But behind the scene, all of them use memory address as the reference. So, instead of copying the data to a new local variable in the function, they send the address of the original data. In C/C++, this is done transparently through pointers, as shown in Sample Code #4-c.

Sample Code #4-c

```
1. void Report(int* data) //receives a pointer now
2. {
3.     printf("data is %d\n",*data);
```

---

[6] You can sync them to a local folder on your computer though, as you do with Google Drive and Dropbox.

```
4.     *data = 0; //reset the data after use
5. }
6. void main()
7. {
8.     int data;
9.     scanf_s("%d",&data);
10.    Report(&data); //sends the address
11.    int y = data; //use the new value after report
12. }
```

In this sample code, the function Report() has a pointer parameter. So, instead of getting a copy of the data, it will have the address of the original data. Lines 3 and 4 have changed to use the pointer. Everything is the same except adding * before the variable name to show that we are using a pointer. The caller changes only one line (10) to send the address of data instead of the data itself. At the end of this code, y will receive the value 0.

Passing parameters by reference (**by-ref**, for short) has two main advantages:

- The original data can be changed if needed.
- No copy is created that saves program memory and time.

The advantage of passing parameters by value (**by-val**, for short) is the simplicity of the code as we don't need to worry about references and the possibility of programming mistakes. Javascript, Python, and most other programming languages support by-ref method automatically based on the data type. The most common approach is that the simple data types such as integer and float are always passed by value, while more complicated ones such as UDTs are passed by reference. I will talk more about this in Chapter 7.

☞ **Key Point:** Passing parameters by value results in local variables that have the given value at the start of the function. These values can change inside the function without affecting anything outside. Passing parameters by reference allows the function to use outside variables directly using the memory address.

☑ **Practice Task:** Typically, C/C++ functions can only return one value. Use by-ref parameters to pass more than one data to a function, and in the function, change their values. See how by-ref parameters can be used to "return" more than one value.

☑ **Practice Task:** Write a function called `Swap()` to swap the values of two numbers using by-ref parameters. For example, if a = 8 and b = 9 before the function call, then calling `Swap(&a,&b)` will result in a = 9 and b = 8. Note that the function needs pointers, and & operator can create a pointer to the variable.

✍ **Reflective Question:** Can you think of a way to do by-val and by-ref parameter passing without explicitly using pointers? C# and many other languages that have no explicit pointer type automatically decide which data type is passed by which method. C# also allows the keyword `ref` to define a parameter as being passed by-ref. Can you imagine how this works and what advantages or drawbacks it has?

6.2 OUTSIDE THE FILE

6.2.1 Multiple Source Files

When we start programming, our code tends to be short and easily manageable within one single source file. But once we start working on bigger projects, managing the code in a single file becomes harder, due to multiple reasons:

- Longer code will result in larger files that are just hard to navigate through.

- Compiling a large file takes a longer time, but if we have multiple smaller files and make a single change, only the affected file has to be recompiled.

- It is a lot easier to collaborate with a team if each person is working on a separate file, which will later be linked together.

- If we plan to reuse our code in another project, it is simpler to just add a small source file that only has the reusable code instead of going through all code in a large file and find the useful parts.

All programming languages allow programmers to divide the code into multiple files. The only trick in doing so is to make sure you can access the data and code in one file from another.

Let's start by seeing how multiple files are used in Javascript. Most examples that we have seen so far included the Javascript code inside the HTML file. For larger projects, it is more practical to have them separate. The Sample Code #5-j shows the Javascript and HTML files used for a simple example of calculating power. Remember that in an example such as $2^3 = 2 * 2 * 2 = 8$, 2 is the base, 3 is the exponent, and 8 is the 2 to the power of 3.

```
Sample Code #5-j
```

Power.js is the first part of our Javascript code that has a function named Power

```
1. function Power(base, exponent)
2. {
3.     var p = 1;
4.     for(var i=0; i < exponent; i++)
5.     {
6.         p = p * base;
7.     }
8.     return p;
9. }
```

Run.js is the second part of our Javascript code that interacts with HTML.

```
1. function Run()
2. {
3.     var base = data1.value;
4.     var exponent = data2.value;
5.     var p = Power(base, exponent);
6.     demo.innerHTML = p;
7. }
```

Main.HTML includes the front end of our program.

```
1. <html>
2. <head>
3. <title>Power Example</title>
4.
5. <script src="Run.js">
6. </script>
7. <script src="Power.js">
8. </script>
```

```
 9.
10. </head>
11. <body">
12. <p id="demo">
13. .
14. </p>
15.
16. Base: <input id="data1">
17. <br>
18. Exponent: <input id="data2">
19. <br>
20. <button onclick="Run()">Calculate Power</button>
21.
22. </body>
23. </html>
```

In the above code, the .JS files no longer include any HTML tags such as <script>, but we expect them to be included in an HTML file. This expected inclusion means they can access the HTML components we expect to have, as we see in Run.js. Note that the Power() function is written in a very general way and can easily be used in another project. But the Run() function assumes we have an HTML page with inputs called data1 and data2, and a paragraph called demo, and that we have another JS file with the function Power(). So, it is less reusable.

The HTML file has two important differences compared to previous examples:

- It shows a full HTML code with elements that are not necessary but commonly used, including

  - <html> and </html> to start and end the whole page

  - <head> and </head> for parts that are not the page data such as title (line 3) scripts (lines 5–8), keywords (not used here), and other "header" information.

  - <body> and </body> for the main content of the page. All previous HTML examples were considered by default as part of a <body>.

- The script resides in the .JS files added using the src attribute of the <script> element.

- Note that the `src` attribute assumes the files are in the same folder. If not, you have to provide a full web address, a.k.a. **Uniform Resource Locator (URL)**.

In Python, using data and code in another source file requires the use of `from` and `import` keywords. For example, if we had Run.py and Power.py, similar to the previous example, then Run.py needs to have the following line before using anything from Power.py:

```
from Power import *
```

This makes everything in Power.py accessible within Run.py. Note that unlike Javascript, where the HTML code determines how and when the JS files are used, for Python projects, you need to identify which Python file is your starting point, as illustrated in Exhibit 6.5 for a Visual Studio project.

In C/C++, using multiple source files is a little more complicated. Just like Python, at the beginning of any C/C++ file, you need to identify which other files you are using. But C/C++ compilers are not designed to just have a reference to an external file and find the resources they need in it. Each individual external function or data has to be separately

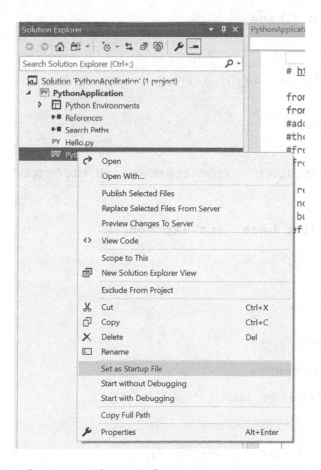

EXHIBIT 6.5 Choosing the Startup File in a Python project.

identified. Sample Code #5-c shows the power example written in two C/C++ files. Similar to Javascript, we have a file that has the Power() function and one that has our main() function with all user interactions. There is no need for a third file here as we don't use the HTML front end.

Note that every C/C++ program needs a function called main() as the starting point. This function has to be unique and only in one of the source files, as shown in Sample Code #5-c.

Sample Code #5-c
Power.cpp

```
1. int Power(int base, int exponent)
2. {
3.     int p = 1;
4.     for(int i=0; i < exponent; i++)
5.     {
6.       p = p * base;
7.     }
8.     return p;
9. }
```

Main.cpp

```
1. #include "stdio.h" //for standard input/output
2.
3. int Power(int base, int exponent);
4.
5. void main()
6. {
7.     int base;
8.     int exponent;
9.     printf("enter base: ");
10.    scanf_s("%d", &base);
11.    printf("enter exponent: ");
```

```
12.    scanf _ s("%d", &exponent);
13.    int power = Power(base, exponent);
14.    printf("power is %d\n", power);
15. }
```

Power.cpp is almost identical to Power.js and has to need to refer to any other file as it doesn't use any external function or data. Main.cpp, on the other hand, needs to identify the function Power(). C/C++ compiler doesn't use external files to look for items, but in line 13, it needs to know what Power() is. For that purpose, prior to the use of the function Power(), we have to let the compiler know what the word Power means. Line 3 is what we call a **function declaration** or **prototype**. It is similar to the **function definition** in Power.cpp but has no body of code. This line tells the compiler that there is a function called Power() that receives two integers as a parameter and returns an integer result. **Compiling** doesn't need any more information when translating Main.cpp to machine code. Power.cpp is also translated to machine code with no problem, and then, these two will be combined into a single program using the second phase of the build process called **linking**, as we saw in Section 3.4.

> ☞ **Key Point:** A function declaration (prototype) is similar to function definition but without the body of code. It shows the name, input type (parameters), and output type (return) for the function, and can be used when the function is defined in another file or later in the same file.
> If function A() uses B() but is defined before B(), the a prototype for B() can be put at the start of the file to avoid compiling errors, as everything has to be defined (or declared) before it is used.

> ✋ **Reflective Question:** Can you recall the two-phase build process with compile and link? It was mainly designed to accommodate multi-file projects. Compile deals with a single file, and link integrates multiple ones. That is why C/C++ compiler needs function declaration, so it knows a function is defined in another file. Why does such a declaration need to specify the return type and parameters, instead of just naming a function as "external"?
> *Hint*: It has to do with proper compiling.

If a file we are using has only one function to share (like Power.cpp), adding a single line is easy. But frequently, we reuse files with many functions. In such cases, it is easier to have all function prototypes put together as a list and then add that list to new source files. A list of function prototypes and other things a file is sharing is called a **header file,** with extension H. In C/C++, the **preprocessor directive** #include is used to read a text file and insert it into the code. #include is almost always used to add a header file to the code.

If the file Power.cpp had multiple functions to share, we could have them all listed in one single header file and then add that to Main.cpp. The header file name can be anything, but it makes sense to have the same name for header and source files that are related.

```
#include "Power.h"
```

We will see more examples of using header files in the next chapters, but for now, it is important to know what they as we cannot use any C/C++ library without them.

☞ **Key Point:** If a source file is using a data or function that is defined somewhere else, then the programmer needs to provide information to the compiler or interpreter of what that item is or where it can be found. In Javascript and Python, you specifically refer to the source files that have what you want to use. In C/C++, you provide a declaration of what you want to use (such as a function prototype), and then, the linker finds it in other files that are being linked together. In Java and C#, the compiler automatically searches through all project source files, so such a declaration is not necessary.

☞ **Key Point:** A C/C++ header file does not include definition (implementation) of functions. It only has a declaration (prototypes) for functions defined in other source files.

☑ **Practice Task:** Add another mathematical operation to Power.cpp above and modify Main. cpp so it can use the new function.

✍ **Reflective Questions:** Why can't we include Power.cpp inside Main.cpp or any other file that needs the Power functions? Is it really necessary to have a header file?

Including the whole Power.cpp in Main.cpp causes all the functions to be compiled twice, once when we are compiling Power.cpp and once when we are compiling Mian.cpp (and potentially more if we include Power.cpp in other files). This will cause repeated functions. Some languages like Java and C# use other methods such as namespaces to identify all files that can share each other's functions, and as such, don't need header files. Can you see how that would work? What are the advantages and drawbacks?

6.2.2 Libraries

After discussing multiple course files, it is easy to understand the notion of libraries. A **library** is a one or a collection of source files created as a reusable code. They provide multiple benefits. Tasks that are hard to do but are frequently used can be done through the use of library functions, so new programmers or those who are not familiar with those tasks don't have to spend time figuring out how to do them.

Libraries can also act as a middleware between applications and some resources such as network, graphics systems, and databases. If those resources change, only the libraries need to be updated. All the applications written to use the library can stay the same. For example, if a library function provides a function `WriteToDatabase()`, it implements how this writing is done, and the application programmer simply calls this function. If later the database changes and needs a different way of writing, the applications stay unchanged (assuming the upgrade maintains compatibility). Such libraries usually provide what is called an **Application Programming Interface (API)**, which is a clear and organized way of accessing their functions and data used by application programmers. As long as the API is maintained, the library can be updated without any need for applications to change.

C/C++ header files are basically the API for the libraries as they list all the functions in the library with input (parameters) and output (return). For example, stdio.h which can be found in the standard include folder for Visual Studio, or any other C/C++ compiler has lines of code such as the followings that describe the function prototypes in `stdio` library[7]:

```
int printf(const char * _Format, ...[8]);
int scanf _s(const char * _Format, ...);
```

These prototypes allow the programmers to know what the right syntax is for using the library functions. They also make it possible for the compiler to determine the proper use of the functions and give errors if functions are not used correctly. For languages such as Python and Javascript that don't have header files, such descriptions are provided in library documentation, and the interpreter can read the right syntax directly from the library source files (PY and JS).

In compiled languages such as C/C++, libraries are usually provided in binary (machine code). This allows the application programmer to use the library code without seeing the source, which may contain trade secrets. For example, all Windows application developers have access to libraries that allow working with the operating system components. But they don't see the source code as it gives away information about the design of the OS, which is a proprietary product. **Open-Source products such as Linux operating system**, on the other hand, provide access to the source code and allow modifying it. However, changes to source code can cause compatibility and support problems.

> ☞ **Key Point:** Libraries are existing code that programmers can use. They provide multiple advantages such as (1) ease of programming, (2) standard way of doing common tasks, and (3) limited access to source code while using the functionality.

SIDEBAR: OPERATING SYSTEMS

Early digital computers were capable of running only one program at a time. Even today, micro-controllers (specialized CPUs designed for control and automation purposes) frequently have a single-task architecture. The common modern computer organization, though, allows multiple tasks (programs or **applications**) to run at the

---

[7] Actual code may look more complicated based on the operating system.
[8] The ... at the end means the function can have a list of other parameters which in this case are the variables used.

same time. This is done through proper scheduling so the CPU switches from one program to another quickly and in a way that the user "feels" the tasks are running at the same time. In cases where the computer has multiple CPUs, they, in fact, run at the same time but on different processors. Scheduling programs and managing their access to computer resources such as disk and memory requires a particular program called **Operating System (OS)**.

Microsoft's Disk Operating System (MS-DOS) was an early operating system for personal computers that was not capable of running programs at the same time. Released initially in 1981, MS-DOS would provide a text-based command prompt to users, start a new program based on user command and, upon the termination of that program, would take control of the system again. While the application program was running, it had full control of the computer, although the OS code was kept in memory in order to (1) take charge after the application ended, and (2) provide functions for common programming tasks (through MS-DOS API).

Windows is another operating system by Microsoft and for personal computers that includes a **Graphical User Interface (GUI)** and allows multi-tasking and, as such, includes various parts for scheduling and resource management. It started as an add-on to MS-DOS in 1985 and with the release of Window NT and Window 95 in 1993 and 1995 became a standalone operating system with two streams which later merged into Windows XP in 2001. Mac OS by Apple is another GUI-based operating system that did not support multi-tasking in its initial release in 1984 but added that feature later on.

Unix is another multi-tasking operating system that has been around since the 1970s and into multiple streams. Currently, the most common variation of Unix is **Linux** for personal computers, but **Android** and the newer versions of **Mac-OS** are also Unix-based.

The primary component of a modern operating system is its **Kernel**, which controls tasks (programs), memory, and disk storage. Networking, security, and user interface are the other important components of the OS.

For more information about operating systems, refer to many great books on the subject such as *Modern Operating Systems* by Tanenbaum and Bos or *Operating System Concepts* by Silberschatz et al.

6.3 GRAPHICS PROGRAMMING

6.3.1 Computer Graphics

Almost all computer systems provide a visual output to interact with users. This visual output can be as complicated as a **Three-Dimensional (3D)** holographic display to a simple set of indicator lights, but the most common is a **Two-Dimensional (2D)** screen. In this book, when talking about visual output, I focus on 2D screens. Note that a 2D screen may be used to display 3D scenes.

The science and technologies required for working with visual output for computers are collectively called **Computer Graphics (CG)**. The screen and other hardware and software required to manage it are usually referred to as the **graphics system**, and the process of determining the final image showed on the display screen is called **rendering**. The in-depth study of various concepts in computer graphics is beyond the scope of this book. But our focus on games and other interactive multimedia applications as a context to explain programming concepts requires some understanding of CG. Two basic concepts in CG are **pixel** and **color**.

A 2D screen consists of a series of points called **pixels** (Picture Elements) that together form the visual output (see Exhibit 6.6). Pixels are organized in a matrix structured as rows and columns. The number of rows and columns defines the **screen resolution (or spatial resolution)**. Higher screen resolution means more pixels, which allow us to either show more objects or show more details of an object. At a lower resolution, we are not able to show details; for example, curves will not look smooth.

As common in Math, the location of a point in a 2D system is identified with two numbers (coordinates) along horizontal (X) and vertical (Y) axes. The X-axis starts at 0 on the left side and increases towards the right. The Y-axis starts at 0 on the top and increases towards the bottom. The positive direction on Y-axis in computer graphics is the opposite of what is common in math literature where the upward direction is considered "increase."

Human vision is based on detecting colors of the light reflected from objects. As such, **color** is another fundamental concept in computer graphics. Just like any other information, colors are coded using binary values. The most common way of representing numeric values for colors in modern computers is the **RGB System**. Since every color can be created as a combination of red, green, and blue, the RGB system defines colors as three integer numbers (RGB channels), corresponding to the intensity of red, green, and blue in that combination. Just like any other computer data, it is important to decide how many bits we allocate to storing these integer values. The number of bits used to show the color

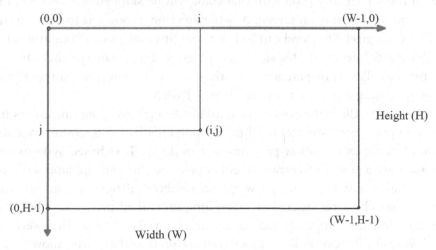

EXHIBIT 6.6   Two-dimension screens and pixels.

determines the total number of colors we can have or the **color resolution**. For example, if we are using only shades of gray (**Gray Scale**[9]) with only 8 bits, then 0 (minimum value) and 255 (maximum value) will mean black and white, respectively, and we will have 256 identifiable shades of gray. For the RGB system, we will have 256 shades for each of the three main colors, and their combinations will determine the final color.

The number of bits will also affect the file size for storing the data, and the time it takes to transfer it (say, from your hard drive to computer RAM). Consider the example of a 100-pixel by100-pixel image. It has $100 \times 100=10,000$ pixels; 4 bytes (standard integer size) per color requires $100 \times 100 \times 4 \times 3 = 120,000$ bytes of data, or roughly 120K.[10] Using 1 byte per color, on the other hand, will need 3 bytes per pixel and $100 \times 100 \times 3=30,000$, or roughly 30K, which makes a much smaller file. One byte, as we saw, allows us to define 256 different shades of each color (levels of intensity). It is argued that the human eye cannot detect much more than that, so having more bits (and more levels between black and full intensity) is of no practical value. As such, the most common system is 1 byte per RGB or **TrueColor**.

☑ **Practice Task:** Open Photoshop, Paint, or any other graphics and image processing program that allows editing colors. Use the Edit Color option and see the dialog box that allows you to pick a color or define it using RGB value. Play with this tool to get a sense of how RGB values change for any color.

While color can be represented using three values for RGB, a pixel may need extra information in addition to color. One of the most frequently used data items for a pixel is its **transparency**. For example, some image file formats such as GIF and PNG include transparent pixels, usually for the background. This is particularly helpful for icons that may need to be placed on different parts of the screen with different colors. Some graphics systems achieve this by allowing a certain color to be defined as transparent. When drawing an image on the screen, any pixel with that color will be skipped, and as such, the color of pixel "behind it" will stay on screen. A better solution, though, is to define a transparency level for each pixel. This level can be from zero (invisible) to a maximum value (fully opaque). Exhibit 6.7 demonstrates such transparency. TrueColor systems can include 4 bytes per pixel to allow transparency. The 4th byte is commonly called the Alpha value, and the systems using this value are referred to as **RGBA**.

Graphics systems allow the control of individual pixels by assigning colors to them. While the computer needs to access all pixels individually, the operating system or the graphics hardware may not allow programmers to do so. **Text-based systems (consoles or terminals)** have a predefined arrangement of pixels to show an alphanumeric character (letters of alphabet, numbers, and a few special symbols). Programs can only choose to show a particular character instead of controlling individual pixels. Text-based systems were dominant in older computers such as mainframes and early PCs. They were not capable of displaying any images. **Full-graphics systems**, on the other hand, allow programs to

---

[9] A shade of gray is a color with the same value for R, G, and B.
[10] 1K is 1024 (2 to the power of 10). So 120,000 is 117.2K.

EXHIBIT 6.7 Transparency. Making the background of an image transparent using alpha values will allow it to be superimposed on top of another with only the foreground object showing. The pixels on the background (or any other pixel) have the alpha value of 0, which makes them transparent.

access any pixel and show different images. While displaying text is part of graphics, it is common to refer to these two types of systems as text-based and graphics-based.

SIDEBAR: GRAPHIC SYSTEMS

The visual display is a common and arguably the default output for modern digital computers, but a screen was not the primary output method for early computers that relied on various paper print-outs to provide information to users. Based on the success of established devices such as television and oscilloscope, computer manufacturers in the 1950s realized the potential of Cathode Ray Tube (CRT) as a display system. Regardless of using text-only or image-based (a.k.a. graphical) outputs, or the recent introduction of flat screens, these visual outputs were controlled by a specific part of the computer commonly referred to as the **Graphics System**. Due to the special needs of displaying and controlling the screen content, these operations are not directly performed by the CPU. Instead, graphics systems include specialized hardware that receives the screen content from the CPU, holds it in their own local memory, and manages the display.

The type of connection to the CPU, supported screen resolution, number of colors, graphics memory (a.k.a. Video RAM), and the type of operations that can be done independent of the CPU are common characteristics that define graphics systems and have evolved over the last few decades. Most commonly, the graphics systems are implemented on replaceable extension cards that attach to the computer's main board through various interface technologies, such as Industry Standard Architecture (ISA), EISA (Extended ISA), or Peripheral Component Interconnect (PCI). Some mainboards, on the other hand, provide their own on-board graphics systems.

While many early computer applications were text-only, graphics systems even back in the 1950s and 1960s did include the ability to show images and perform typical graphics operations. For example, Spacewar was a 2D graphics game developed for the DEC PDP-1 computer in 1961.

The original IBM PC was released in 1981 with a Color Graphics Adapter (CGA) card. Shipped with 16 Kbytes of video memory, CGA could use colors from a palette of 16 colors (4 bits per pixel) in text modes. Only four of those colors could be used simultaneously in graphics mode that supported the resolutions of 320×200 or 640×200. CGA was followed by Enhanced Graphics Adapter (EGA) in 1984. EGA increased the maximum resolution to 640×400, 640×480, and 720×540, and allowed all 16 CGA colors to be used at the same time.

Video Graphics Array (VGA) was released in 1987 with 256 Kbytes of video memory. It introduced 16-color 640×480 and 256-color 320×200 resolution that had fully redefinable palettes. Each color in the system palette could be defined using 24-bit RGB values, which meant programs could change the system palette and show images with palette-based file formats such as GIF. VGA cards used a 15-pin analog connector for communication with the monitor, which is still in use. Super VGA (SVGA) cards, released in the late 1980s and early 1990s, increased the maximum resolution to 800×600 and 1024×768, which became common desktop resolutions till the introduction of HDMI and other high-definition standards. SVGA cards also made it possible to directly use RGB values (16 or 24 bits) for each pixel, instead of selecting a color from the palette.

In addition to increasing the resolution and color depth, graphics cards evolved by providing extra functionality that allowed programs to pass some responsibility to the graphics card. For example, the VGA card supported the split-screen feature, which meant the program could divide the screen into two sections, each corresponding to a different part of the video memory. The starting point of each part could easily change by the program, which would result in screen content shifting quickly or completely switching from one image to another. Such a hardware operation was much faster than the alternative "software" method that required the program to change all the pixels to replace an image on the screen. This introduced the notion of hardware-supported graphics, which lead to the introduction of the Graphics

Processing Unit (GPU), which were programmable components on the graphics card that are now commonly used to perform complicated operations at a much faster speed than the software (CPU-based) alternative.

For more information on graphics systems and computer graphics in general, see great books such as *Computer Graphics: Principles and Practice in C* by James Foley et al. and *Fundamentals of Computer Graphics* by Marschner & Shirley.

Understanding that the graphics system is mainly manipulating pixels, it is easy to see that basic graphics operations are

- Defining the graphics screen in terms of width and height (resolution) and color bits/bytes per pixel (color depth)

- Reading a pixel (getting the color value)

- Writing a pixel (setting the color value)

- Drawing **primitive shapes** (such as circle and rectangle)

- Displaying an image (copying Image data to a location on the screen which is writing to a group of pixels)

- Displaying text (similar to an image but with predefined shapes).

> **Key Point:** Graphics systems allow programmer access to individual pixels on screen, so they can display any shape. A graphics screen is defined by the number of pixels and the color information per pixel.

> **Key Point:** RGB system is the common way of identifying color in computers. It includes 1-byte values for each of the three main colors (red, green, and blue) and optionally a transparency/alpha value.

6.3.2 Graphics in Javascript

Many programming languages, especially older ones, do not come with any built-in ability to perform graphics operations. They come with standard libraries that are limited to text manipulation on terminals. Examples of such languages are C/C++ and Python. Both these languages require extra and non-standard libraries to do graphics programming. The non-standard nature of these libraries results in two important issues:

- In order to use a particular graphics library, it has to be made available for the target platform.

- Developers and end users have to install the library on their computers.

So, it is possible that a developer creates a game on Windows using C/C++ and OpenGL (one of the most popular graphics libraries). But it cannot be compiled and made available on another operating system because that library is not available for it. Even other Windows users may not be able to play the game because they failed to install the library properly, or it was not compatible with other programs on their computer.

Javascript, on the other hand, was designed to interact with HTML pages. So, right from the start, it has the ability to work with GUI elements. With the introduction of HTML 5, a specific element, called **canvas**, was added that allows direct pixel manipulation. Canvas is a blank graphics screen where the code can perform all the operations mentioned earlier. Sample Code #6-j shows a simple Javascript code that uses canvas to perform graphics operations.

Sample Code #6-j

```
 1. <html>

 2. <head>

 3.

 4. <script>

 5. var canvas;

 6. var context;

 7. var image;

 8. function Init()

 9. {

10.     canvas    = document.getElementById("Demo");

11.     context   = canvas.getContext("2d");

12.     context.fillStyle = '#00ff00';

13.     image     = new Image();

14.     image.src = "star.png";

15. }

16. function DrawPoint()

17. {

18.     context.fillRect(10,10,1,1);

19. }

20. function DrawStar()
```

```
21. {
22.    context.drawImage(image,40,40);
23. }
24. </script>
25.
26. </head>
27.
28. <body onload="Init()">
29. <p>Canvas to fill:</p>
30. <canvas id="Demo" width="450" height="300" style="border:1px
       solid #ff0000">
31. Your browser does not support the HTML5 canvas tag.
32. </canvas>
33.
34. <p><button onclick="DrawPoint()">Draw a point</button> </p>
35. <p><button onclick="DrawStar()">Draw a star</button></p>
36.
37. </body>
38. </html>
```

The above code describes an HTML page and is divided into two parts: <head> and <body>. As is common, the Javascript code is in the <head> and <body> shows the general structure of the page. Let's start with the <body> and see how the graphics screen is created.

Line 30 uses the <canvas> element to define a 450 × 300 area identified with the name "Demo." See how the canvas element has attributes id, width, height, and style. The style attribute defines the border of the canvas, a solid line with a one-pixel thickness, and a red color. It is common to identify colors with the RGB value in hexadecimal. Remember that in Hex, a digit can go from 0 to F (15). FF (Hex) is 255 (all 1's in a single byte). #FF0000 has two Hex digits for each of red, green, and blue, and shows a full red. A number like #009900, on the other hand, means a shade of green, #FFFFFF is white, and #000000 is black.

There is a text between <canvas> and </canvas>, which will not be shown if the browser supports the canvas element. Using an older browser with no support for HTML5, the canvas element will not be recognized, and we will see the text. The <body> also has two buttons to call two Javascript functions and the onload attribute (line 28) that calls a third Javascript function to initialize.

The Javascript code starts by defining three global variables that can be used in all functions. These are the canvas, context (a programming interface for canvas), and an image. The Init() function initializes these three variables. Standard Javascript functions are used to initialize canvas and context. Image is a user-defined type (or object) from a Javascript library. On line 13, you see how a variable of this type is created by a call to its constructor function, as in Sample Code #2-j.

A context has a fillRect() function for filling a rectangle on the screen with the selected color. The Init() function selected the color green (line 12), and the DrawPoint() function fills a rectangle of size 1 with this color. The four parameters of fillRect() functions are X and Y coordinates of the top-left of the rectangle, plus the width and height of the rectangle. So, the following code would draw a rectangle with the top-left corner at X=50 and Y=100, with a width of 10 and a height of 20.

```
context.fillRect(50,100, 10,20);
```

The drawImage() function copies an image to screen at the given point. Just like the rectangle, the position of an image is identified by the position of its top-left corner.

☞ **Key Point:** The canvas element in HTML-5 allows Javascript programs to use a graphics screen.

☑ **Practice Task:** In Sample Code #6-j, create a new function called ClearScreen() that uses fillRect() to clear screen to white color. You need to change the fillStyle (line 12) to white (#ffffff). Add a button to call this function.

☑ **Practice Task:** Create new functions called MoveLeft() and MoveRight() to move the star image to the left and right.
   Hint-1: You need to clear the screen before moving the objects.
   Hint-2: You need variables to hold the current location of the star (remember Programming rule #1).
   Hint-3: You need to use the above variables to draw the star instead of hard-coded values on line 22.

✋ **Reflective Question:** What is the main advantage of HTML-based graphics applications?
   Hint: Compare HTML development with Python or C/C++. What do we need in either case? What makes HTML development easier for both programmers and viewers? Get back to this question at the end of this chapter.

6.3.3 Graphics in Python

Standard Python does not come the ability to perform graphics operations. But there are many third-party libraries that add such abilities. The main problem with using these libraries is that they follow different Application Programming Interface (API). It means that while they more or less follow the same logic, function names, syntax, and some capabilities are not the same. For example, one of the most common graphics libraries for Python is PyGame.[11] It has to be downloaded and properly installed on a computer in addition to Python itself. You need to make sure to use proper and compatible versions of Python and PyGame. Then, in order to use this library in any Python file, you need to add the following line:

```
import pygame
```

If you run a Python program, you will see the standard console (text-based interface). Just like Javascript, the first thing we need to do in order to perform graphics operations is to define a graphics screen. PyGame includes the following functions corresponding to the canvas element in HTML5/Javascript:

- `pygame.init()` to initialize PyGame and create the graphics screen as a new window

- `pygame.display.set_mode()` to define/change the screen resolution

- `pygame.display.set_caption()` to give a title to graphic screen window.

Other Python graphics libraries provide similar functionality but in different forms. For example, Graphics.py[12] is a simple library developed as a single Python file that you can add to your projects. It was created by John Zelle for his book, *Python Programming: An Introduction to Computer Science*. Sample Code #6-p shows the basic operations of creating a graphics window and drawing a point and displaying an image.

Sample Code #6-p

```
1. from graphics import *

2. def gtest():

3.     win = GraphWin('Demo', 200, 150)

4.

5.     point = Point(40,40)

6.     point.draw(win)
```

---

[11] http://www.pygame.org
[12] https://mcsp.wartburg.edu/zelle/python/graphics.py

```
7.
8.     star _ location = Point(100,100)
9.     star = Image(star _ location,"star.gif")
10.    star.draw(win)
11.
12.    win.getKey()
13.
14.    win.close()
15.
16. gtest()
```

The above code has to be written in a new Python file. As we discussed before, this source file and Graphics.py have to be in the same folder, and then, line 1 allows our new source file to use functions in Graphics.py. The first function we use is `GraphWin()`, which is the constructor for a Python object (UDT) representing the graphics screen as a separate window. The function returns the object to be stored in the variable `win`. The next lines, similarly, create objects for a point and an image and draw them on the screen. The program then waits for any key before it closes.

You notice that while the concepts are the same, PyGame and Graphics.py use different syntaxes. This causes the programs written with different libraries to not be compatible. For example, a game written using Graphics.py cannot be integrated with another written with PyGame. Programmers need to decide which library to use, based on their preferences and needs, and make sure that it is available to their team and users. Examples in this book use Graphics.py due to its simplicity.

☞ **Key Point:** Standard Python functions do not support graphics operations. Third-party libraries such as PyGame and Graphics.py can be used to add graphics support, but they will cause compatibility issues.

☑ **Practice Task:** In Sample Code #6-p, replace line 12 with a loop that repeats 10 times. In each iteration, move the star shape by 10 pixels to the right and then wait for a keypress.
  Hint: In Graphics.py library, any object has a function that allows it to move. `star.move(10,0)` will move the star to the right by 10 pixels. So, you don't need to worry about erasing the screen.

6.3.4 Graphics in C/C++

Just like Python, standard C/C++ also has no support for graphics operations. The number of options to do graphics in C/C++ can be overwhelming and confusing. Over the decade that these languages have been around, many groups have developed graphics libraries to be used by C/C++ programmers. To develop professional and high-quality programs, there are highly optimized, feature-rich, and popular ones such as OpenGL (https://www.opengl.org, available on almost all platforms) and DirectX (https://docs.microsoft.com/en-us/windows/win32/directx, for Microsoft platforms only). But these are probably more suitable for experienced programmers. To increase usability, they are written for C but can also be used in C++, even though C++ features are not used.[13]

On the other hand, there are C/C++ graphics libraries that are simpler to use, although they usually have fewer features. For example, OpenGL and DirectX allow complicated 3D graphics, but many simpler graphics libraries are limited to 2D. Examples of the libraries with an easier interface for new programmers are

- OpenFrameworks (http://www.OpenFrameworks.cc)
- Cinder (http://libcinder.org)
- Marmalade (http://www.madewithmarmalade.com)
- Cocos2d-x (http://www.cocos2d-x.org)
- Allegro (https://www.allegro.cc)
- SDL or Simple DirectMedia Layer (https://www.libsdl.org).

In my programming classes, I used Allegro for a long time. I found the new version 5 more optimized but less friendly for new programmers. So, recently, I have moved to **SDL** for starting C/C++ programmers and **OpenFrameworks** for intermediate ones. In this book, we follow the same idea. In this and the next two chapters, C/C++ examples are based on SDL. In Chapter 9, I introduce **OOP** and **OpenFrameworks**, which is a more advanced library and uses C++ features such as OOP.

Both SDL and OpenFrameworks need to be installed on your computer. See the Companion Website for instructions on how to do that. To make using SDL even easier, I have created a set of functions that I collectively refer to as **SDLX**. These functions are implemented in SDLX.cpp, and their prototypes are listed in SDLX.h. These files can be found on the Companion Website and have to be added to your project. Any of your source files that use SDLX has to start with the following line:

```
#include "SDLX.h"
```

Please refer to Section 6.2 for information on how to use multiple files in your C/C++ project, and to the Companion Website for how to set up SDL and create a

---

[13] Remember that C++ is a superset of C.

Visual Studio project for SDL/SDLX applications. All SDL/SDLX samples in this book assume that

- You have Visual Studio on your computer.

- You have installed SDL.

- You have created an SDL project.

- You have added SDLX.cpp and SDLX.h to the project.

- You have included SDLX.h in all your source files.

Sample Code #6-c demonstrates how to do simple graphics in C/C++ using SDL. Similar to the previous examples in Javascript and Python, it includes creating a graphics window, accessing a single pixel, and drawing an image. Exhibit 6.8 shows the output of this code.

Sample Code #6-c

```
1. #include "SDLX.h"
2. int main(int argc, char *argv[])
3. {
4.      //initialize SDL
5.      SDLX _ Init("Demo", 640, 480);
6.
7.      //draw single pixel
8.      SDLX _ PutPixel(NULL, 10, 10, 255, 0, 0);
9.
10.     //draw line of pixels
11.     for (int i = 15; i< 50; i++)
12.        SDLX _ PutPixel(NULL, i, 15, 255, 0, 0);
13.
14.     //draw rectangle of pixles
15.     for(int j=20; j < 50; j++)
16.        for (int i = 20; i< 50; i++)
17.             SDLX _ PutPixel(NULL, i, j, 255, 0, 0);
18.
```

```
19.    //load and draw bitmap
20.    SDLX_Bitmap* img = SDLX_LoadBitmap("Hello.bmp");
21.    SDLX_DrawBitmap(img, 200, 200);
22.
23.    //render to screen
24.    SDLX_Render();
25.
26.    //wait for user input
27.    SDLX_GetKey();
28.
29.    return 0;
30. }
```

The function SDLX_Init() creates a graphics window with a given size and title (line 5). The function SDLX_PutPixel() can draw a pixel on any SDLX_Bitmap. It writes on screen if the bitmap parameter is NULL (line 8). Note how combining SDLX_PutPixel() with a single or double loop allows us to process lines or rectangular regions of pixels. Every operation that needs to be repeated can go inside a loop. If we want to draw pixels at the same Y value but with varying X values, then inside the loop we have a constant Y but use the loop index (i) as the X (line 12). The whole loop for X can be repeated

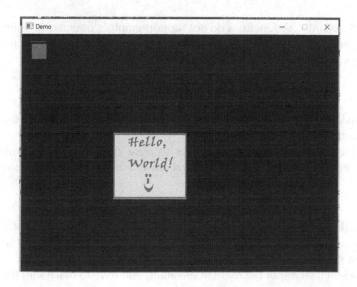

EXHIBIT 6.8   Output of the simple graphics program with pixels, lines, and images.

inside another loop for Y, resulting in the outer loop index (j) to be used as value for Y in
SDLX _ PutPixel() (line 17). The double loop to process an image (or an area of it) is
commonly used in **Image Processing**. We will see more examples of it in Chapter 12.

SDLX _ PutPixel() can use SDLX _ Color structure instead of separate RGB values:

```
SDLX _ Color white;
white.r = white.g = white.b = white.a = 255;
SDLX _ PutPixel(NULL, 10, 10, &white);
```

SDLX _ GetPixel() does the opposite operation, i.e., reading from bitmap to a SDLX _
Color structure:

```
//read the pixel
SDLX _ GetPixel(NULL, 10, 10, &white);
//turn green component to full
White.g = 255;
//write the new value
SDLX _ PutPixel(NULL, 10, 10, &white);
```

The functions SDLX _ LoadBitmap() and SDLX _ DrawBitmap()are to perform the
basic operations on the image files, and SDLX _ GetKey() waits for any key to be pressed
before the program ends.

☞ **Key Point:** Standard /C++ functions do not support graphics operations. Third-party libraries such as OpenGL and DirectX can be used to add graphics support, but they will cause
compatibility issues. Examples in this book use SDL and OpenFrameworks.

☑ **Practice Task:** Similar to Javascript example, modify the Sample Code #6-c to stay in a
loop and move the shape.
   Hint: Use the function SDLX _ Clear () to clear the screen in each iteration, then
change the coordinates (needs new variables), redraw, and then wait for keyboard input.

☑ **Practice Task:** Use SDLX _ PutPixel(), SDLX _ GetPixel(), and SDLX _ Color to
load a bitmap and change its colors (for example, increase redness by 10 for all pixels), and
then, draw it to screen.
   Hint: When you increase or decrease RGB values, you have to make sure they stay within
0–255 range.

🖐 **Reflective Question:** What is the main advantage of HTML-based graphics applications compared to Python or C/C++?

6.3.5 Command-Line Parameters

There are two lines in Sample Code #6-c that require special attention. They are lines 2 and 24. In all previous C/C++ examples, the main() function had no parameters. While this is the more common usage, it is not always the case. A C/C++ program can start with a set of parameters. For example, a database program may start with a parameter that is the name of a storage file, or a chat program may start with the address of a chat server on the Internet. These parameters are called **Command-Line Parameters.**, due to their usage as options when running the program from a command line (the command prompt on a console, as shown in Exhibit 6.9). In Visual Studio, where you run the program from inside the IDE, you may provide the command-line parameters in the Project Properties.

If a program uses command-line parameters, then they will be passed as parameters to main(). The format is standard and includes two parameters: argc is the number of parameters on the command line (2 for the example shown in Exhibit 6.9), and argv is a list including the parameters as text ("dir" and "/AD" for Exhibit 6.9). The list is defined as an array which we discuss in the next chapter. None of the examples in this book use the command-line parameters, but SDL requires the main() to be defined with that option.

```
Command Prompt                                          —    □    ×

Microsoft Windows [Version 10.0.17763.805]
(c) 2018 Microsoft Corporation. All rights reserved.

C:\Users\aliar>cd Documents/temp

C:\Users\aliar\Documents\temp>dir /AD
 Volume in drive C is Local Disk
 Volume Serial Number is 1A76-E077

 Directory of C:\Users\aliar\Documents\temp

2018-11-07  09:53 AM    <DIR>          .
2018-11-07  09:53 AM    <DIR>          ..
2016-12-15  11:00 AM    <DIR>          img
2017-11-06  05:49 PM    <DIR>          New folder
               0 File(s)              0 bytes
               4 Dir(s)  684,118,618,112 bytes free

C:\Users\aliar\Documents\temp>
```

EXHIBIT 6.9   Command-line parameters. /AD is a parameter for dir program that tells to show only directories.

6.3.6 Rendering

Line 24 of the Sample Code #6-c deals with rendering, which is creating the information corresponding to what is being displayed on the screen (all pixel values). This information is stored on a special piece of hardware called **Graphics Card** or **Video Card**. This sub-system of a computer is usually separate from the **Mother Board**, the electronic circuit that holds the CPU and main computer memory (a.k.a. System RAM). The Graphics Card has its own memory for storing display data, which is called Video RAM or VRAM. This memory is not directly accessible by the CPU, and so the graphics operations that involve updating VRAM by CPU are quite slow compared to the operations that involve process-ing data on regular memory.

The slow access speed can become a major issue in applications like games that require many objects to visually change at high speed (for example, hundreds of objects in a scene at a rate of 60 frames per second). Even fast changes to the screen can cause flickers and other visual disturbances that reduce the quality of user experience. To speed up graph-ics operations and reduce issues such as flickering, most graphics systems use a technique called **Double Buffering**. While the content to be shown on screen is stored in VRAM as a primary or front buffer, a copy of it, called secondary or back buffer, is used to perform all graphics operations. Once a video frame is ready, the computer will copy it from the secondary buffer to the primary.

When the secondary buffer is created in system memory, the process is called software double buffering. It still requires copying data from system memory (RAM) to video mem-ory (VRAM), but only one VRAM access is required for any video frame, which makes rendering faster. More advanced graphics systems provide a feature called **Hardware Acceleration**. A special processing unit on the graphics card called **Graphics Processing Unit (GPU)**, which can access the VRAM directly, performs many tasks such as double buffering, loading images, and processing them directly in VRAM. As such, graphics pro-cessing will be done much faster. Many modern programs are written to check if a GPU is available and, in that case, make use of it. Programming and using GPU is beyond the scope of this book.

SDL/SDLX functions are written to use GPU if available; otherwise, use a software dou-ble buffering. `SDLX _ Render()` function in line 24 copies this buffer to screen once the video frame is ready.[14]

Note how C/C++ programs tend to be more complicated compared to Javascript and Python equivalents. This is due to additional features they provide that allow faster and more efficient programs but comes at the cost of programming complexity.

6.3.7 Simple 2D Game

In the following chapters, I will discuss more advanced concepts in programming that allow us to write better and more complicated code and solve more difficult problems. But what you have learned so far in this chapter is enough to make a simple 2D game.

---

[14] SDLX allows programmers to add their own software buffer by passing a value of `true` as the fourth parameter to `SDLX _ Init()`. This can be helpful when there is no hardware acceleration.

The design and development of games can be quite extensive, with many topics that are beyond the scope of this book. Despite the complexity and diversity of design, there are two fundamental ideas that can be applied to almost any game, and understanding them helps us design and develop interesting games with the basic knowledge that we already have: Game Objects and Game Code Structure.

6.3.7.1 Game Objects

Each game is made of a series of objects or entities. While these can be quite different in how they look and what they do, they fit into one of these categories as far as movements are concerned:

- Objects that are stationary, so have no movement on their own. They may still be picked up by players or be "destroyed" by the game.

- Objects that move or act on their own (or in fact by the game). An example of this is what we call Non-Player Character (NPC), although non-character items such as a bullet you fire or a flowing river can also be in this category.

- Objects that are controlled by the player (Player Character or Avatar).

Once we know how to represent an image on the screen, working with game objects is simply to change them by changing different aspects of that image. This can be either the image itself (as it happens in the case of animated characters, for example) or changing its location (for moving objects). More advanced operations on an object involve scaling and rotation, but in the end, they all go back to (1) create or manipulate an image, and (2) show it at a particular location, which are the basic operations we learned in the previous sections.

All these happen within the context of a **game world** or **levels**, which are background objects that define the environment and rules that control the game.

6.3.7.2 Game Code Structure

Realizing that a simple game is a collection of three types of objects helps us visualize the game and its code as we discussed in previous chapters. The main entities, as shown in Exhibit 6.10, are:

- An all-encompassing game that includes and controls all the others

- Player object

- Active non-player objects

- Passive non-player objects.

In previous chapters, we saw examples of code that followed a linear and transactional structure, **Input-Process-Output**. In such a structure, the program starts by reading some

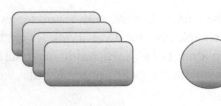
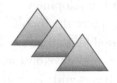

Active Non-Player Objects   **Player Object**   **Passive Non-Player Objects**

Game

EXHIBIT 6.10 Visualizing a Game with three types of objects.

input information; then, it processes them and finally creates output before it terminates. We also saw the programs that followed the **Command Processor** structure, where a continuous and repeated interaction with the user happens until the terminating condition happens (usually an explicit input, or command, from the user). The command processor generally involves two parts:

- Initialization that happens once

- Main loop where the commands are received and processed continuously.

Games and other interactive programs follow a structure similar to the Command Processor, but due to the graphical nature, rendering becomes an important part of the program. The modified game code structure can be described using the algorithm of Sample Code #7-a that deals with the basic objects listed above.

```
Sample Code #7-a
  1. What We Have (Data)
  2.   Background
  3.   Player Object
  4.   Active Objects
  5.   Passive Objects
  6. What We Do (Code)
  7. Initialize
  8.    Load assets (multimedia data)
  9.    Initialize variables
 10. Main Loop
```

```
11.     Update variables

12.          Receive input from user

13.          Receive input from other sources

14.          Modify variables

15.     Draw objects

16.          Create a buffer

17.          Draw objects from far to near

18.          Render Buffer for one frame

19.     Wait for the time for the next frame
```

Understanding and following this code structure is essential in writing efficient and manageable games, regardless of the gameplay. Just like the Command Processor, the program has two parts: initialization and the main loop. Note that there may be many other "loops" in the program for various reasons, and that is why we refer to this as the main loop.

Initialization deals with two important tasks:

- It loads all the audio/video assets (multimedia data). In most of our examples in this book, this is limited to simple images. Remember from Chapter 1 that loading is transferring data from permanent storage (disk) to working memory (RAM). This is a time-consuming process. When we are creating visual outputs in a continuing loop, time delays can be very dangerous as they can affect the timing of what happens on screen. As such, loading assets almost always happen at the start, and that is why you see the message "Loading…" when you start a game or at the start of a new level. Game programmers avoid loading assets, while the user interaction is happening.

- Other than assets, there are many variables that control different aspects of the game, such as the location of objects, scores, and remaining lives. All these need to be initialized prior to actual gameplay.

Once every data is properly initialized, the main loop starts. Games and many other graphics programs (for example, media players and animations) create their visual output as continuously changing **frames**. Each frame is a snapshot of the game world. To give an illusion of being animated and alive, the frames have to be updated very quickly. While traditional movies have standard frame rates of 25–30 frames per second (FPS), modern movies and games can have much higher rates. The process necessary for updating a frame for games involves updating all the variables that define the game and rendering the game world using the updated values. This corresponds to one iteration of the main loop.

The game's main loop starts by checking all inputs (generally from the user but also other devices that may generate input information, for example, timers and network connections). Then, based on these inputs all the game variables will be updated. For example, pressing the Right key can increment the X coordinate of the player object, or an active object's location may change based on time.

Once all variables are updated, a new frame will be redrawn at new locations and states. It is important to conceptually and practically separate updating and drawing. Objects can affect each other, so we should not draw anything until all updates are done. Separating update and draw parts also makes the code more manageable as two different parts of the code can be responsible for these two actions. The drawing process depends on the rendering system being used, but in general, we should remember that all objects are being drawn on the same screen. Anything that is drawn first will be overwritten by object drawn later at that location. So, those objects that are supposed to be upfront (**foreground object**s) should be drawn later than the ones behind (**background**).

The last part of the main loop deals with timing. To create the visual effect of smooth movement, it is important that all frames are created at the same time intervals. For example, to have 60 FPS the program needs to create one frame every $1000/60 = 33$ ms. So, at the end of each iteration, there has to be the right amount of delay before we start the next iteration. We will discuss timing issues in more detail later.

> ☞ **Key Point:** Typical 2D or 3D games use a common program organization that includes an initialization plus the main loop itself divided into the update and draw sections.

> ✋ **Reflective Questions:** What is the main advantage of separating Update and Draw parts of the main loop? Do you see any drawbacks?

6.3.7.3 Example
Sample Code #7-c shows a simple game code in C/C++. The code doesn't offer any exciting gameplay, but it has all the essentials that we can expand and turn into our first cool game. We will do that in the next chapter, where we also discuss how to implement this game in Javascript and Python (it only requires adding support for keyboard input to Sample Codes #6-j and #6-p).

Sample Code #7-c

```
1. #include "SDLX.h"

2. #include "stdlib.h"

3.

4. int main(int argc, char *argv[])

5. {

6.     //initialize
```

```
7.
8.     //init SDL library and graphics window
9.     SDLX_Init("Simple 2D Game", 640, 480);
10.
11.    //load asset
12.    //player data
13.    SDLX_Bitmap* imgPlayer = SDLX_LoadBitmap("Player.bmp");
14.    int xPlayer = 0;
15.    int yPlayer = 0;
16.    //enemy data
17.    SDLX_Bitmap* imgEnemy = SDLX_LoadBitmap("Enemy.bmp");
18.    int xEnemy = 200;
19.    int yEnemy = 200;
20.    //prize data
21.    SDLX_Bitmap* imgPrize = SDLX_LoadBitmap("Prize.bmp");
22.    int xPrize = 400;
23.    int yPrize = 400;
24.
25.    SDLX_Event e;
26.    bool quit = false;
27.
28.    //main loop
29.    while (!quit)
30.    {
31.        //update
32.
33.        //keyboard events
34.        if (SDLX_PollEvent(&e))
35.        {
```

```
36.                    if (e.type == SDL_KEYDOWN)
37.                    {
38.                        //press escape to end the game
39.                        if (e.keycode == SDLK_ESCAPE)
40.                            quit = true;
41.                        //update player
42.                        if (e.keycode == SDLK_RIGHT)
43.                            xPlayer++;
44.                        if (e.keycode == SDLK_LEFT)
45.                            xPlayer--;
46.                        if (e.keycode == SDLK_UP)
47.                            yPlayer--;
48.                    if (e.keycode == SDLK_DOWN)
49.                        yPlayer++;
50.                    }
51.                }
52.            //update enemy
53.            xEnemy += rand() % 7 - 3;
54.            yEnemy += rand() % 7 - 3;
55.

56.            //collision detection
57.            if (xPlayer == xPrize && yPlayer == yPrize)
58.                xPrize = 1000; //off screen
59.            if (xPlayer == xEnemy && yPlayer == yEnemy)
60.                quit = true;
61.

62.            //draw
63.

64.            //First clear the renderer/buffer
```

```
65.          SDLX_ Clear();
66.

67.          //Draw the objects
68.          SDLX_ DrawBitmap(imgPrize, xPrize, yPrize);
69.          SDLX_ DrawBitmap(imgEnemy, xEnemy, yEnemy);
70.          SDLX_ DrawBitmap(imgPlayer, xPlayer, yPlayer);
71.

72.          //Update the screen
73.          SDLX_ Render();
74.

75.          //pause
76.          SDLX_ Delay(1);
77.        }
78.

79.      //clean up
80.      SDLX_ End();
81.

82.      return 0;
83. }
```

The game involves three objects (similar to visualization in Exhibit 6.10): a player, a prize that player needs to pick up, and an enemy with small random movements that player has to avoid. For each of these, we have a bitmap and a pair of XY coordinates. Remember the first rule of programming discussed earlier in this book. As soon as we define our data, we create variables for them (lines 11–23). Then, we answer the three main variable-related questions (Three HOW Questions, 3HQ) we saw earlier:

- How is the variable initialized?
- How is the variable changed?
- How is the variable used?

The answer to the first question goes to the **Initialization** part of the game, while the other two are in the **Main Loop** (line 28), itself divided into the **Update** and **Draw** sections (starting at lines 31 and 62, respectively). This is the basic structure of almost all games. While it is not required to organize your code in this way, it is extremely helpful as it allows better modularization.

- Initialization is where we define and load all our important data. These are variables that will be used throughout the program. Other parts of the code may still define variables, though, but they are usually local and for use within a limited scope of the program, such as temporary data, counters, and loop index.

- Main Loop is the part of the code that repeats. Each iteration of the loop corresponds to the creation of one output frame or one small increment in game time.

  - The Update section receives input from the user and calculates the new values of the game data.

  - The Draw section generates a new frame based on the new data values. Separating Draw from Update allows us to use different modules for each of these tasks without affecting the other.

In the update part of the loop, we have the following tasks:

- Change the location of the player if keys are pressed (lines 36–49). Note how we detect the keypress event and change location accordingly. In every frame, if the key is pressed, the object moves by one pixel (or more if we want faster movements).

- Change the location of the enemy automatically (lines 53 and 54). Note how we create a random number between 0 and 6, and then subtract 3 to create a random number between −3 and 3. This makes the enemy to "vibrate" around its location.

- Detect if the player has hit the prize, and if it has, make the prize disappear or "picked up" (lines 57 and 58).

- Detect if the player has hit the enemy, and if it has, end the game (lines 59 and 60).

The separation of Update and Draw parts allows us to organize them into different modules that can later be replaced without affecting the other parts. For example, when using game engines like Unity and Unreal, the programmers don't even see the Draw part as it is implemented by the engine. We will see more examples of such modularization in the next chapters.

Note that we are no longer using the SDLX _ GetKey() function. Instead, the function calls SDLX _ PollEvent() with a pointer to an SDLX _ Event variable (line 31). SDLX _ GetKey() pauses the program while waiting for a key to be pressed. This may work in programs that are not supposed to do anything when waiting for user input. But a game has many actions happening even when the user is not interacting (remember the objects that move by the game). SDLX _ PollEvent() checks to see if an event has happened (returns true or false).

Events are actions originated from outside the program, such as a user input through a mouse and keyboard, the arrival of a network message, or a timing signal. A common method for the programs to respond to events is to receive them as a message from the operating system. These messages are written in a list called an **event queue**, and the program can read them at any time. In SDLX, the data type `SDLX _ Event` holds all the data for an event, and `SDLX _ PollEvent()` reads one event from the queue and returns the information in `SDLX _ Event` variable, which is passed by reference (as a pointer). This part of the code is commonly referred to as **Event Handler**.

☞ **Key Point:** Events are actions that can happen outside the routine instructions in the program. Event handlers are functions that respond to events that affect the program.

☑ **Practice Task:** The Sample Code #7-c assumes a collision has happened when the coordinates of two objects are the same. This is not very practical because objects hit each other as soon as they touch. To make this a little more realistic, change the conditions in lines 57–60 to include a range.

For example, (`xPlayer>xEnemy-5 && xPlayer<xEnemy+5`) means that `xPlayer` is within a 5-pixel distance from `xEnemy`.

☑ **Practice Task:** Modify the code to have two enemies.

HIGHLIGHTS

- UDTs allow the programmer to manage data more efficiently.

- Enumerators assign limited numeric values to textual labels.

- Structures and classes group together related data.

- Pointers are memory addresses.

- Passing a parameter to a function by value means initializing a local variable in the function to the same value as a variable outside the function.

- Passing a parameter to a function by-reference means allowing the function to access the original data using its address.

- If a function or data is defined outside the source file, the programmer needs to provide information to the compiler about that item, such as its type, parameters, and where it is defined.

- Graphics systems allow programmer access to individual pixels on the screen, so they can display any shape.

- A graphics screen is defined by the number of pixels and the color information per pixel.

- RGB system is the common way of identifying color in computers. It includes 1-byte values for each of the three main colors (red, green, and blue) and optionally a transparency/alpha value.

- Graphics operations are usually supported through standard or third-party libraries.

- Typical 2D or 3D Games use a common program organization that includes an initialization plus the main loop itself divided into the update and draw sections.

- Events are actions that can happen outside the routine instructions in the program. Event handlers are functions that respond to events that affect the program.

END-OF-CHAPTER NOTES

A. Things I Should Mention

- While using pointers may seem scary and complicated, it is well worth it to learn about them even if you are using languages like Python, C#, and Java that don't explicitly support pointers. Understanding value types and reference types, and references in general, is impossible if you don't know pointers. If you are using C/C++, then there is probably no escape from pointers, anyway!

- I started graphics programming back in the time of MS-DOS PCs. My bible was the *EGA/VGA: A Programmer's Reference Guide*, a book about the PC graphics cards. I will never forget the joy of writing a program that used the VGA card's split-screen and shifting features to create a digital real-time signal monitoring application.

- Game development helps you explore many programming concepts. It is also a good way to explore various interactive technologies and computing concepts such as Artificial Intelligence, Networking, and Data Science. A quick way to introduce yourself to game development is to attend the Global Game Jam, a 48-hour game design, and development event that happens annually around the world.

B. Self-test Questions

- How does an enumeration work?

- What does a C/C++ structure do?

- What is a pointer?

- How do passing parameters by-ref and by-val compare? What are their advantages?

- What is the primary role of a C/C++ header file?

- What are the two main concepts in computer graphics that define a graphics system?

- What is a canvas element in HTML?

- What are the common graphics operations?

- What is the basic program structure of a computer game?

C. Things You Should Do

- Design a simple 2D game using GameMaker or a similar tool, and then, write the code for it in Javascript, Python, and C/C++.

- Attend a Global Game Jam event in your area.

  - http://globalgamejam.org

- Check out these web resources:

  - https://ggjnext.org/curriculum (introductory educational material for game design and development)

  - https://www.coursera.org/learn/game-development

  - https://simpleprogrammer.com/started-game-development/

  - https://www.tutorialspoint.com/cprogramming/c_pointers.htm

- Take a look at these books:

  - *Computer Graphics: Principles and Practice in C*, by James Foley et al.

  - *Fundamentals of Computer Graphics*, by Marschner & Shirley.

  - *Understanding and Using C Pointers: Core Techniques for Memory Management*, by Richard M Reese.

D. Reflect on the Experience of Reading This Chapter

- What did you expect from this chapter before reading it?

- What was it about, and what did you learn?

- What tasks did you perform, and what difficulties did you face?

- How did you feel about the material and tasks presented in this chapter?

- How can you improve your learning experience?

- How do you see this topic in relation to the goal of learning to develop programs?

- How comfortable do you feel with basic programming concepts and development environments? You may want to review previous chapters before moving forward as we are starting a new part of the book, and examples start to be more complicated.

- How comfortable do you feel with the pace of the material? Feel free to spend more time on the parts you have difficulty with and make sure you do all the tasks.

- Are the notions of pointers and references clear to you? Does it make sense to use a pointer? Where?

- Can you see how game development can be a good way of learning programming? Does this approach seem reasonable to you?

CHAPTER 7

Modularization of Data

```
                    Topics
              • User-defined Types
                   • Arrays
              • Text processing
      • Games and Applications with Data Modules
```

```
   At the end of this chapter, you should be able to:
    • Define and use arrays and structures for related data items
         • Create loops to process array members
       • Use arrays to process text or strings of characters
        • Use string processing library functions
  • Combine array and structures in games and other applications
```

OVERVIEW

In previous chapters, I introduced **Structured Programing (SP)** as a common programming paradigm based on three important concepts: **modularization, selection,** and **iteration**. SP is a more recent development than some other programming paradigms that may seem more complicated, such as **Object-Oriented Programming (OOP)** that I discuss in Chapter 9. Through an emphasis on selection and iteration, SP introduced a better organization of code. Still, it shares the notion of modularization with OOP, while almost all programming paradigms now follow the concepts of selection and iteration as the basic ways of organizing code.

Modularization is a key concept in computer programming and is common between many design processes. As we saw for LEGO® example in the Introduction, it is a common design approach to assemble elements into modules and simpler modules into more complicated ones. This helps us manage our design more efficiently as (1) we deal with details only when needed and (2) we reuse modules. For example, an arm LEGO module can be attached to a character without worrying about how the arm was made, and it can be used in different characters.

Two important questions when dealing with modularized designs are what goes into a module and how a module can be used by others. These define the **content** and the **interface** of modules, respectively. By allowing the content to include other modules, we create a **hierarchy**, where at low levels, we have small modules such as functions, and at higher levels, we have system architectures that define the big components of a large software. By establishing standard interfaces, we create an **Application Programming Interface (API)** that allows programmers to use the modules without knowing how their contents are implemented.

So far, we have seen the use of functions as the primary method of creating modules of code in a program. In this chapter, we are going to discuss the methods for creating modules of data so that we can manage our information more efficiently. We have already talked about user-defined types, and in the next section, we see how they can be used as modules of data. Then, I will introduce another type of data modularization called **Arrays**, and we will see how these two methods together can help us organize our data.

7.1 USER-DEFINED TYPES AS MODULES OF DATA

In Chapter 6, I introduced the notion of User-Defined Type (UDT) as a collection of other data types. They allow programmers (the users of a programming language) to define new entities in their programs that are more complex. For example, while the red component of a pixel color can be represented using a single integer value, the overall color is more complex as it is a combination of 4 values (red, green, blue, and potentially alpha). If you look at SDLX.h or SDLX.cpp, you will see that SDLX _ Color is defined as such a combined type:

```
struct SDLX _ Color
{
    unsigned char r;
    unsigned char g;
    unsigned char b;
    unsigned char a;
};
```

Similarly, an image is an even more complex item that includes many pixels and other information such as dimensions and encoding. The type SDLX _ Bitmap puts together some of that information. Looking at the Sample Code #7-c in the previous chapter, you can see that using such combined types makes programming a lot easier. For example, the function SDLX _ PutPixel() no longer needs individual RGBA values as four integer parameters, and we can simply pass a single variable of type SDLX _ Color. Creating and using complex data will be a lot easier if we can refer to it as a single variable instead of manipulating the elements one by one.

Let's consider the individual objects in the simple game we made at the end of the last chapter. Each of these objects had three pieces of information: shape, X, and Y. Following the idea of using user-defined types as modules of data, it makes sense to define a new type that has all the information for any of our game objects.

```
struct GameObject
{
    SDLX_Bitmap* shape;
    int x;
    int y;
};
void DrawObject(GameObject obj)
{
    SDLX_DrawBitmap(obj.shape, obj.x, obj.y);
}
```

☛ **Key Point:** User-defined types (structures and classes) combine related data items with different types and names. These data items can have different types and roles but are generally needed together.

☑ **Practice Task:** Think of a book or a student as modules of data. What members will they include? What operations can be performed on them?

✋ **Reflective Questions:** Can user-defined types be used to manage multiple global variables, multiple parameters to a function, or multiple results from a function? How do you see them, in general, as a way of managing data?

All SDLX functions use pointers to pass bitmap data by reference. We defined our shape member in GameObject as a pointer to a bitmap, which is what we receive from the SDLX _ LoadBitmap() function. All functions that needed a bitmap also use pointers (reference). As we saw in the previous chapter, the primary advantage of using a memory address (reference or pointer) is that we don't make a copy of the data as a parameter to a function. This saves time and memory, and also allows the function to change the original data.

While the DrawObject() function above does not need to modify the original data (the object being drawn), using a by-ref method is more efficient and compatible with other functions calls. Note that if we have a pointer to a structure, we use -> instead of the "."

symbol to access members. This is only in C/C++, though. Javascript and Python don't use pointers explicitly. The language decides which types are sent by reference and which ones by value.

```
void DrawObject(GameObject* obj)
{
    SDLX_DrawBitmap(obj->shape, obj->x, obj->y);
}
```

Using the new type and its related function, operations such as drawing will look a lot simpler:

```
GameObject player;

player.shape = SDLX _ LoadBitmap("Player.bmp");

player.x = player.y = 0;

DrawObject(&player);
```

There are a few things that you should pay attention to in the above code:

- We still need to initialize the new variable (player).
- When accessing the members of a struct (or any other combined type), we use a ".". symbol to connect the combined variables to its members.
  - If we have a pointer to the combined variable, then we use the -> symbol instead, as in DrawObject() function, which receives a pointer to the game object.
- If passing a variable to a function that needs pointers, we use the & operator. See how player is passed to DrawObject().

Sample Code #1-c shows the full code for the simple game written using the new GameObject type.

Sample Code #1-c

```
1. #include "SDLX.h"

2. #include "stdlib.h"

3.

4. struct GameObject

5. {

6.     SDLX _ Bitmap* shape;
```

```
 7.      int x;
 8.      int y;
 9. };
10.
11. void DrawObject(GameObject* obj)
12. {
13.         SDLX _ DrawBitmap(obj->shape, obj->x,obj->y);
14. }
15.
16. //main function
17. //needs to be in this format because of SDL
18. int main(int argc, char *argv[])
19. {
20.       ///////////////////////
21.       //initialize
22.       ///////////////////////
23.
24.       //init SDL library and graphics window
25.       SDLX _ Init("Simple 2D Game", 640, 480);
26.
27.       //load asset and initialize objects
28.       GameObject player;
29.       player.shape = SDLX _ LoadBitmap("Player.bmp");
30.       player.x = player.y = 0;
31.       GameObject enemy;
32.       enemy.shape = SDLX _ LoadBitmap("Enemy.bmp");
33.       enemy.x = enemy.y = 200;
34.       GameObject prize;
35.       prize.shape = SDLX _ LoadBitmap("Prize.bmp");
```

```
36.        prize.x = prize.y = 400;
37.
38.        SDLX_Event e;
39.        bool quit = false;
40.        //main loop
41.        while (!quit)
42.        {
43.            ///////////////////////
44.            //update
45.            ///////////////////////
46.
47.            //keyboard events
48.            if (SDLX_PollEvent(&e))
49.            {
50.                if (e.type == SDL_KEYDOWN)
51.                {
52.                    //press escape to end the game
53.                    if (e.keycode == SDLK_ESCAPE)
54.                        quit = true;
55.                    //update player
56.                    if (e.keycode == SDLK_RIGHT)
57.                        player.x++;
58.                    if (e.keycode == SDLK_LEFT)
59.                        player.x--;
60.                    if (e.keycode == SDLK_UP)
61.                        player.y--;
62.                    if (e.keycode == SDLK_DOWN)
63.                        player.y++;
64.                }
```

```
65.          }
66.          //update enemy
67.          enemy.x += rand() % 7 - 3;
68.          enemy.y += rand() % 7 - 3;
69.
70.          //collision detection
71.          if (player.x == enemy.x && player.y == enemy.y)
72.              quit = true;
73.          if (player.x == prize.x && player.x == prize.y)
74.              prize.x = 1000; //off screen
75.
76.          ////////////////////////
77.          //draw
78.          ////////////////////////
79.
80.          //First clear the renderer/buffer
81.          SDLX_ Clear();
82.
83.          //Draw the objects
84.          //in order: background to foreground
85.          DrawObject(&prize);
86.          DrawObject(&enemy);
87.          DrawObject(&player);
88.
89.          //Update the screen
90.          SDLX_ Render();
91.
92.          //pause
93.          SDLX_ Delay(1);
```

```
94.        }
95.
96.        //clean up
97.        SDLX _ End();
98.
99.        return 0;
100. }
```

The new type (GameObject) and the related function (DrawObject()) are not part of SDL or SDLX. Those libraries are general-purpose and for any graphics operation. The new types and functions we are discussing here are for our specific application (a game) and are defined in the source files we create for our project. As a general rule, do not modify the content of library files or other third-party files you are reusing in your project. Such modifications will cause the loss of compatibility as your version of the library is no longer what others are using. The only exception is when you are contributing to new versions of that library.

☞ **Key Point:** User-defined types are usually accompanied by related functions.

☑ **Practice Task:** Imagine we wanted the enemy to continually move in a random direction. What new data members do we need to add to the module? What operations?

Hint: We can have speed values in X and Y, and they can be positive or negative, for example, a random number between -1 and 1.

7.1.1 Collision Detection

In Sample Code #1-c, we detected an enemy hit (or picking up a prize) is the player, and the other objects are at the exact same location. If you recall, we identify the location of objects (images) using the location of their top-left corner. But objects are not one pixel and occupy more space than that one point. So, they can hit each other even if their top-left points are not at the same coordinates. It is also very unlikely and inconvenient for players to expect such a point-on-point hit, as you probably have noticed when playing the game.

In computer graphics and games, the task of deciding if two objects have hit each other is called **Collision Detection**. This task is essential in many games and can be done in a variety of methods. What I've done in Sample Code #1-c is probably the simplest to implement but least useful. Let's consider this problem and find better solutions using general problem-solving approaches we know.

One of the most useful problem-solving techniques that we discussed in Chapter 2 is an analogy. In real life, two objects don't need to be at the exact same locations to hit. They need to be close enough. So, what is important is the distance between them. Similarly, in our game, we can detect a collision if the distance is less than a certain value. To determine that value, we can use trial-and-error, another common problem-solving technique.

The code required to do collision detection based on distance is shown as follows:

```
//assume obj1 and obj2 are two GameObject variables
float distance = pow(obj1.x-obj2-x, 2)+ pow(obj1.y-obj2-y, 2);
distance = sqrt(distance);
if(distance < 10)
{
        //do whatever needed if there is a hit
}
```

The functions pow() and sqrt() above calculate power and square root. They are defined in math.h, which is a standard C library. Other programming languages such as Javascript and Python have similar functions. This code calculates the distance using the Pythagorean theorem that says the length of the hypotenuse (the side of a triangle opposite the right angle) is equal to the square root of the sum of the squared values of the other two sides, as illustrated in Exhibit 7.1.

If you want to not use any library function, you may revise the above code:

```
float distance2 = (obj1.x-obj2-x)*(obj1.x-obj2-x)

                + (obj1.y-obj2-y)*(obj1.y-obj2-y);

if(distance2 < 100)

{

        //do whatever needed if there is a hit

}
```

Here, we calculate the power of 2 (square value) by multiplying the values by themselves. We also skip the square root calculation and compare the distance as squared values (100 instead of 10, if the distance is less than 10, then distance squared is less than 100).

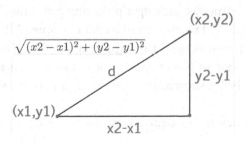

EXHIBIT 7.1 Calculating distance.

There are other methods for collision detection that are more accurate. For example, the distance check is an estimate, and depending on the size of the object and direction of movements may not work. Another approach is to see if any of the four corners of one object are inside the other object (which is a rectangle). Many professional games also use **Collision Box** or **Collision Sphere** for 3D objects, which is an extension of the rectangle check to three dimensions. We will see some of these methods throughout the book.

☞ **Key Point:** Collision Detection is the process of detecting if two objects have hit each other. It is essential in most games and can be done with different methods. One of the most common ones is to compare the distance between objects with a pre-defined value.

☑ **Practice Task:** Try a different collision detection method. See the next sample code for an example.

👋 **Reflective Questions:** Which one of collision detection methods makes more sense to you? Do you see their advantage and disadvantages?

7.1.2 Vanishing Prize: How to Make Objects Disappear

In Sample Code #1-c, we saw how to write a code that lets the player "pick up" a prize. We achieved this by moving the prize off the screen. This method works in simple games but in more complex ones can cause some problems:

- Many games have a larger "world" than what shows on screen.

- In many situations, we want the object to disappear and then re-appear.

- It is also common to make an object gradually disappear or have a certain level of transparency.

Thinking about the case of "vanishing prize," our initial solution to the problem of how to make the prize object vanish was to move it away. This makes sense because it is what we do when playing a board game; we pick up a piece and put it off the board. This is a solution using an analogy, a method we discussed back in Chapter 2. Back then, we also talked about other methods such as divergent thinking and redefining the problem. Using these two methods, we can see that an object appearing at a location is not just about where the object is but also about it being shown. So, while an option is to move the object, the alternative is not to draw it.

We start by defining a new goal (requirement) for our program:

Make a prize disappear after it is picked up by turning it invisible.

The Draw section of our game code is responsible for showing (displaying or drawing) the objects. We can simply skip the line for displaying the prize if it is picked up. Based on our first rule of programming, any information (such as an object being picked up) needs to be stored in a variable. We can define a new variable for each prize object that shows their state (picked up or not), but if you remember the primary motivation behind creating the GameObject structure, we wanted to keep all information related to an object together. So, it makes sense to add a new member to that structure.

```
struct GameObject
{
    SDLX_Bitmap* shape;
    int x;
    int y;
    int visibility;
};
```

While at this time, our player or enemy doesn't need to disappear, having this information for all objects is helpful for future expansion of the game.

We could also use a bool variable to make visibility a binary (false/true) data, but using an integer can allow us to have different levels of transparency. It can be between 0 and 100 (to show a percentage) or 0 and 255 (to correspond to alpha values in graphics systems). If the graphics library we are using does not support alpha/transparency levels, then we can still use the integer value as a binary data; 0 means invisible and non-zero means visible.

Once we have the information about visibility as a variable in our code, we can use the variable-related 3HQ (Three HOW Questions) to see how it changes the program. I will do that in Javascript to also demonstrate how our C/C++ game translates to Javascript.

☞ **Key Point:** Adding a visibility member to GameObject is an example of solving a problem through rethinking it (divergent thinking and redefining the problem). While picking up a score may initially mean "moving it," it can also be achieved by making it disappear.

☑ **Practice Task:** Implement visibility in C/C++.

☝ **Reflective Questions:** How can you solve other problems using this "rethinking" approach? How about collision detection?

7.1.3 Simple Game in Javascript

Sample Code #2-j shows how to implement a vanishing prize feature (and our simple game) in Javascript. It also demonstrates how to define the main loop in a Javascript game, how to get keyboard input, and how to calculate the distance between two objects for collision detection. As mentioned earlier, we see that our sample codes are starting to get lengthier. So, it is important to pay attention to details, understand different parts of the code, and establish a proper foundation before we move forward. Let's take a look at this code and analyze those different parts. The logic (as we will see) is the same regardless of the language, but the syntax will vary.

The program is embedded in an HTML file as before. The HTML code is quite short. The <head> section includes the script (starting at line 4), and the <body> section (starting at line 99) calls the Init() function and creates the canvas and a button to restart the game by moving the player to the initial locations.

The GameObject module is created using a constructor function (line 10), as opposed to the struct keyword in C/C++, as we saw earlier in Chapter 6. The keyword this in the function shows a data member. The data members are all initialized to a default value. We will see later how structures (and classes) can have a constructor function for similar initialization.

The program logic is very similar to C/C++, and the changes we have made in order to implement the visibility feature can be similarly applied to the C/C++ code. We discuss these changes below. Here is the Javascript code:

Sample Code #2-j

```
 1. <!DOCTYPE html>

 2. <html>

 3. <head>

 4. <script>

 5. var c;

 6. var ctx;

 7. var player;

 8. var enemy;

 9. var prize;

10. function GameObject()

11. {

12.     this.shape = new Image();

13.     this.x = 0;

14.     this.y = 0;
```

```
15.          this.visibility = 100;
16. }
17. function DrawObject(obj)
18. {
19.      if(obj.visibility != 0)
20.           ctx.drawImage(obj.shape, obj.x, obj.y);
21. }
22. function Distance(obj1, obj2)
23. {
24.      var dx = obj1.x - obj2.x;
25.      var dy = obj1.y - obj2.y;
26.      var d = Math.sqrt(dx*dx + dy*dy);
27.      return d;
28. }
29. function Init()
30. {
31.      // initialize graphics screen
32.      c = document.getElementById("myCanvas");
33.      ctx = c.getContext("2d");
34.      //initialize objects
35.      player = new GameObject();
36.      player.shape.src = "Player.png";
37.      enemy = new GameObject();
38.      enemy.shape.src = "Enemy.png";
39.      enemy.x = enemy.y = 200;
40.      prize = new GameObject();
41.      prize.shape.src = "Prize.png";
42.      prize.x = prize.y = 400; prize.visibility = 100;
```

```
43.      //set main loop
44.      setInterval(MainLoop, 100);
45.      //set event handler for keyboard input
46.      window.addEventListener('keydown', KeyInput);
47. }
48.
49. //main loop for the game
50. //calls Update and Draw
51. function MainLoop()
52. {
53.      Update();
54.      Draw();
55. }
56. function Update()
57. {
58.      enemy.x += Math.floor(Math.random() * 7)-3;
59.      enemy.y += Math.floor(Math.random() * 7)-3;
60.      if(Distance(player,enemy) < 5)
61.           Reset();
62.      if(prize.visibility != 0)
63.      {
64.           if(Distance(player,prize) < 5)
65.               prize.visibility = 0;
66.      }
67. }
68. function Draw()
69. {
70.      ctx.clearRect(0,0,c.width,c.height);
71.      DrawObject(enemy);
```

```
72.        DrawObject(prize);
73.        DrawObject(player);
74. }
75. function Restart()
76. {
77.        player.x = player.y = 0;
78. }
79. function KeyInput()
80. {
81.        switch (event.keyCode)
82.        {
83.        case 37: // Left
84.               player.x --;
85.               break;
86.        case 38: // Up
87.               player.y --;
88.               break;
89.        case 39: // Right
90.               player.x++;
91.               break;
92.        case 40: // Down
93.               player.y ++;
94.               break;
95. }
96. }
97. </script>
98. </head>
99. <body onload="Init()">
100. <p>Canvas to fill:</p>
```

```
101. <canvas id="myCanvas" width="640" height="480"

102. style="border:1px solid #d3d3d3;">

103. Your browser does not support the HTML5 canvas tag.</
     canvas>

104. <p><button onclick="Restart()">Restart</button></p>

105. </body>

106. </html>
```

The visibility feature is implemented using the following changes that correspond to our 3HQ: how to initialize, how to use, and how to change every data item:

- A new data member added to the GameObject (line 15) and initialized to 100, although any non-zero value works for now.

- The DrawObject() function (line 17) checks if the object is visible and draws it only if visibility is not zero. Later, we will see how we can draw with different levels of transparency.

- Collision with the prize is checked only if the prize is visible (line 62). This is important because the prize object is still there at the previous location, but we don't want the game to do anything with it (for example, get scores if we hit the prize, as we see in the next section).

- If there is a collision, the visibility for the prize is set to zero (line 65).

One of the tricky parts of this code is the function Init() that sets up event handlers, introduced at the end of Chapter 6. I explain this function in the next section with more information about events.

> ✋ **Reflective Question:** How does the 3HQ method relate to the notion of requirement analysis?

7.1.4 Events Revisited

While the HTML code provides a front end (GUI) for the game, it is important to have some actions happening in the background even if there is no user interaction. The most obvious example of such actions is moving the enemy. In general, we want the game to update and create new frames at the rate we set. In C/C++, this was done through a while loop and a delay element. The Javascript code, embedded in HTML, is written as specific actions in response to things that happen on the HTML page. The browser, or any software that renders the page, implements the main loop to deal with user interaction, and the Javascript code is not supposed to implement another continuing loop.

In the case of HTML/Javascript, we set up the process of updating and drawing frames by defining a function that is called automatically at certain intervals. This method provides more precise timing and is also available in C/C++, as we will see later. A continuously running loop was necessary in C/C++ and Python programs to keep them running, but in Javascript, the HTML page maintains that continuity and the scripts only respond to events.

Recall that **events** are actions originated from outside the program, such as a user input through a mouse and keyboard, the arrival of a network message, or a timing signal. In order to respond to any of these actions, for example, to read user input through the keyboard, part of the program works as an **event handler**. A program can handle events in two common ways:

- **Event Queue:** The operating system creates a list of event messages for each program (called event queue). For example, every time that the user does a mouse click on a program window, a mouse click event will be put on the list. The program can then read its event queue and perform the required actions. The SDL/SDLX code that I introduced in Chapter 6 and use throughout the book uses this approach.

- **Callback Functions:** The program can define certain functions as event handlers and let the operating system know about them. These functions will then be called automatically when an event happens. Sample Code #2-j uses callback functions to handle time-based and keyboard events.

The Init() function calls a system function named setInterval() to create a **timer**, a timing-based event that happens at a given interval (line 44). Two parameters are provided: the name of a function to be called and the time interval in milliseconds. I have defined a function called MainLoop (as it replaces the main loop) to act as a timer callback and set the interval to 100. The main loop (line 51) itself is made of two parts, Update to move the enemy and do collision detection and Draw to render a frame, similar to C/C++ code. Note that to organize better, I have created separate functions for these two parts. Something we can and will do in C/C++ code as well. I will show an example of using timers in C/C++ later in Chapter 8.

Note the use of Math functions random() and floor() on lines 58 and 59. The random() function creates a random number between 0 and 1, which is then multiplied by a number to scale it to a different range (here 0–6). The floor() function removes the non-integer part of the data.

The init() function also calls addEventListener(), which is a standard Javascript function for any system window (line 46) and defines callback (event handler) functions for mouse and keyboard events. The parameters are the names of the event and our event handler function. The event name (keydown) is standard, but the event handler function can be called anything. In our case, I have called it KeyInput(), which is very similar to what the C/C++ code does (line 79) and simply changes the player location based on user input. Note that there is no drawing done in the KeyInput() as all our drawing will be performed together in the Draw() function.

☞ **Key Point:** Events are actions that can happen outside the routine instructions in the program. Events can be handled using an event queue or callback functions.

7.1.5 Distance and Collision

As mentioned before, detecting collision by comparing the location of the two objects is not very practical. In Sample Code #2-j, I have created a function (line 22) to calculate the distance between two objects, as shown in Exhibit 7.1.

The `Distance()` function is used in `Update()` (lines 60 and 64) to detect collision between player and enemy or player and prize. The minimum distance of 5 pixels is considered for a collision. This value is somewhat arbitrary. You may hard-code it based on the size of the objects, or have it configurable in run-time through a user or file input, or program it to be determined at run-time based on the object sizes.

☑ **Practice Task:** Detect the collision based on the distance from the center of the objects.

☑ **Practice Task:** Detect the collision based on the size of the objects.

☑ **Practice Task:** Detect the collision if any of the corners of the object-1 is inside the object-2.

✍ **Reflective Questions:** How do Javascript, Python, and C/C++ compare in terms of ease of game programming? Do you see the similarity of concepts? Do you have a preference, and why?

7.1.6 Lives and Scores in Python

Similar to transparency (or visibility), we may also need other information about our objects. Imagine, for example, not wanting to end the game as soon as the player hits an enemy. Instead, we have a certain number of lives, and the player loses one after each enemy hit. The game then ends when the player runs out of lives. Another example is to have multiple prizes and give player scores every time we pick one.

Sample Code #3-p demonstrates how to implement this feature (and our simple game, so far) in Python. The modules are defined at the top of the code. Similar to the Javascript version, we have created functions for drawing the object, calculating the distance, and also initialize, update, and draw parts of the game. `MainLoop()` function creates a loop and calls `Update()` and `Draw()`.

Just like the Javascript version, we are using global variables to share data among various functions. The global variables are defined on lines 142–148, where the program starts after all module definitions. Unlike Javascript and C/C++, Python does not explicitly define variables and creates a variable on the first use. As such, if we use a global variable inside a Python function, it will assume we are creating a local variable with the same name. Python programs use the keyword `global` to identify the variable is defined globally, and there is no need to create a new one. We have used this keyword in `Init()`, `Update()`, `Draw()`, and `MainLoop()` functions to identify the global variables they are using.

The Python code is shown below, and the implementation of movements, collision detection, and lives and score are discussed after the code.

Sample Code #3-p

```
1. # graphic library
2. from graphics import *
3. # for timer and sleep functions
4. import time
5. #for movements
6. import math
7. import random
8.
9. ##################################################
10. # Game Object modules
11. ##################################################
12.
13. class GameObject:
14.     #location
15.     x = 0
16.     y = 0
17.     #speed
18.     sx = 5
19.     sy = 5
20.     #appearance
21.     shape = Circle(Point(0,0), 20)
22.     visibility = 100
```

```
23.
24. def DrawObject(obj):
25.     if obj.visibility != 0 :
26.         dx = obj.x - obj.shape.getAnchor().x
27.         dy = obj.y - obj.shape.getAnchor().y
28.         obj.shape.move(dx,dy)
29.     else :
30.         obj.shape.undraw()
31.
32. def Distance(obj1, obj2):
33.     dx = obj1.x-obj2.x
34.     dy = obj1.y-obj2.y
35.     d = math.sqrt(dx*dx + dy*dy)
36.     return d
37.
38.
39. ####################################################
40. # INIT part of the game
41. # things that are done only once at the start
42. ####################################################
43. def Init():
44.     #global variables
45.     global quit
46.     global win
47.     global player
48.     global enemy
49.     global prize
50.
51.     # create graphics screen
```

```
52.      win = GraphWin('Simple 2D Game', 640, 480)

53.

54.      # init objects

55.      player = GameObject()

56.      player.shape = Image(Point(player.x,player.y), "Player.gif")

57.      enemy = GameObject()

58.      enemy.x = 200

59.      enemy.y = 200

60.      enemy.shape = Image(Point(enemy.x,enemy.y), "Enemy.gif")

61.      prize = GameObject()

62.      prize.x = 400

63.      prize.y = 400

64.      prize.shape = Image(Point(prize.x,prize.y), "Prize.gif")

65.      #draw all objects (THIS IS NEEDED FOR GRAPHICS.PY)

66.      prize.shape.draw(win)

67.      enemy.shape.draw(win)

68.      player.shape.draw(win)

69.

70. ####################################################

71. # UPDATE part of the main loop

72. # Game logic should go here

73. ####################################################

74. def Update() :

75.      #global variables used here

76.      global quit

77.      global player

78.      global enemy

79.      global prize

80.      global score
```

```
81.      global life
82.      #keyboard
83.      dx = 0
84.      dy = 0   #remember that positive y is down
85.      key = win.checkKey()
86.      if key=='Escape':
87.          quit = True
88.      if key=='Up':
89.          dy -= 1
90.      if key=='Down':
91.          dy += 1
92.      if key=='Right':
93.          dx += 1
94.      if key=='Left':
95.          dx -= 1
96.      #update player location
97.      player.x += dx*player.sx
98.      player.y += dy*player.sy
99.      #update enemy location
100.     enemy.x += randint(-3,3)
101.     enemy.y += randint(-3,3)
102.     #collision detection
103.     if Distance(player,enemy)<5 :
104.         life -= 1
105.         if life == 0 :
106.             quit = True
107.     if prize.visibility != 0 and Distance(player,prize) < 5 :
108.         prize.visibility = 0
109.         score += 1
```

```
110.
111. ##########################################################
112. ## DRAW part of the main loop
113. ## Rendering code goes here
114. ##########################################################
115. def Draw() :
116.     #global variables used here
117.     global player
118.     global enemy
119.     global prize
120.     #drawing
121.     DrawObject(prize)
122.     DrawObject(enemy)
123.     DrawObject(player)
124.
125. ##########################################################
126. # Main game Loop
127. ##########################################################
128. def MainLoop() :
129.     #global variables used here
130.     global quit
131.     global win
132.     #start the loop
133.     while not quit:
134.         Update()
135.         Draw()
136.         time.sleep(0.05)
137.     #end the game
138.     win.close()
```

```
139.
140. ###########################################################
141. # Full game
142. # program starts here
143. ###########################################################
144. #global variables
145. quit = False
146. win = None
147. player = None
148. enemy = None
149. prize = None
150. score = 0
151. life = 3
152. #function calls
153. Init()
154. MainLoop()
```

7.1.7 Movements and Collision

The player movement in Python is done in a way very similar to Javascript and C/C++ using arrow keys. The logic for enemy movement is also the same but Python implementation looks a little different due to the availability of randint() function (lines 100 and 101) that gives us an integer number in the given range without the need for the operations we had in Javascript and C/C++:

```
C/C++:              enemy.x += rand()%7-3;

Javascript: enemy.x += Math.floor(Math.random() * 7)-3;

Python:             enemy.x += randint(-3,3)
```

☑ **Practice Task:** Write the randint() function in Javascript and C/C++ using the code we already have.

Note that Graphics.py library does not require drawing the objects on every frame as it doesn't refresh the screen. It only draws an object when it is moved. As such, our `DrawObject()` code compares the location of the object and the location of its shape, and moves the shape if there is any difference. It also takes visibility into account, similar to Javascript, but if the object is not visible, then it will be "undrawn" or erased.

7.1.8 Life and Score

Two global variables are created to hold the information for life and score. They are initialized right away: `life` to a maximum value and `score` to zero.

The `life` value is decreased every time we hit an enemy. The `score` is increased every time we hit a prize. Note that hitting a prize that is already picked up and is not visible should not give us more scores.

Life is used to end the game when the value has reached zero, and the player has lost. Similarly, the game can end when we have collected all prize objects, and `score` has reached a maximum value.

☑ **Practice Task:** Add another prize, and end the game when `score` is 2.

☑ **Practice Task:** Add all the new features to the C/C++ version of the game. You will find the code in later examples.

7.2 ARRAYS AS MODULES OF DATA

7.2.1 Arrays

So far, we have seen the use of user-defined types to create modules of complex data made of related items. The members of a structure/class can be of any type and can potentially have different roles in the program, but they are all related to a particular part, such as a game object. There are many cases, though, that we have a series of data items of the same type and with the same role. Imagine the list of a student's grades or your friends' list on social media. This way of combining data allows us to introduce a different type of modularization.

Let's revisit the familiar example of entering a series of numbers and calculating their average, as shown in Sample Code #3-c below. This program receives 10 numbers and calculates their average. For simplicity, some details, such as the function that holds this code and #include line, are removed.

Sample Code #3-c

```
1. int number;
2. int sum = 0;
3. for(int i=0; i < 10; i++)
```

```
 4. {
 5.     printf("enter a number: ");
 6.     scanf_s("%d", &number);
 7.     sum += number;
 8. }
 9. float avg = sum / 10.0;
10. printf("average is %f \n", avg);
```

Now imagine that these numbers are the grades for a student. It makes sense for us to store all these grades in addition to the average. According to the Golden Rule of Programming #1, if we want a piece of information to be stored and used in our program, we need to make a variable for it. The variable number holds the input grade for the student. But as you can see, this variable is overwritten every time we enter a new value. As such, we are not storing our "older" data when a "new" one is entered.

It is intuitive to assume we need 10 variables to store 10 data items (grades):

```
int grade1;

int grade2;

...

int grade10;
```

If you try to use these new variables in the program, you will notice a major issue. The loop will not work anymore as we don't have the "same" code to repeat (the names change). So, we have to write the same operation ten times but for ten different variables:

```
printf("enter a grade: ");

scanf_s("%d", &grade1);

sum += grade1;

printf("enter a grade: ");

scanf_s("%d", &grade2);

sum += grade2;

...

printf("enter a grade: ");

scanf_s("%d", &grade10);

sum += grade10;
```

This means that our program will be much longer and harder to write or read. A typical database program, for example, may include hundreds and thousands of numbers that need to be entered and processed. A simple loop can do that in a few lines. But with different names, these lines have to be repeated multiple times, instead of using a loop.

There is a second problem that is probably less obvious but can be equally challenging to deal with and that is naming. When dealing with a large number of data, it is difficult to name them individually.[1] As our example shows, all these data items are performing a similar task, and numbering is probably a better approach than naming. This is the inspiration for the concept of arrays.

An **array** is a single data item since it is defined once and with one name, but it also represents multiple data items since it consists of more than one element. It is a **module of data** aimed at grouping a set of data items that are related and have a similar type and role in the program. In our previous example, the grade variables are integer numbers representing student grades, and they are all treated the same way by the program (entered by the user and added to a sum variable).

In C/C++, an array is defined using the [ ] pair after the name of a variable that holds the number of data items that are grouped together as a module and represented by that variable:

```
int grades[10];
```

The definition follows the standard for all variables: a type followed by a name. While the definition of a single variable ends there (int number;), the array has the part that defines the size. Just like single variables that can be initialized right away (int number=0;), we can initialize an array immediately:

```
int grades[] = {100, 90, 83}; //array of three members
int numbers[5] = {100, 90, 83}; //5 members. The last 2 are
    zero
```

Accessing the array has to be by referencing a particular member which is done through an **index value**:

```
grades[0] = 100;
grades[1] = 95;
grades[2] = 0;
```

The index values are placed inside the [ ]. Note that the first member of the array is at index 0, and the index value for the last member is the array size minus one. Instead of a constant

---

[1] It is possible in some programs to use a loop and dynamically create new names, but that process is complicated and not possible in all languages.

number, we can use any variable or operation for an index, as long as they have a value which is an integer greater than or equal zero and less than the array size:

```
int n = 1;
grades[n] = 0;
grades[n+1]=0;
for(int i=0; i < 3; i++) //loop over all grades
        grades[i] = 100; //all grades set to 100
```

☞ **Key Point:** Arrays are collections of related data of the same type.
Arrays and user-defined types are similar because they are both modules of related data. Arrays and UDTs are different because

- Array members have to be of the same type.
- Arrays members have no name and are accessed using the array name and an index number.
- Arrays members usually have the same role in the program.

☑ **Practice Task:** Similar to the example above, use a loop to initialize an array to random numbers.
   Hint: `rand()%101` will give you a random number between 0 and 100.

☑ **Practice Task:** Use a loop to print the values of array members.
   Hint: you can use `grades[i]` as an int variable in any function.

✍ **Reflective Question:** Do you think by now you have an understanding of when to use UDTs or arrays to modularize data?

The syntax for defining and using an array in most programming languages, including C/C++, Java, C#, Javascript, and Python, is similar but with some small variations. Let's rewrite our program using arrays in multiple languages to explore the syntax and usage.

Sample Code #4-c

```
1. int grades[10];
2. int sum = 0;
```

```
3. for(int i=0; i < 10; i++)
4. {
5.       printf("enter a grade: ");
6.       scanf_s("%d", & grades[i]);
7.       sum += grades[i];
8. }
9. float avg = sum / 10.0;
10. printf("average is %f \n", avg);
```

In C/C++, dealing with arrays is very straight-forward:

- When creating an array, we use a name followed by [ ] that holds the size of the array. So, `int number[10]` means to define an array called number with a size of 10, i.e., with 10 **grades**.

- Once the array is created, we access each member of the array using the name and an **index** inside the [ ]. So, `number[i]` means the (i + 1)th member of the array `number` (Index 0 is the first, and so on). Except for this notation, an array member is used exactly the same way as any other variable (see lines 6 and 7, for example).

The array size in C/C++ has to be a constant number, not a variable. Any value that is constant, or known to the programmer at the time of writing the code and compiling it (known as **compile-time**), is called **Static**. Variables are **Dynamic** as they can change when we run the program (known as **run-time**). Languages such as Java and C# use a similar syntax for creating arrays but are less restrictive. They can have arrays created using a variable as size, so the size of the array depends on the value of that variable when we run the program. C/C++ allows the creation of dynamic arrays using pointers. We will discuss this later in this chapter. For now, keep in mind that when defining an array, you need to give a constant size. If you don't know how many members you will need, then use a size big enough to cover the maximum possible need. For example, the grades for a class can be done as in the following code, assuming we can't have more than 100 students. Remember that the program cannot go beyond that size.

```
int grades[100];
```

As we see in Sample Code #4-c, arrays are commonly used with loops because iterations allow simple access to the members of an array performing the same action. The loop variable (`int i` in this example) can easily be used as the index to the array. For example, now that we have all the user data, we can continue the code and show if each grade was below or above average (see Sample Code #5-c, below, line 12).

☑ **Practice Task:** Write a simple C/C++ program that creates an `int` array of size 10 and initializes all members to 0.

Change the initial value to a random number between 0 and 10.

Change the initial value to integer numbers from 1 to 10. (*Hint*: member at index 0 is 1, and so on)

Sample Code #5-c (the first 10 lines are the same as #4-c)

```
1. int grades[10];

2. int sum = 0;

3. for(int i=0; i < 10; i++)

4. {

5.        printf("enter a grade: ");

6.        scanf _ s("%d", & grades[i]);

7.        sum += grade[i];

8. }

9. float avg = sum / 10.0;

10. printf("average is %f \n", avg);

11.

12. // now we can process each grade

13. for(int i=0; i < 10; i++)

14. {

15.      if(grades [i] > avg)

16.            printf("grade %d: %d, above average\n",

17.                 i, grades [i]);

18.      else if(grades[i] < avg)

19.            printf("grade %d: %d, below average\n",

20.                 i, grades[i]);

21.      else

22.            printf("grade %d: %d, average\n",

23.                 i, grades[i]);

24. }
```

☞ **Key Point:** Arrays are commonly processed using loops, where the loop index (or a number based on it) is used as an array index.

✺ **Reflective Question:** Can you think of a case when we shouldn't use a loop to access array members?

Arrays are modules of data that work with a series of items of the same type. This makes databases a good candidate for using arrays, as they are made of similar elements such as all the grades of a student or all employees of a company. Note that in these two examples, the grades are simple integer values, but employee is a more complex data type. In a later section, we will talk about arrays of such UDTs.

Keep in mind that the array index starts at 0, and the last valid index is the size of the array minus one. So, an array of size 10 (with 10 members) can have index values of 0–9. Using a value outside this range will access a memory location outside the array. Depending on where that location is, this operation may access uninitialized memory, another variable in your program, or even memory that is not available to your program. The latter case can cause your program to crash.[2]

☞ **Key Point:** Array index values start from 0, so the last member of an array with N members has an index value of N − 1.

☑ **Practice Task:** Write a program that creates an int array of size 10 and fills it with odd numbers (1, 3, 5, etc.)
Hint: if i is 0, 1, 2, etc., then 2*i is 0, 2, 4, etc., and 2*i+1 is 1, 3, 5, etc.

7.2.2 Arrays in Javascript and Python

The use of [ ] after the array name is a common syntax in almost all languages. This is similar to using ( ) after a function name and is an easy way to identify arrays. Using an index value inside the [ ] to access a member of the array is also another common syntax.

The method of creating arrays, on the other hand, can be different from one language to another. Sample Codes #5-j and #5-p show how to create and use arrays in Javascript and Python.

Sample Code #5-j

```
1. var number = [];

2. var sum = 0;
```

---

[2] Operating system allocates memory to each running program in your computer. We cannot know how this allocation happens, so the next byte after an allocated array may not be accessible to your program.

```
3. for(var i=0; i < 3; i++)
4. {
5.      number[i] = prompt("enter a number");
6.      sum += Number(number[i]);
7. }
8. text.innerHTML = sum/3;
```

Sample Code #5-p

```
1. number = []
2. sum = 0
3. for i in range(3):
4.      print("enter a number: ")
5.      x = int(input())
6.      number.append(x)
7.      sum = sum + number[i]
8. print(sum/3)
```

Note that in both Python and Javascript, a newly created array has a zero-length (line 1 of both programs above). We need to explicitly call the `append()` function to add a new member to the Python array (line 6). Javascript makes it easier as the appending is done automatically and implicitly when we use an index (line 5). Both are more flexible than C/C++ when you have to specify the array size in the code (statically), and there is no built-in way to append new members. Creating a C/C++ array with a length that is calculated at run-time (dynamically) or extending the array size will be discussed in Chapter 9.

> ☞ **Key Point:** In Javascript and Python, arrays are created with a zero-length, and members can be added dynamically (at run-time). C/C++ arrays are created with a fixed-length set statically (at the compiling time).

7.3 EXAMPLES OF USING ARRAYS

7.3.1 Linear Search

In Sample Code #5 (c, j, and p), we saw how to use an array to calculate the average as a very common task we perform on database elements. Another common task that is frequently performed on data is search. This is the operation we perform when we enter a keyword into a search engine to find web pages with related content. There are multiple algorithms for searching through different data types, but for now, let's consider the simplest example.

Imagine that we have a series of numbers corresponding to students' grades and want to find all the A-level grades. Schools don't follow a standard way of defining letter grades, so let's assume:

A. 80–100

B. 70–79

C. 60–69

D. 50–59

E. Below 50

Our goal is to find the students with A-level grades and create a new list (array) of these top students so that in the future, we can use it (for example, to give the awards). We can break this down into the following steps:

1. Browsing through data (accessing all grades)

2. Finding the items we are interested in (comparing to 80)

3. Doing something with the selected data items (saving in a new list if 80+).

These steps are common among any search task. Like any other complex task, multiple steps and sub-tasks may feel overwhelming. So, to start, we set our goal to simply print/display the A-level grades, but later we will add step 3 to create a new list. The algorithm for steps 1 and 2 is pretty straight-forward, as shown in Sample Code #6-a.

Sample Code #6-a

```
Data:
        Arrays of grades
        Threshold for A-level
Code:
    1.          Create array
    2.          Initialize array
    3.          Get Threshold
    4.          Search Loop over array
    5.              If grade greater or equal to threshold
    6.                  Print grade
```

The C/C++ code is shown in Sample Code #6-c.

Sample Code #6-c

```
1. #define SIZE 20

2. //create array

3. int grades[SIZE];

4. //initialize array

5. for (int i=0; I < SIZE; i++)

6. {

7.      grades[i] = rand()%101;

8. }

9. //get threshold

10. int threshold=90;

11. //search loop

12. for(int i=0; i < SIZE; i++)

13. {

14.      if(grades[i] >= threshold)

15.          printf("%d\n", grades[i]);

16. }
```

Remember that in C/C++, we need to identify the array size with a constant number. Later, I will talk about how to create arrays with a variable as size, but for now, let's simply add a #define line and have a hard-coded size of 20. Using the defined SIZE still creates static arrays, but at least if we need to change the arrays size, we only have to modify one line of code. When using static arrays, we set the size to the maximum value we think we may need.

To initialize the array, we could ask the user to enter or read from a data file. But for now, we simply simulate the data using the rand() function that generates a random number. It is defined in "stdlib.h" so it has to be included at the top of our source file. Also, keep in mind that %101 means the remainder of division by 101, which is between 0 and 100, assuming that is the range for our grades. The rest of the code is fairly clear and similar to the previous sample code.

Note that we have two loops, one to create and one to process the data. While in this simple example, we could combine them and do everything in one loop, it is easier to manage the code if it is divided into clear parts. More sophisticated search methods may

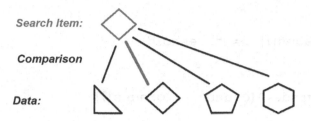

Search Item:

Comparison

Data:

EXHIBIT 7.2   Visualization of search program.

require the whole data to be created before the search can begin. For example, to make the search more efficient and fast, many algorithms start by sorting the data.

Now that we have successfully identified the A-level grades, we need to store them somewhere. According to the Golden Rule #1, any information needs to have a variable. But as of now, our code has no variable showing only the A grades. Exhibit 7.2 shows the visualization of this search problem with two arrays: one for all data and one for only the A-level grades. We also have a variable that holds the number of A-level grades. While the size of the original array is hard-coded 20, the number of A-level grades is dynamic and needs to be a variable.

Sample Code #7-c shows the modified code with a new array to hold the A grades.

Sample Code #7-c

```
1. #define SIZE 20
2. //create arrays
3. int grades[SIZE];
4. int gradesA[SIZE];
5. //initialize array
6. for (int i=0; i < SIZE; i++)
7. {
8.     grades[i] = rand()%101;
9. }
10. //get threshold
11. int threshold=90;
12. //search loop
13. int countA = 0;
14. for(int i=0; i < SIZE; i++)
```

```
15. {
16.     if(grades[i] >= threshold)
17.     {
18.         gradesA[countA] = grades[i];
19.         countA++;
20.     }
21. }
```

The first difference is adding a new array (line 4). Even though it is unlikely that all grades are A, such a case is possible, so we need to create an A list that has the same size.

The next addition to the code is a variable that holds the number of A-level grades (line 13). Note that this variable is not the size of the A array, as the array has to be created with maximum possible size.

Lastly, line 18 shows how to copy data from one array to another. The loop index can be used as the index to the grades array as we are going over all members of that array, one by one, with each iteration. The gradesA array, on the other hand, does not increment with each iteration. It only moves to the next member if we find an A-level grade. The countA variable plays a double role here:

- It is the number of A-level grades we have found so far (that is why it is initialized with zero).

- It is the index for the next location in the array to be used.

The role that countA is playing is similar to what the loop index does in lines 6–8. We created an empty array (grades) in line 3. The members of this array are not initialized, so they can have any random value (you should not assume they are zero). The initialization loop is basically a series of "adding" values to the array and filling up the empty array. As we are filling the array up, at any iteration, i is the next available (non-empty) location in the array. At the start, i is zero means that we have no initialized value, and the first member of the array (grades[0]) is available to receive data (it's empty). Then, i becomes 1, which means grades[0] is no longer empty, and grades[1] is ready to receive data, and so on. The variable countA does the same thing for gradesA except that it cannot increment at every loop. It has to be increased only when we find a match for the search and copy it from grades to gradesA. So, in the search loop (lines 14–20), the index for grades increases every iteration (loop index i), but the index to gradesA doesn't.

> ☑ **Practice Task:** Write a program that creates an array, and initialize it to random numbers. Then, define a new array and copy the values of the first array to the second array.
>
> *Hint*: You need a loop to copy array members one by one.

7.3.2 Ground Levels in Games

In many computer games or other applications, there is a need to simulate the ground level, which is not always flat. While there can be very realistic terrains, in many cases such as 2D platformer games, the ground is a series of flat areas at different levels, as shown in Exhibit 7.3. To move the characters properly and allow them to jump and fall, and stop them when reaching a big bump of ground level, it is necessary for the program to know what the ground level is at any X coordinate. I will discuss the algorithm for implementing jumps and falls using gravity in Chapter 8. For now, I only focus on the ability to determine the ground level.

Depending on the shape of our ground, there are different methods to detect the ground level. For example, the ground in Exhibit 7.3a has a very random shape that cannot be determined algorithmically. As such, the common method used in such cases is a **HeightMap**, an image similar to what we have in Exhibit 7.3c that has no details but shows the height. This image is created by game artists and loaded by the program in addition to

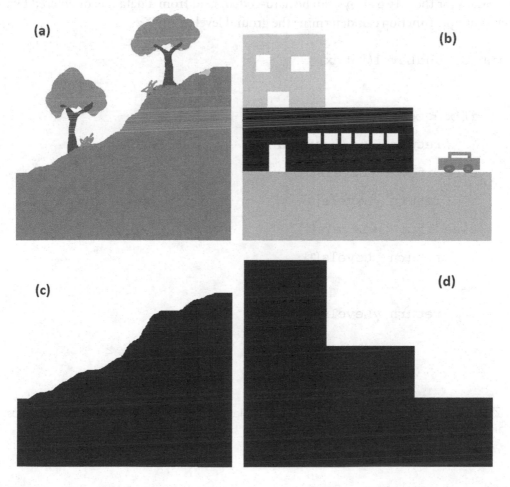

EXHIBIT 7.3   Ground levels. (a) Very realistic and random, (b) piece-wise flat, (c) heightmap for (a), (d) heightmap for (b).

the realistic image of the game world, 3a. The program reads the heightmap and, for every X value, scans the pixels from the top until it reaches the first back one. The Y coordinate of that pixel is the ground level (height) for that X. Assuming that the map has a width of, for example, 1000, the Y coordinates (height values) will be stored in an array[3]:

```
int groundLevels[1000];
```

The ground level in Exhibit 7.3b and d, on the other hand, doesn't need a big array like the one in Exhibit 7.3a and c. Since the ground is flat, we only need to know at which X values the level changes and what is the new Y value. Assuming that we have a maximum of, for example, 3 bumps, the data can be stored in two small arrays:

```
int xLevels[3];

int yLevels[3];
```

The values for these two arrays can be hard-coded, read from a data file, or entered by the user. A simple function can determine the ground level at any X:

```
int GroundLevel(int x)
{
    if(x < xLevels[0])
        return -1; //invalid
    else if(x < xLevels[1])
        return yLevels[0];
    else if(x < xLevels[2])
        return yLevels[1];
    else
        return yLevels[2];
}
```

☑ **Practice Task:** For a larger number of ground levels, revise the above code, and use loops and a variable number.

---

[3] In Chapter 12, I will talk about image processing and show an example of a game using heightmaps.

7.3.3 Plotting Data

Processing data through search operations is a very common task which we reviewed in the previous section. Another common task when dealing with data is visualization or the graphical representation, and plotting a curve is one of the most common ways to visualize data. Plotting data uses one array of variables as horizontal X values and another as vertical Y in a two-dimensional coordinate system. Each pair of XY values identifies a point of the plot. For example, if we are showing a plot for daily temperature, days of the week are X values, and temperatures are the Y ones, as shown in Exhibit 7.4.

Let's consider a simple program to plot data. Imagine a store owner who has five types of products and wants to show the number of items sold vs. the price. The idea is to see if more expensive items are selling in lower quantities. Here is our algorithm:

```
Data:

    Two arrays for price and sales

Code:

    Initialize graphics window

    Initialize arrays X (price) and Y (sales)

    Loop i=1 to 5 (or 0 to 4)

        Put pixel at X[i],Y[i]
```

Sample Code #8-c shows the implementation of this algorithm using SDL/SDLX library.

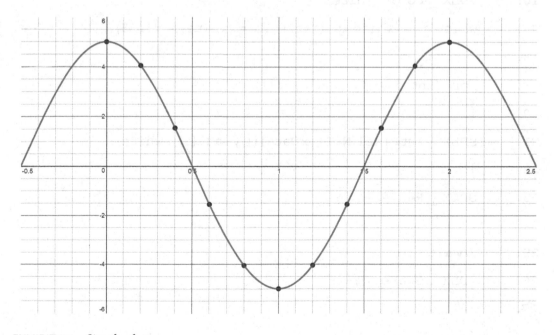

EXHIBIT 7.4   Simple plot.

Sample Code #8-c (SDL project)

```
1. int main(int argc, char *argv[])
2. {
3.      //init SDL library and graphics window
4.      SDLX _ Init("Plot: Sales vs Price", 640, 480);
5.
6.      //data
7.      int xData[] = { 230, 120, 80, 350, 100 };
8.      int yData[5];
9.      yData[0] = 100;
10.     yData[1] = 300;
11.     yData[2] = 400;
12.     yData[3] = 50;
13.     yData[4] = 350;
14.
15.     //plot
16.     SDLX _ Color white;
17.     white.r = white.g = white.b = white.a = 255;
18.
19.     for (int i = 0; i < 5; i++)
20.     {
21.     SDLX _ PutPixel(NULL,xData[i],yData[i],&white);
22.     }
23.     SDLX _ Render();
24.
25.     //wait
26.     SDLX _ GetKey();
27.
```

```
28.    //clean up
29.    SDLX _ End();
30.
31.    return 0;
32. }
```

Note how in the above code, I have illustrated two methods for initializing the arrays: X array (xData) is initialized in one line (7) by providing all data inside a pair of { }. There is no need to identify the array size as the program can count the number of data. Y array (yData), on the other hand, is created as an empty array first, and then, each member has received a value.

We use SDLX _ PutPixel() function to draw points on a bitmap, which will then be rendered to screen once all drawings are done. If the bitmap is NULL, then the pixel will be written to the screen.

☑ **Practice Task:** In the above sample code, replace PutPixel() with drawing a small image at that location.

7.3.4 Scaling and Translation

You probably noticed that the X and Y values in this example were conveniently chosen to fit into our screen (640 by 480). The data you use in different projects is most like not designed to fit into your screen, though. If you plot a Y variable that is between 0 and 5, for example, then you almost won't see it as all points in a 5-pixel-thick band on top of the screen. In many cases, we will need to stretch or shrink the data. You may have done something like that on your TV screens when displaying content with different aspect ratio and size, in order to make it "fit into your screen." This process is called **Scaling**.

Scaling is done through multiplying X and Y data values by a scale factor, which is the ratio of the display range to the data range. For example, if your data is between 0 and 5, and you want to show it between 0 and 400, then the scale factor is (400−0)/(5−0) = 80. So, we will multiply all Y values with 80. Pay attention that if you are using a "typed" language like C, then the calculations have to be done using a non-integer type, even if the data is finally stored in an integer variable.

Sample Code #9-c shows the same plot as #8-c but assumes a different data range that needs to be scaled. We assume prices (X) are in 0-$20 range, and we want to show them in 0–600 on the screen. The sales data (Y) are in 0–50 customers range, and we want to show them in 0–400 range on the screen. Lines 15 and 16 define the scale factors, while the SDLX _ PutPixel() uses the data multiplied by those factors.

Sample Code #9-c (SDL project)

```
1. int main(int argc, char *argv[])
2. {
3.        //init SDL library and graphics window
4.        SDLX _ Init("Plot: Sales vs Price ", 640, 480);
5.
6.        //data
7.        int xData[] = { 10, 8, 5, 15, 6 };
8.        int yData[5];
9.        yData[0] = 10;
10.       yData[1] = 30;
11.       yData[2] = 40;
12.       yData[3] = 5;
13.       yData[4] = 35;
14.
15.       float xScale = 400.0/20; //20
16.       float yScale = 600.0/50; //30
17.
18.       //plot
19.       SDLX _ Color white;
20.       white.r = white.g = white.b = white.a = 255;
21.
22.       for (int i = 0; i < 5; i++)
23.       {
24.           SDLX _ PutPixel(NULL,
25.               xScale*xData[i],
26.               ySCale*yData[i],
27.               &white);
28.       }
```

```
29.     SDLX _ Render();

30.

31.     //wait

32.     SDLX _ GetKey();

33.

34.     //clean up

35.     SDLX _ End();

36.

37.     return 0;

38. }
```

Scaling is an operation that takes us from the original data range to screen range by affecting the size; if our data range is from −10 to 10 (that is 20), it has to be stretched to fill a screen with a width of 800 pixels. Other similar operations are **Translation** (or **Shifting**) and **Rotation** that moves the object and change the orientation, respectively. In a translation, the size (of the data range) stays the same, but a data range of −10 to 10 may change to 0–20, for example. Rotation, on the other hand, works with 2D and 3D shapes and changes the way they are aligned with axes. All these operations are required when we use data of any type (from a range of weather values to a 3D animated character) in a new environment. Scaling, Translation, and Rotation are generally called **Linear Transformations**. While scaling is done through multiplication, translation is done by adding a factor to data. Both can be applied to X- or Y-axis independently. Rotation, on the other hand, is a 2D or 3D operations. These transformations move us from one coordinate system (or space) to another, as shown in Exhibit 7.5.

The combination of scaling and translation creates an operation that can be shown using formula $F = aX + b$ where X is the data, a is scaling factor, and b is the translation factor. F is the new, transformed value. This is an example of a mathematical function.

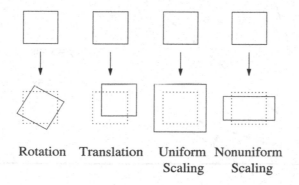

Rotation   Translation   Uniform   Nonuniform
                           Scaling   Scaling

EXHIBIT 7.5   Translation, scaling, and rotation as a change in coordinate systems.

Mathematical Functions are operations that receive one or more input (independent variable, commonly shown with X) and create an output (dependent variable, commonly shown with Y). The dependent variable is a "function of" the independent variable and sometimes shows as Y = F(X) or "Y equals to F of X." You can see that they are very similar to programming functions with parameters and return values.

A very common group of functions are **Linear Functions** represented with the general formula Y = F(X) = aX + b where a and b are the slope and offset values for a line. They are called linear because if we plot the data (**dependent variable vs. independent variable**), the result will be a line, as shown in Exhibit 7.6 with different values for a and b. Linear functions belong to a group of functions called **Polynomial**, where Y is defined as the sum of terms with varying powers of X such as $Y = aX^2 + bX + c$ (second-degree polynomial) and $Y = aX^3 + bX^2 + cX + d$ (third-degree polynomial).

Linear functions, and polynomials in general, play an essential role in various scientific, artistic, and technical fields. For example, the movement of an object at constant speed is a linear function of time where the slope is the object's speed, and offset is its initial location. This is particularly helpful when we want to simulate a moving object in our program and need to know where the object is at any time. A movement with accelerating speed, on the other hand, is a second-degree polynomial function of time. I will discuss this later when I add jumping and falling to our simple game.

☞ **Key Points:** Mathematical Functions
- Mathematical Functions calculate the value of a dependent variable based on an independent variable.
- Linear Functions have a line plot.
- Linear Transformations change the size, location, or orientation of objects, but not the general shape.

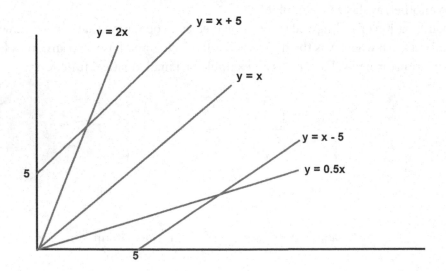

EXHIBIT 7.6   Plotting lines.

Let's consider the problem of creating and plotting a linear function. Using our common algorithmic approach, we identify our data elements first. They include the independent (X) and dependent (Y) variables but also parameters a and b. We need to create a variable in our program for each of these, but remember that our X values simply start from 0 and increase by 1 to the end of our plot area (screen). We can think of our plot data as an array of Y values where X values are the index to the array. Our code will be very similar to #9 without the need for xData or scaling.

Sample Code #10-c (SDLX project. Partial code)

```
1. float a = 1;

2. float b = 0;

3. int y[600];

4. for (int i=0; i < 600; i++)

5.     y[i] = a*i + b;

6. for (int i=0; i < 600; i++)

7.     SDLX _ PutPixel(NULL, i, y[i], &white);
```

You notice that the result doesn't look quite like what we saw in Exhibit 7.3. The reason is that in typical math systems, the Y dimensions increase as we go up. But in computer graphics, it is the opposite. Also, there are no negative values. To fix this, we need to do scaling and shifting as we saw before.

Imagine you want to draw a plot similar to Exhibit 7.3 in which X and Y both range from −100 to 100. The total range for both dimensions is 200, (100 − (−100)), so assuming a screen size of 640 × 480, the scale factors are

xScale = 640 / 200 = 3.2

yScale = 480 / 200 = 2.4

We also need an offset of −100 to translate (shift) the numbers from screen space to the intended range so that the pixel 320/240, which is in the middle of the screen, will be translated to 0/0 or the center of coordinate system for the data. Sample Code #11-c shows the code for this version of the plot. The variables a and b are the line parameters. yData holds 640 Y values in the data space (−100 to 100) range. The index to the array corresponds to X values but in the screen space. In order to use the linear function to calculate Y (line 25), we first need to transform the X values from screen space (0–640) to data space (−100 to 100). This is done in line 24 by scaling and then shifting. Drawing the pixels is done in line 35. Similarly, we need to transform the Y values from the data space to screen space (opposite of what we did in line 24). This is done by shifting and then scaling in line 34.

Note that the positive Y dimension on the screen is downward. To fix this, we can always subtract the Y value from the height of the screen, as shown in line 35.

Sample Code #11-c (SDLX project. Partial code)

```
1. //data
2. float xScale = 640.0 / 200;
3. float ySCale = 480.0 / 200;
4. int   xyShift = -100;
5.
6. float a = 1;
7. float b = 0;
8. int yData[640];
9. for (int i = 0; i < 640; i++)
10. {
11.     float x = i / xScale + xyShift;
12.     yData[i] = a*x + b;
13. }
14.
15. //plot
16. SDLX _ Color white;
17. white.r = white.g = white.b = white.a = 255;
18. for (int i = 0; i < 640; i++)
19. {
20.     int y = (yData[i] - xyShift)*ySCale;
21.     SDLX _ PutPixel(NULL, i, 480-y, &white);
22. }
```

☑ **Practice Task:** In the above code, try changing the values of a and b and see how the line changes.

☑ **Practice Task:** Another common mathematical function is sine, shown as Y = sin(X). The sine function is defined in trigonometry for an acute angle (less than 90') in a right triangle (one with a 90 angle). The sine of an angle is a cyclical curve, as shown in Exhibit 7.5.

7.3.5 Text Processing

Strings of text can be considered arrays of single characters. While some languages have a specific data type for text (usually called a **string**), in C/C++, we simply use an array of char data types.

Recall that char data type is a 1-byte integer number if we are using the ASCII standard. While it can hold any values, it is common to use it for character codes. So, for example, the following lines will print characters A and B.

```
char cData = 'A';
printf("%c ", cData);
cData = 66; //character code for B
printf("%c", cData);
printf("%d", cData);
```

You notice that the operator ' ' returns the code for a character. It is different from " " that holds a string (series of characters ending with zero, NULL). Also, note that the char value can be used just as a number as in the last line that uses %d instead of %c.

A string is created as an array of characters. The following code creates an empty array, initializes its members with the word Hello, terminates the string with a zero so the program knows where the end of string is, and then prints the text using %s.

```
char text[50];
text[0] = 'H';
text[1] = 'e';
text[2] = 'l';
text[3] = 'l';
text[4] = 'o';
text[5] = 0;
printf("%s", text);
```

Alternatively, we could not specify the array size and initialize it in one line. The program automatically allocates six bytes (the length of the string plus the terminating zero) and adds all the data to the string. Of course, in this case, we are limited to that size and cannot use longer text while in the other method, we can have text up to 49 characters (keep in mind that we need one byte for terminating zero).

```
char text[] = "Hello";
```

☞ **Key Point:** Text strings are arrays of characters.

Char arrays or strings can be used just like any other arrays, but C/C++ has special functions to help with text processing defined in **string.h**. Sample code #12-c demonstrates some of these functions.

Sample Code 12-c (needs `stdio.h` and `string.h`)

```
1. #include "stdio.h"

2. #include "string.h"

3. void main()

4. {

5.       char name[50];

6.       printf("Enter your name: ");

7.       scanf_s("%s", name, 50);

8.       printf("%s\n", name);

9.       if (strcmp(name, "Ali") == 0)

10.          printf("Hello, Ali!\n");

11.      else

12.      {

13.          printf("Unknown user!\n");

14.          strcpy_s(name, 50, "Ali");

15.          int length = strlen(name);

16.          printf("%s, %d\n", name, length);

17.      }

18. }
```

Some of the most commonly used C/C++ functions to process strings are

- `printf()` and `scanf()` can write and read text using the %s option.
  - Notice the _s at the end of `scanf()`. This makes a more recent and secure version of the function that limits the input data to a particular length. For example, in this example, the string has a maximum size of 50, and so the last parameter is the maximum length (line 7).
- To strings can be compared using `strcmp()`. It returns zero if the parameters are equal (line 9). Note that strings in C/C++ are case sensitive.

- A string can be copied to another using `strcpy()`. Just like scanf and any other function that receives data into a string, the _s identifies the secure version that requires a maximum length (line 14).

- The length of a string (not including the terminating zero) can be determined using `strlen()`, as shown in line 15.

☑ **Practice Task:** In Sample Code #12-c, add a password. Define two names and a password for each. When the user enters a name, compare to the defined names, and then, ask for a password and compare that too.

7.4 COMBINED DATA MODULES

7.4.1 Simple Database

In Sample Code #5 (c, j, and p), we saw how to use arrays to hold student grades and then use that array to calculate the class average and identify the stronger and weaker students. This is what happens in one course, but students usually take multiple courses, and the school needs to do something similar based on all their grades. Imagine a simple case where each student has three classes, and we want to create a simple database program that stores all student data and identifies weak students (below average) based on their GPA (Grade Point Average, or simply the average of all grades). For simplicity, let's assume we need only the following information about each student:

- Grades

- Student ID

- Name

- GPA.

In database terminology, each one of these information items is called a **field**. The collection of all the fields for one person (or other unit) is called a **record**. A real school database may require more fields, but for now, this will suffice. Based on what we have learned so far, you can probably guess that a UDT can collectively represent each student, and an array of such students can be the basis of our database. Let's start with only one student and make sure we have the code to process that data. Sample Code #13-p shows such a simple program in Python.

Sample Code #13-p

```
1. class Student :
2.      name = ""
3.      ID = 1000
4.      grades = []
```

```
5.        GPA = 0
6.
7.  def CalculateGPA(student) :
8.        sum = 0
9.        for i in range(3) :
10.            sum += student.grades[i]
11.       return sum/3
12.
13. def Report(student) :
14.       print(student.ID)
15.       print(student.name)
16.       for i in range(3) :
17.            print(student.grades[i])
18.       print(student.GPA)
19.
20. #Program starts here
21. student = Student()
22. student.name = "Ali"
23. for i in range(3) :
24.       print("enter a grade: ")
25.       x = int(input())
26.       student.grades.append(x)
27. student.GPA = CalculateGPA(student)
28. Report(student)
```

The main parts of this code are

- The Student class that combines all the information for a student and itself has an array for grades.

- The CalculateGPA() function which is similar to the average calculation in Sample Code #5 except that instead of an independent variable grades is now a part of student.

- The `Report()` function prints all the members of `Student`. Note that accessing an array requires setting up a loop.

- The rest of the program simply creates an instance of Student, initializes the grades, and then calls the above functions.

Lines 8–29 of the Sample Code #13-c show the same program in C/C++. Of course, creating just one student is far from even a simple database. Let's assume that the minimum requirements for this database are

1. Hold information for each student. Allow up to 3 courses per student and up to 20 characters per name.

2. Hold information for up to 30 students.

3. Allow users to enter the name and grade data.

4. Automatically calculate GPA and assign an ID starting with 1000.

Requirement #1 is already taken care of by the Student structure and its related functions. Requirement #2 will need an array of students. Line 33 of Sample Code #13-c defines such an array. Note how we use labels and `#define` preprocessor directive in C/C++ to make the program easier to read and change.

```
#define NUM_COURSES      3     //number of courses per student
#define NUM_STUDENTS     30    //number of students
#define NUM_CHARACTERS   20    //number of name characters (max)
```

Seeing numbers like 3, 30, and 20 in the code is not very informative, but the above labels make much more sense when trying to understand (or remember) what the code does. Also, if we later decide to change, for example, the number of courses, using a label we only need to change one line as opposed to looking in the code for all occurrences of the number 3 and see if they are actually the number of courses and need to be changed.

> ☞ **Key Point:** Hard-coded numbers in the program are not ideal as (1) they can make the code hard to understand, and (2) they are hard to change if used multiple times. Labels such as those created by `#define` in C/C++ are better alternatives.

> ✋ **Reflective Questions:** #define doesn't have to be used for labeling a number. It simply means a text equals another text. What other uses can you imagine for this ability? How can it be used to define a short text to be used as a replacement for a long one, or a more meaningful name for a strangely named library function?

Once the primary data elements of our program are identified, we only need to create a loop (line 36) and repeat the process for all students (lines 46–55). We could use a for loop and repeat for a definite number of times (<DS>NUM _ STUDENTS</DS>). Instead, and for flexibility, I have a while loop (line 36) that ends if

- We reach the max number (count < NUM _ STUDENTS), or

- The user specifies that no more student is needed (line 39–43).

Note that the C/C++ break keyword ends the current loop and goes immediately to the first line after the loop, which in this case, is the end of the program. Another useful C/C++ keyword is continue, which ends the current iteration of the loop and goes to the top (checking the condition).

Sample Code #13-c

```
1. #include "stdio.h"

2. #include "string.h"

3.

4. #define NUM _ COURSES     3

5. #define NUM _ STUDENTS    30

6. #define NUM _ CHARACTERS 20

7.

8. struct Student

9. {

10.    int ID;

11.    char name[NUM _ CHARACTERS];

12.    float grades[NUM _ COURSES];

13.    float GPA;

14. };

15.

16. int CalculateGPA(Student s)

17. {

18.    float sum = 0;

19.    for (int i = 0; i < NUM _ COURSES; i++)

20.        sum += s.grades[i];
```

```
21.    return sum / NUM_COURSES;
22. }
23. void Report(Student s)
24. {
25.    printf("ID=%d , Name=%s\n", s.ID, s.name);
26.    for (int i = 0; i < NUM_COURSES; i++)
27.       printf("%f\n",s.grades[i]);
28.    printf("GPA=%f\n\n", s.GPA);
29. }
30.
31. void main()
32. {
33.    Student students[NUM_STUDENTS];
34.    int count = 0;
35.
36.    while (count < NUM_STUDENTS)
37.       {
38.       //check if the user wants a new student
39.       int x;
40.       printf("New student? (1-YES, 2-NO) ");
41.       scanf_s("%d", &x);
42.       if (x != 1)
43.          break;
44.
45.       //main process for each student
46.       printf("enter new student's name: ");
47.       scanf_s("%s", students[count].name, NUM_CHARACTERS);
48.       students[count].ID = 1000 + count;
49.       for (int i = 0; i < NUM_COURSES; i++)
```

```
50.            {
51.                printf("enter grade: ");
52.                scanf _ s("%f", &students[count].grades[i]);
53.            }
54.            students[count].GPA = CalculateGPA(students[count]);
55.
56.            Report(students[count]);
57.
58.            count++;
59.        }
60. }
```

The Sample Codes #13-p and #13-c define a class (Python) or structure (C/C++) that corresponds to a Student. They then create a single variable or an array of those types to demonstrate how to work with this UDT. The main point that deserves special attention in these examples is how we create an array of Student data. Arrays can be made of any type, including Student. Once the array is defined, any member can be accessed using the index value (students[i]). Any member of the array is a Student and has its own members that can be accessed using a "." symbol. Anything that we could do with one Student can be done similarly for any member of students array.

Combining UDTs and arrays is a very common practice as it allows us to use a set of data items of similar type (array), and each of them made of related members of different types (class/structure). In the next example, we will see how we can use this to improve our simple game code.

☑ **Practice Task:** Expand the Sample Code #13-p to include similar functionality as #13-c.

☑ **Practice Task:** Write the Javascript version of Sample Code #13-p and 13-c.

✍ **Reflective Questions:** Arrays of any type are usually used to group members with a similar role in the program. It is, then, common to perform similar operations on these members, which is typically done by loops. Does the connection between arrays and loops make sense to you? When repeating operations using loops, it is important to note what needs to be repeated, and what needs to be done only once at the start or end. Can you think of cases when we need to deal with some members of the arrays differently, and as such, they have to be excluded from the loop?

7.4.2 2D Game with Arrays

In games, just like many other applications, we frequently use a large number of similar items; there are only four ghosts attacking Pac-Man,[4] but many Pac-Dots that are controlled by the same logic. From enemies to environmental objects, we always have such entities with repeated looks and functionalities. Arrays offer a perfect way to manage these items together, while UDTs (class or structure) can collect all the data each one of them needs.

Recall the Sample Code #2-p and #2-c that defined the GameObject module with location and shape. In previous sections, we added visibility to it. Sample Code #14-c shows the C/C++ program that uses an array of such objects and places them at random locations on the screen. The player can then move around and "pick up" the dots (prizes). We have also added two new members to GameObject to control the speed of movement in X and Y directions (lines 8–16). These speed values are used for the player movement in the Update part of the code (lines 74–88) and allow faster movement:

```
player.x += player.sx;
```

Instead of

```
player.x ++; // same as +=1
```

The code combines the use of arrays with all the new features we added in previous sections. These features include better collision detection, visibility, and data for score and life, and their implementation is almost precisely like Javascript and Python.

The new Distance() function is added (line 22) and used for detecting collision with enemy and prize (lines 95 and 103, respectively).

The new visibility variable is added to the GameObject structure (line 15) and used in DrawObject() (line 17). It is set upon collision with a prize (line 105).

The score and life variables are added and initialized (lines 59 and 60) and then changed upon collision. They are used to end the game (lines 94–108). Note the use of NUM _ LIVES and NUM _ PRIZES (lines 5 and 6) to avoid hard-coding numbers in multiple places in the code. This allows us to easily change these numbers without going through the code.

The use of arrays in this sample code (and in most other cases) is very straight-forward. If we have a program that works for a single variable (for example, a prize object) and we want to expand it to multiple similar objects using an array, we follow these steps:

- Change the variable to an array (line 49).

  - Single: GameObject prize;

  - Array: GameObject prize[NUM _ PRIZES];

---

[4] See sidebar on Computer Games History.

- Identify the parts of the code that need to be repeated. Move them inside a loop with the array size, and change all single variables to an array member with proper index.

  - Single:
    - `prize.shape = SDLX_LoadBitmap("Prize.bmp");`
    - `prize.x = rand() % 400;`
    - `prize.y = rand() % 400;`
    - `prize.visibility = 100;`
  - Array:
    - `for (int i = 0; i < NUM_PRIZES; i++)`
    - `{`
      - `prize[i].shape = SDLX_LoadBitmap("Prize.bmp");`
      - `prize[i].x = rand() % 400;`
      - `prize[i].y = rand() % 400;`
      - `prize[i].visibility = 100;`
    - `}`

See how this is done for collision detection (lines 101–110).

Sample Code #14-c

```
1. #include "SDLX.h"        //for SDL and SDLX
2. #include "stdlib.h"      //for rand()
3. #include "time.h"        //for time()
4.
5. #define NUM_LIVES      3
6. #define NUM_PRIZES 5
7.
8. struct GameObject
9. {
10.     SDLX_Bitmap* shape;
11.     int x; //location
12.     int y;
```

```
13.     int sx;  //speed
14.     int sy;
15.     int visibility;
16. };
17. void DrawObject(GameObject* obj)
18. {
19.     if(obj->visibility != 0)
20.         SDLX _ DrawBitmap(obj->shape, obj->x,obj->y);
21. }
22. float Distance(GameObject* obj1, GameObject* obj2)
23. {
24.     int dx = obj1->x - obj2->x;
25.     int dy = obj1->y - obj2->y;
26.     float d = sqrt(dx*dx + dy * dy);
27.     return d;
28. }
29.
30.
31. int main(int argc, char *argv[])
32. {
33.     ////////////////////////
34.     // Initialize
35.     ////////////////////////
36.
37.     //init SDL library and graphics window
38.     SDLX _ Init("Simple 2D Game", 640, 480);
39.     srand(time(NULL));
40.
41.     //game objects
```

```
42.     GameObject player;

43.     player.shape = SDLX _ LoadBitmap("Player.bmp");

44.     player.x = player.y = 0;

45.     player.sx = player.sy = 3;

46.     GameObject enemy;

47.     enemy.shape = SDLX _ LoadBitmap("Enemy.bmp");

48.     enemy.x = enemy.y = 200;

49.     GameObject prize[NUM _ PRIZES];

50.     for (int i = 0; i < NUM _ PRIZES; i++)

51.     {

52.         prize[i].shape = SDLX _ LoadBitmap("Prize.bmp");

53.         prize[i].x = rand() % 400;

54.         prize[i].y = rand() % 400;

55.         prize[i].visibility = 100;

56.     }

57.

58.     //other variables

59.     int score = 0;

60.     int life = NUM _ LIVES;

61.     SDLX _ Event e;

62.     bool quit = false;

63.

64.     // Main Loop

65.     while (!quit)

66.     {

67.         /////////////////////

68.         // Update

69.         /////////////////////

70.
```

```
71.          //keyboard events
72.          if (SDLX _ PollEvent(&e))
73.          {
74.               if (e.type == SDL _ KEYDOWN)
75.               {
76.                    //press escape to end the game
77.                    if (e.keycode == SDLK _ ESCAPE)
78.                         quit = true;
79.                    //update player
80.                    if (e.keycode == SDLK _ RIGHT)
81.                         player.x += player.sx;
82.                    if (e.keycode == SDLK _ LEFT)
83.                         player.x -= player.sx;
84.                    if (e.keycode == SDLK _ UP)
85.                         player.y -= player.sy;
86.                    if (e.keycode == SDLK _ DOWN)
87.                         player.y += player.sy;
88.               }
89.          }
90.          //update enemy
91.          enemy.x += rand() % 7 - 3;
92.          enemy.y += rand() % 7 - 3;
93.
94.          //collision detection
95.          if (Distance(&player, &enemy) < 5)
96.          {
97.               life--;
98.               if (life == 0)
99.                    quit = true;
100.         }
```

```
101.          for (int i = 0; i < NUM_PRIZES; i++)
102.          {
103.              if ((prize[i].visibility != 0) && Distance(&player,
     &prize[i]) < 5)
104.              {
105.                  prize[i].visibility = 0;
106.                  score++;
107.                  if (score == NUM_PRIZES)
108.                      quit = true;
109.              }
110.          }
111.
112.     ////////////////////////
113.     // Draw
114.     ////////////////////////
115.
116.     //First clear the renderer/buffer
117.     SDLX_Clear();
118.
119.     //Draw the objects (in order: background to
     foreground)
120.     for(int i=0; i< NUM_PRIZES; i++)
121.         DrawObject(&prize[i]);
122.     DrawObject(&enemy);
123.     DrawObject(&player);
124.
125.     //Update the screen
126.     SDLX_Render();
127.
```

```
128.            //pause
129.            SDLX _ Delay(1);
130.     }
131.
132.     //clean up
133.     SDLX _ End();
134.
135.     return 0;
136.}
```

☑ **Practice Task:** Use similar code and move enemies.

SIDEBAR: MILESTONES IN COMPUTER GAMES HISTORY

In 1952, Alexander Shafto "Sandy" Douglas, a British Ph.D. student at the University of Cambridge, created a graphical computer game called *OXO* on a mainframe computer as part of his thesis research on human–computer interaction. *OXO* was quite simple and would simulate a tic-tac-toe game on a 35 × 16 monochrome screen. *OXO* and a checkers game created Christopher Strachey around the same time are the first examples of computer games.

In 1962, Steve Russell created *Spacewar!*, a simple space combat computer game and the first video game that could be installed and played on multiple computers.

In 1967, Sanders Associates, Inc. invented a multiplayer, multi-program game system called *The Brown Box* that could connect to a TV as the visual output. While it wasn't a commercial success, one of its games inspired the design of *Pong*. Released in 1972 by Atari, *Pong* was the first commercially successful video game available on a dedicated game console.

In 1977, Infocom released *Zork*, a text-based interactive fiction game that inspired many other games in that genre. Late 70's and early 80's also saw the release of some other influential games such as *Pac-Man* and *Space Invaders*.

The game industry had rough years in the early to mid-80s but started to recover with the release of the Nintendo Entertainment System (NES) in North America. It was followed by Nintendo Gameboy and the 16-bit Sega Genesis in 1989.

In the early 90s, games such as *Doom* and *Myst* introduced and popularized 3D worlds, multimedia content, story-based gameplay, and reusable game engines.

In the early 2000s, more game consoles such as Xbox and PlayStation 2 came to the market, and games such as *SimCity* and *Sims* introduced game AI while those such as *Half-Life* advanced the graphics and storytelling in games.

Later in 2006–2010, Nintendo Wii and Microsoft Kinect introduced players to new ways of interacting with games that were based on physical activity, while the release of smartphones created the new category of mobile games.

With the advances in Augmented and Virtual Reality, the 2010s saw the popularity of AR/VR games such as *Pokémon Go* and games made for Oculus Rift, HTC Vive, and other immersive headsets.

7.4.3 Animated Objects

In previous examples, we combined arrays and UDTs by defining arrays of complex items such as `Student` or `GameObject`. Another way of using these two data modules is to have new types that include arrays. For example, in Sample Code #13-c and #13-p, if the student has 5, 10, or more courses, then it is not convenient to have a separate variable for each. An array of grades will be a more efficient solution. Another interesting example is adding an animated object to our simple games such as Sample Code #14-c.

Animation simulates the effect of moving objects or any change over time. Any motion picture, be it live-action or animated, is a series of pictures (frames) that, once played back, gives the illusion of continuous motion or change. Exhibit 7.7 shows a few frames that together can make an animated character who is running. That is what we call an animation sequence.

In live-action films, one photo of the whole scene is taken for every frame. In computer-generated animations and games, for every frame, each of the moving objects needs to change its shape and so has its own separate animation sequence. Depending on what is happening, for every frame, some of the objects go through their animation sequence. For example, if the character in Exhibit 7.5 is used in a scene that involves running, for every frame, we use one of the running shapes, and we keep advancing to the next shape as long as running continues. Since running can continue for a long time and our animation sequence is not unlimited, we can simply restart from the beginning. That is why the animation sequence is sometimes called the **animation cycle** and is designed in a way that the last frame leads to the first one so it can be used in a loop. Once running ends and character is standing still, it won't change its shape or plays an idle animation if it has one.

In Sample Code #14-c, and our previous game programs, we saw that each game object has a few essential data items, including the coordinates and its shape. An animated object has a series of shapes, as shown in Exhibit 7.5. The game code decides which one to use based on what the character is doing at that frame. In order to figure out how this affects our program, we follow our data-centered approach; we start by identifying what new information we have and then what we have to do with that data. Let's ask ourselves these questions:

EXHIBIT 7.7   Animated character.

- How to store the information for more than one shape? Any new information needs new variables, but a series of similar variables that have the same role in our program are good candidates to be defined as an array. So, we will have an array of shapes (animation frames).

- How does the program know which frame of the animation to use at any time for the character shape? We start with the first frame and then move to the second, then third, and so on. But we need to remember which one we rendered, so we can decide which one is next. Again, following our first rule of programming, we need to define a variable (call it `currentFrame`, for example) that is pointing to the first frame to start and advances as we move forward.

- What do we do when we reach the end of the animation? Remember that in the end, we need to restart the animation cycle. We can know that we have reached the end when the current frame is the last frame. This, again, needs to be a piece of explicit information in our program by defining a variable that shows the number of frames, so we can compare the current frame to it and see if we have reached the end.

To implement these features, the following changes have to be made in Sample Code #14-c:

- `GameObject` structure has an array of frames and two variables for the number of frames and the current frame. Note that for compatibility with other parts of the program, we keep the variable `shape` and add `frames[]`. This allows the structure to be used for non-animated objects without any change.

```
#define NUM_FRAMES 3
struct GameObject
{
    SDLX_Bitmap* shape;
    int x;  //location
    int y;
    int sx; //speed
    int sy;
    int visibility;
    //animation
    SDLX_Bitmap* frames[NUM_FRAMES];
    int numFrames;
    int currentFrame;
};
```

- The initialization code defines animation frames for the player object. The member `shape` is still used in `DrawObject()`, so it is initialized to the first frame.

```
GameObject player;
player.frames[0] = SDLX_LoadBitmap("r0.bmp");
player.frames[1] = SDLX_LoadBitmap("r1.bmp");
player.frames[2] = SDLX_LoadBitmap("r2.bmp");
//player.shape = SDLX_LoadBitmap("Player.bmp");
player.shape = player.frames[0];
player.numFrames = 3;
player.currentFrame = 0;
```

- A new function is added to advance the animation. Notice that it doesn't do any actual drawing and is limited to the update part of the code. Drawing is still done in the same way as before using `shape`, to maintain compatibility.

```
void AdvanceAnimation(GameObject* obj)
{
    obj->currentFrame++;
    if (obj->currentFrame == obj->numFrames)
        obj->currentFrame = 0;
    obj->shape = obj->frames[obj->currentFrame];
}
```

EXHIBIT 7.8   Game with background and moving characters. The character on the left shows its background, but the character on the right has a transparent background.

- The update code advances the animation when the player is moving. As long as we hold down the arrow key, the shape will change on each frame.

```
//update player
if (e.keycode == SDLK_RIGHT)
{
    player.x += player.sx;
    AdvanceAnimation(&player);
}
```

7.4.4 Transparent Pixels and Background Image

As we discussed in Section 6.3, graphics systems use an alpha value to control the transparency of pixels. Another method for achieving transparency is to choose a key color as transparent. This color (usually the background color of the image) will not be rendered, and as such, the image can be drawn on top of another without the background causing an issue. This is particularly important for images that represent a game character as they move over background images, as shown in Exhibit 7.8.

Some image file formats such as PNG and GIF allow transparency, while others such as BMP and JPG don't include any transparency information. The Simple DirectMedia Layer (SDL) library (introduced in Chapter 6 and used in our C/C++ graphics examples) only supports BMP images.[5] To achieve transparency, SDL offers the use of Alpha values that can be set for any pixel or choosing a key color as transparent for any image. To demonstrate this, we can add a background image to the game and see how the background pixels for the player image can be rendered transparent.

---

[5] SDL_Image extension supports other image formats but is not discussed in this book.

In the initialize part of the code, we can add a new image that is our background image (same size as the game screen):

```
SDLX _ Bitmap* bg = SDLX _ LoadBitmap("Sky.bmp");
```

Then, we replace clearing the screen with drawing this background. Note that the background image covers the whole screen, so there is no need to "clear" the screen before rendering a new frame.

```
//SDLX_Clear();
SDLX_DrawBitmap(bg, 0, 0);
```

Now, we specify the background color of our animated frames (green) as transparent, right after we load them:

```
player.frames[0] = SDLX _ LoadBitmap("r0.bmp");

player.frames[1] = SDLX _ LoadBitmap("r1.bmp");

player.frames[2] = SDLX _ LoadBitmap("r2.bmp");

//RGB value of background set as transparent

for(int i=0; i < NUM _ FRAMES; i++)

    SDLX _ SetTransparentColor(player.frames[i], 0, 128, 0);
```

☑ **Practice Task:** Change the animation code so that the shape goes back to the "idle" shape when not moving.

HIGHLIGHTS

- UDTs and arrays are two common methods of modularizing data.
- UDTs and arrays allow programmers to combine different data elements into one module. User, in this case, refers to the programmer, the user of the language.
- UDTs (structures and classes) are collections of related data members that have different roles in the program, can have different types, and use different names.
- Arrays are collections of related data members that have the same role in the program and the same type and are accessed using an index value.
- Modules of data are usually processed using modules of code (functions) specially written for them.
- Array index values start from zero.

- Arrays are usually processed through loops.

- Strings of text are arrays of characters.

- Linear mathematical functions have a linear XY plot.

- Linear transformations are scaling, rotation, and translation.

- Games follow a common organization: an initialization part and a main loop (itself divided into update and draw sections).

- Animation is an array of frames (images).

END-OF-CHAPTER NOTES

A. Things I Should Mention

- The terms UDT, structure, and class are used frequently in this chapter and the rest of this book. By the time we get to Chapter 9, they will be more clear, but for now, you can assume that they are the same thing: a collection of related data items of, potentially, different types.

- The notion of array and UDT as modules of data is important in understanding modularization and software design in general.

- Using the term UDT can be confusing as the "user" refers to the "programmer." A programing language defines certain standard types, and those who use it (programmers) can define their own.

B. Self-test Questions

- What is a UDT?

- What are the differences between a UDT and an array? What are the similarities?

- In C/C++, how do you show the first member of an array, called mydata?

- If an array has ten members, what is the index of the last one?

- A database is made of fields and records. How do you implement them using UDTs and arrays?

- If you have three enemies in a simple game, what is the advantage of using an arrays compared to three separate variables?

- How can we use arrays to show an animated object?

C. Things You Should Do

- You now know almost anything you need to code cool 2D games. So, get your artist friends and make a good-looking one! You can make the art assets yourself too.

- Check out these web resources to learn more about databases:

  - https://www.coursera.org/lecture/sql-data-science/introduction-to-databases-XO9Ak

  - http://www.w3schools.com/sql (SQL is a language designed for working with databases.)

D. Reflect on the Experience of Reading This Chapter

- What did you expect from this chapter before reading it?

- What was it about, and what did you learn?

- What tasks did you perform, and what difficulties did you face?

- How did you feel about the material and tasks presented in this chapter?

- How can you improve your learning experience?

- How do you see this topic in relation to the goal of learning to develop programs?

- How comfortable are you with the concepts of UDT and array?

- Does the notion of modularization of data using these two make sense to you?

- Can you think of how software programs you use may have such modules?

- Can you combine them to make complicated data modules?

CHAPTER 8

Modularization of Code

Topics

- Using functions as modules of code in Structured Programming
- Introduction to software design

At the end of this chapter, you should be able to:

- Define and use multiple functions with various parameters and return types
- Design a modularized program using functions
- Design a modular code organization for games
- Design a function that is reusable
- Apply the principle of data hiding for deep self-sufficient functions

OVERVIEW

Learning programming starts with writing a few lines of code that show the result of a single operation on the screen. But later, it can advance to complicated programs that manage all transactions of a banking system or a multiplayer online game. Just like using LEGO®, the key to successfully designing software is the concept of modularization: we start with simple but highly versatile and reusable modules; then, we built new modules that are more complicated from smaller modules in a hierarchical way.

I introduced arrays and user-defined types (classes or structures) as basic modules of data, and functions as the basic module of code. Groups of these modules can together define **Software Components**, things such as a Game Renderer, Physics Engine, Voice Recognition, or Network Communication. These components can be used in different programs if they are made in a reusable way. They can also be grouped together in certain patterns to define **Software Architectures**, which are high-level arrangements of different components. For example, one program may have a tiered architecture with three tiers (layers): front-end (all modules related to interaction with the user, including both input and output), back-end (behind the scene operations), and storage (persistent data). Another system may use a distributed client–server architecture, where many programs running on different user computers (called **Clients**) connect and get service from a program running on a known central computer (called **Server**). Web applications usually combine a tiered architecture with a client–server one.

When we design smaller programs, we may not think of reusable components and high-level architectures, but as our programs grow in complexity, those issues become important, and it becomes essential to decide on an architecture, then the components, and finally smaller modules.

As you can see, modularization is the thread that connects many different concepts in software development, and it starts with basic and simple modules of data and code. As we move forward in the book, I will introduce some commonly used components and architectures, but the detailed study of those subjects is within the scope of another book on advanced software design. In this book, we focus on creating modules of code and data, and small components and programs made with them. After a review of how to define modules of data in Chapter 7, in this chapter, I talk more about creating modules of code, functions. Our goal is to see how to design functions that are reusable and manageable, that is to say, we can easily use them in another program because they have general purposes, and we can easily integrate them together because they are designed to allow modularity and distribution of work among different modules.

When you start programming, your goal is probably to "get your code to work." As you move forward to become a professional software developer, you realize that there is more to software development than writing code that works. It is hard to define what a good software is, but there are many features that are commonly associated with it that are mostly related to proper design to make using and reusing the code easier. In this book, I refer to these as **Manageability** and **Reusability** and aim to show how they can be achieved through proper modularization.

In this chapter, I revisit functions and see how we can create reusable and manageable modules of code and data. I demonstrate this through a few examples that also introduce some simple yet commonly used architectures. I continue discussing modularization in the next part of this book by introducing Object-Oriented Programming (OOP) that allows us to combine data and code into a single module.

8.1 FUNCTIONS REVISITED

If you have ever assembled a LEGO object, you are familiar with the concept of modular design. A complicated LEGO robot, for example, consists of parts each made of small LEGO pieces. The designer of this LEGO robot has thought about pieces that can be reused to create different parts and objects, and also about parts that can make building a complicated object a more manageable process through assembling part-by-part. Functions allow modularization of code in a similar fashion.

8.1.1 Using Functions to Build Modular Programs

Imagine you are writing a program that

- Reads an input image file

- Processes the data by searching for areas of the image that show a face

- Recognizes the known faces using a database

- And then displays an output that identifies the found faces and the name of the person if there was a match.

This task, called Face Recognition, is becoming more popular with advances in computer hardware and software. Face recognition is now used as a convenient and arguably secure way to authenticate users. It is part of a broader technology called Computer Vision that allows computers to process images and perform actions such as object and motion detection and recognition.

A typical face recognition program performs multiple tasks following an algorithm that can look like the Sample Code #1-a.

Sample Code #1-a

```
1. Get a filename

2. Read the file

3. Prepare image data

4. Detect regions that resemble a face

5. Compare with database

6. Initialize graphics system

7. Display output with faces identified

8. Add a description for recognized faces
```

The algorithm is fairly readable; we can easily and quickly get a sense of what this program does and what the main parts are. But based on what we have experienced so far, it is not hard to imagine that each part of this algorithm requires many lines of code. An actual program doing face recognition may have hundreds or thousands of lines of code. It is not so easy to read the code and know what it is doing without scrolling through pages of code and trying to relate different parts. In addition to readability, it is not easy to make changes in the code as we again have to search through many lines to find out where we need to make a change and how it can affect other parts. We refer to the ease of understanding, maintaining, and revising a program as **Manageability**.

Using functions, as modules of code, makes programs more manageable. Imagine if your code for the face recognition program looks like the Sample Code #1-c.

Sample Code #1-c (partial)

```
1. void main()
2. {
3.     GetFilename();
4.     ReadFile();
5.     PrepareImage();
6.     DetectFace();
7.     CompareDatabase();
8.     Init();
9.     DisplayFace();
10.    DisplayFaceData();
11. }
```

Looking at the main() function of this C/C++ program (even without any comments), you can easily know what it does and what modules are involved. Using the tools provided by your Integrated Development Environment (IDE), or by just searching, you can find any of the functions and see how they perform their tasks. One can imagine that if those functions are too complicated, they will also be broken down into smaller modules. So, at any time looking at the code for one function, it is fairly short, and we can easily know what it is doing and which parts it has. These functions can be grouped into multiple files and worked on by multiple people independently for final integration.

This modular structure makes the program more manageable because it is easy to read and to change. But it is also easier to use the code later. Imagine you have to write another program that looks for cars in an image or a home security system that needs to detect a human intruder body or any type of motion. This program shares many parts with the face recognition one. Looking through lines of code to find how we can reuse them can be hard, but having general-purpose and self-sufficient functions makes it easy to simply copy them to another program. We refer to this feature as **Reusability**.

For a software module to be reusable, it has to be

- General-purpose, so it can be used in multiple cases

- Self-sufficient, so the outside code doesn't have to be written in a certain way to work with the module.

Consider a simple example for our GetFilename() function above. It can be written like this:

Sample Code #2-c

```
1. char name[100]; //global data
2. void GetFilename()
3. {
4.     cout << "enter filename for face recognition: ";
5.     cin >> name;
6. }
7.     void main()
8. {
9.     GetFilename();
10.         cout << name;
11. }
```

The function receives an array to be filled with the name that the user provides. It seems a very reasonable module that takes care of user input. If, in the future, we would like to change the way we get the filename (for example, read it from a database), we don't need to change anything in the rest of the code and simply access the module in charge.

On the other hand, if we want to use this module in another program, then we will have to make changes in it because it is **customized** for face recognition (see the message in line 4). Such change can be inconvenient (in more complex cases), or impossible (if we don't have access to source code and use this module as a library machine code), and will result in the code not being reusable. An alternative could be what we have in Sample Code #3-c.

Sample Code #3-c

```
1. char name[100]; //global data
2. void GetFilename(char message[])
3. {
4.     cout << message;
5.     cin >> name;
6. }
7. void main()
8. {
9.    char message[] = "enter filename for face recognition: ";
10.    GetFilename(message);
```

```
11.    cout << name;
12. }
```

Another issue that this code has is its dependence on a global variable. If you want to use the module, you need to know what **dependencies** it has and make sure that (1) they are also copied to target program, and (2) the target program is written in a way that uses these dependencies (for example, that global variable). Such a module is not self-sufficient. Another alternative can be what is shown in Sample Code #4-c where name is another parameter passed to the function.

Sample Code #4-c

```
1. void GetFilename(char name[], char message[])
2. {
3.     cout << message;
4.     cin >> name;
5. }
6. void main()
7. {
8.     char name[100];
9.     char message[] = "enter filename for face recognition: ";
10.    GetFilename(name, message);
11.    cout << name;
12. }
```

You may have noticed that the variable name is modified in the function, and the change is affecting its value in the caller (main() function). We have talked about passing parameters by value (a copy of data) and by reference (a pointer to original data). In C/C++, arrays are always passed by reference because the name of an array is a pointer to the beginning of it. So, in the above example, name is a pointer and equal to &name[0] (address of the first member). I will talk about this relation between arrays and pointers in detail later when I discuss dynamic arrays.

☞ **Key Point:** Manageability and reusability are two important features of a good software design.

☑ **Practice Task:** Rewrite some of your previous programs to make the functions more reusable.

👋 **Reflective Questions:** *Can you see the above features in other forms of design? Why are they important? If you customize a design for a particular task, it may be more manageable, but less reusable. How do you approach such a trade-off? Good modules are both manageable and reusable, but we may have priorities for either feature.*

8.1.2 A Modular Command Processor

In Chapter 4, I introduced command processor as a very common structural pattern in programs. It involves receiving a command (input) and then performing various actions based on the value of the command. Simple examples of a command processor, like what we had in Chapter 4, are easy to write as a single module. But the need for reusability and manageability in larger programs makes a command processor a good candidate to demonstrate modularization.

While every command processor may have different commands and tasks, there are common tasks in all programs that follow such a structural pattern. Sample Code #5-a shows the algorithm for a typical command processor.

Sample Code #5-a

```
Loop
    GetCommand
    Switch Command
            Case X: ProcessX
            Case Y: ProcessY
            Case Z: ProcessZ
```

GetCommand is a clear candidate for being a separate module, just like GetFilename in the earlier example. Process modules (X, Y, and Z) are also good candidates since the related processes may be needed in other programs or other places in this program, without the need for user interaction. For example, if command X is for printing a report, automatically running it may be necessary when a problem happens.

With manageability and reusability in mind, Sample Code #5-p shows an example of such a modular command processor in Python. The program is a simple command-line (text-based, non-graphics) calculator.

Sample Code #5-p

```
1. #functions to process commands

2. def GetCommand():
```

```
3.      Command = int(input("enter a command: 0-exit, 1-add,
    2-subtract, 3-multiply, 4-divide: "))

4.      return Command

5.

6. def GetParameter():

7.      N = int(input("enter a number: "))

8.      return N

9.

10. def Add():

11.      N1 = GetParameter()

12.      N2 = GetParameter()

13.      print(N1+N2)

14.

15. def ProcessCommand(Command):

16.      if Command == 0:

17.          return False

18.      elif Command == 1: #add

19.          Add()

20.          return True

21.

22.      #subtract, multiply, and divide commands go here

23.

24.      else:

25.          return True

26.

27. def MainLoop():

28.      result = True

29.      while result == True:

30.          Command = GetCommand()
```

```
31.          result = ProcessCommand(Command)
32.
33. #main code
34. MainLoop()
```

The starting point of this program is line 34; the first line that is not in any function. It simply calls our high-level module called MainLoop(). When looking at this function, it is easy to understand what the program does: it creates a loop and keeps calling GetCommand() and ProcessCommand() until the latter returns False. Notice that ProcessCommand() itself uses another module called Add() that performs the actual process, and Add() uses GetParameter() to receive input data for the operation.

The hierarchy of modules for this program is shown in Exhibit 8.1. Most of these modules are reusable. While MainLoop() and GetCommand() can be used in any other command processor, GetParameter() is useful for any program that needs user input. Later, we see that ProcessCommand() can be written in a more flexible and reusable way without hard-coding the command values and process modules.

> ✋ **Reflective Questions:** Do we need to have two separate functions GetCommand() and GetParameter()? What are the advantages and disadvantages of separating these two modules?

8.1.3 What Is a Good Function?

In my early years as a programmer, I received an important and influential advice from one of my managers. He said: "If you can't see the full code for a function on screen, the function is too long. Break it down into smaller ones."

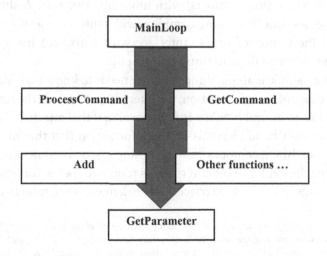

EXHIBIT 8.1 Modules in the simple calculator.

Of course, he didn't mean it literally. Screen size is by no means a clear unit of program length, and the right size of a software module can never be decided by rigid rules. What he said was meant to be a reference to the general guideline of making smaller modules so that they are more manageable.

The notion of building a software system as a collection of small modules goes back to the Unix operating system. Unix is an operating system developed mainly by Ken Thompson and Dennis Ritchie at the Bell Labs research center. It dominated the computer industry in the 1970s and 1980s, and its development is tightly related to the development of C programming language, which was designed by Dennis Ritchie to make utility programs for Unix. This operating system still plays a significant role in the industry in the form of Linux, which is a free, open-source version of Unix running on personal computers. The design of the Unix operating system had a significant influence on computer software, primarily as an example of a robust multi-user operating system, but also through a series of guidelines for software design collectively referred to as Unix Philosophy. One of the first of these guidelines was "write programs that do one thing and do it well," and another famous one was "small is beautiful."[1]

The notion of making small and simple software modules has been frequently used by many programmers to create reusable and manageable software modules. The idea makes sense in many ways; longer code is harder to read and maintain, and more likely to be suitable for only one application as it probably does more things and such a combination is less likely to be useful in many cases. But not all software designers and programmers favor the Unix Philosophy.

One of those who oppose this philosophy is Don Norman, who is the author of the widely respected book, *Design of Everyday Things* (1988, revised 2013). In his 1981 article titled "The truth about Unix: The user interface is horrid," Norman criticizes Unix for its non-user-friendly interface. John Ousterhout, in his book, *A Philosophy of Software Design* (2018), uses the notion of the interface and defines **Deep vs. Shallow Modules**. A deep module, which Ousterhout favors, is one that has a simple small (narrow) interface but provides the user (in this case programmers) with more functionality. A shallow module, on the other hand, does less but has a more complex (wide) interface, as shown in Exhibit 8.2. Ousterhout shares the notion of simple interface with Unix, but his approach may not match the small-module and do-one-thing philosophy.

The **Interface** of a module is all the things the user needs to know and deal with in order to use a module. In the case of a single function, the interface can include the return type, parameters, internal dependencies, and how the function works if that affects using it. The depth of the function, on the other hand, depends on the information that the function holds inside or all the things it does. In Ousterhout's view, designing a good module means decreasing the width and increasing the depth. In other terms, his proposed metric for how good a function is, when it comes to information vs. interface, can be defined as the ratio of depth to width.[2]

---

[1] See *The Art of Unix Programming* by Eric Steven Raymond for a discussion of the Unix Philosophy, which, according to Raymond, "boils down to one iron law, the hallowed 'KISS principle'" – Keep It Simple, Stupid!

[2] Of course, Ousterhout does not say that this ratio is the only thing that make a module "good." Interestingly enough, his current research is on Granular Computing, which is the study of running software as a collection of very small tasks.

Function A	Function B	
Interface	**Interface (many parameters)**	
Implementation	**Implementation**	
(many actions)		

EXHIBIT 8.2 Deep vs. shallow functions. Function A is deep (few parameters with substantial actions) and Function B is shallow (a large set of parameters for a simple but configurable action).

There is no doubt that a function that needs only one parameter and based on that takes care of multiple things is more convenient to use. The GetFilename() function in Sample Code #2-c has a higher depth to width ratio since it has no parameters for the same task, compared to #3 and #4. On the other hand, the version of this function in Sample Code #4-c is more reusable because it can be customized. There is no single rule that can be proved right and used in all cases. As a software developer, you need to be aware of these notions and use them appropriately in different cases. The experience will tell you when reusability and the possibility of customization are more important than the convenience of performing multiple tasks in one simple call.

The need to deal with various criteria suggests the usefulness of a hierarchical approach where the interface (width) and information (depth) vary depending on how low on the hierarchy we go. Hierarchies define roles that can affect the level of detail each item or person has to deal with. A corporate manager has the big picture and makes top decisions that affect the whole company but may not know how an actual task like accounting is done. Lower-level employees, on the other hand, perform very specific tasks but have detailed information about them. This brings us to the concepts of information hiding and abstraction that are essential in modular design.

☞ **Key Point:** Interface of a module is what the users need to know about it. It does not include what is "inside" the module. While some software designers are in favor of simple short modules, others believe good modules have simple interfaces but can be long and complicated inside.

✍ **Reflective Questions:** Does the debate about good design make you confused? What are your own thoughts? There is no silver bullet when it comes to software design. Embrace the uncertainty and try to adapt to each context and adjust your solution based on the specific requirements.

8.2 INFORMATION HIDING AND ABSTRACTION

The C/C++ function getch()[3] reads one character from the keyboard and returns it. This function is much simpler than scanf() or Python's input(), which can read different types of text input and also return in various types. We can imagine that the implementation of getch() is fairly straight-forward; it checks the keyboard buffer (where all entered keys are stored in a first-come-first-serve format and stored character by character) and reads the first (oldest) character. The scanf() function, on the other hand, needs to do more processing in order to detect the data type and prepare the results. One can imagine that this involves local variables that deal with the type and size of the data, and operations that are more complicated than getch(), such as determining when to how to read the data.

Despite the need for many details in order to implement the above functions, the information about those details is not required by the user (the programmer who is using these functions). They are used as **black boxes**, and all the programmer needs to know is how to call them (parameters) and what to receive (return value). The information we need to use a software module is usually referred to as the **interface**. In addition to the interface, the module holds more information internally about what it does and how (its **implementation**), but such information is not visible or needed by the user. This arrangement is called **Information Hiding** and plays an important role in modular software design. The notion of deep function is related to information hiding as it implies that a good function hides many details from the programmer to make the function easier to use.

The concept of information hiding is not limited to one single function. More complex modules, even whole software libraries, are designed to hide details from programmers and only expose a simple and easy-to-use interface. When using C/C++ standard library, for example, we don't know how they are implemented and what internal data and code they use. In fact, the implementation of these standard libraries is probably different in Visual Studio compared to another compiler/IDE that provides its own version. They all perform the same action from an external point of view, though, following the black box idea, and do not bother the programmer with the burden of knowing how things are done. All programmers need to know is the Application Programming Interface (API) for the library, which is the set of descriptions for all modules, as seen in header files (explained in Section 6.2).

In addition to the convenience of not needing to deal with details, information hiding provides another, more important, advantage; the internal implementation of the module can change without affecting the programs using it. This can happen every time that you install an upgrade to your programming library or port your code to a new platform, without the need to modify your code.

While information hiding is an important concept and a goal in many cases, there are situations in which the programmers may need to have access to the details of a module.

---

[3] The functions getch(), getche(), getc(), and getchar() are all for reading one character. They differ in the way they echo the character n screen and the need for the Enter key. Visual Studio now requires the use of _getch() and _getche() as new versions.

Such a need can result in a software design problem caused by conflicting goals. A solution to this problem is to have a module that exposes details and higher-level modules that use the detailed ones and hides their information. Sample Code #6-c demonstrates such an approach for the example of GetCommand() from Sample Code #5-p. The logic is the same for C/C++ and Python.

Sample Code #6-c

```
1. struct Command
2. {
3.      char text[20]; //for example "exit"
4.      int  code;          //for example, 0
5. };
6. Command commands[10];
7. int numCommands;
8.
9. void GetAvailableCommands()
10. {
11.     //numCommands has to be less than 10 (array size)
12.     printf("enter number of commands: ");
13.     scanf_s("%d", &numCommands);
14.     //loop for all commands
15.     for (int i = 0; i < numCommands; i++)
16.     {
17.         printf("enter command text: ");
18.         scanf_s("%s", commands[i].text, 20);
19.         printf("enter command code: ");
20.         scanf_s("%d", &commands[i].code);
21.     }
22. }
23. void ShowAvailableCommands()
24. {
```

```
25.      for (int i = 0; i < numCommands; i++)
26.      {
27.          printf("%d - %s\n",
28.                  commands[i].code, commands[i].text);
29.      }
30. }
31. int GetUserCommand()
32.      {
33.          int c;
34.          printf("Enter command code: ");
35.          scanf_s("%d", &c);
36.          return c;
37. }
38. int GetCommand()
39. {
40.          If(numCommands == 0)
41.              GetAvailableCommands();
42.          ShowAvailableCommands();
43.          int c = GetUserCommand();
44.          printf("%s\n", commands[c].text);
45.          return c;
46. }
47. void main()
48. {
49.          int c = GetCommand();
50.          //the rest of the program
51. }
```

The main() function (line 47) starts by calling GetCommand(). For the sake of this example, it doesn't matter what we do with this command. As a programmer using the GetCommand() function, we may interact with it in different ways:

1. We may not care at all how it works. In that case, the other functions in this sample code will be ignored. We know they are needed to get a command, but we don't deal with them. If the ability to "get a command" is provided by a library, we don't need to see the source code or have any more information.

2. We may need to know what steps the function will go through in order to, for example, prepare the user. In that case, we need to see the code for GetCommand() function itself (line 38) but no need to see how the other functions work. We will know that getting a command from user involves

   a. Initializing the list of available commands if necessary (if the number of commands is zero)

   b. Showing a menu (list of available commands)

   c. Showing a message and getting the current command.

3. Finally, we may need to know exactly how the data for available commands are created and stored, which requires looking at the rest of the code and the Command type.

To allow programmers to deal with the information at the level they need, the program is structured at three different levels of detail. Creating a higher-level construct that hides some details is usually referred to as **Abstraction**, and the higher-level modules are said to form **Abstraction Layers**. Such layers hide details at different levels. For example, computer networks follow a layered model[4] in which parts of the system that deal with actual data transfer need to have functions dealing with the type of connection (wireless vs. wired, etc.). In contrast, the operating system may group the data based on which application is using it and regardless of how it is transferred, and the application programs can ignore all those details and organize their data based on their needs. Each layer of this model provides a different level of abstraction.

Abstraction can also refer to the process of creating and using **Abstract Modules**, which are conceptual definitions that have the general properties of a certain module type without actual implementation. For example, list is an abstract concept that defines the properties of a data module made of a series of elements, combined with operations to add, remove, access, and search for those elements. Lists are implemented differently in different languages or even programs written in the same language. But all these implementations have those common properties. Such abstract concepts, used as the basis of actual program modules, are called **Abstract Data Type (ADT)**. Some programming languages allow the definition of abstract modules directly in the program (not just as a concept). They define a basic set of properties without showing how it is implemented, and are commonly called **Interfaces** or **Abstract Classes**. I will talk more about all these abstract modules in the next part of the book when I discuss OOP.

---

[4] Open Systems Interconnection (OSI) model.

In the remainder of this chapter, we will review some examples to demonstrate the ideas of modular design and information hiding through the definition of manageable and reusable functions.

> ☞ **Key Point:** Information hiding is a basic principle of software design that deals with exposing programmers to only what is necessary. Abstraction can be one way to achieve this.

8.3 MODULAR DESIGN

8.3.1 School Database

A simple database program for a school needs to perform various actions and process data related to not just students but also courses, classrooms, and many other items. While the detailed design for real-world software to manage a school can be too complicated as an example here, I try to introduce the main aspects of such a system and consider them in our design.

Using our data-centered approach, we start by identifying the main data items. Exhibit 8.3 shows these items grouped into three main categories as they are related to student, course, or classroom.

Each of these three groups can be represented with a user-defined type, and the program can use three arrays of those types to store the data.

- Student
  - Identifying data (name, ID number)
  - Courses registered
  - Grades.
- Course
  - Identifying data (title, ID number)
  - Students
  - Classroom
  - Schedule (time/date for weekly class).

EXHIBIT 8.3   Main data items for school program.

- Classroom

    - Identifying data (address, ID number)

    - Capacity

    - Schedule (list of all timeslots and courses assigned to them).

Notice that each of the three main data types has references to others. For example, student data includes an array of ID numbers for all the registered courses. There is no need to store the entire course data for each student. Course data exists only in one place under the course array, but all students or classrooms will have references to these courses. A similar method will be used to allow cross-referencing between data items.

Once our data items have been identified, using our three variable-related "how" questions (3HQ), we decide what operation are necessary for our program:

- **How do we initialize the values?**

    - The program may assume that the classrooms are fixed. We can either initialize them hard-coded or allow some flexibility and read them from a data file. They all have an empty schedule at the start.

    - The program can start with no courses or students, so the user interactively defines new courses. An empty array of maximum size can be created for courses and students.

- **How do we change the values?**

    - The user can create new students and courses through a menu of commands. For each new student or course, the identifying data will be provided by the user.

    - The user can schedule a course by selecting a classroom and an empty timeslot.

    - The user can register a student in a course if it has not reached maximum capacity.

    - The user can enter a grade for the students in each of their registered courses.

- **How do we use the values?**

    - The list of registered courses for each student will be used when entering grades to see if the student is registered in the course for which the user wants to enter a grade.

    - The capacity of a course is used when registering a student.

    - The capacity of a classroom is used when assigning a course to see if the room is big enough.

    - The classroom schedule is used when scheduling a course to see if the room is available.

    - All data items are used for reporting.

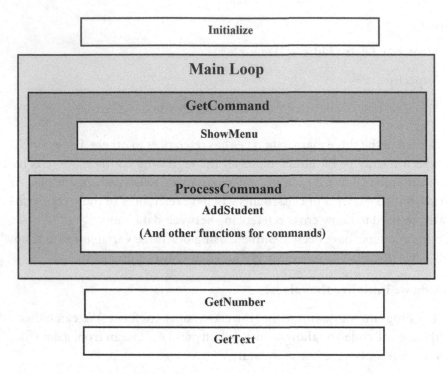

EXHIBIT 8.4   Functions for school program.

You see that by answering the 3HQ for all data items, we have a clear picture of what our program needs to do. This process is related to what I called the **Requirement Analysis** in the Introduction, as it defines the behavior of the program. It is also part of the **Design** as it deals with how the program should be structured (variables, etc.). You can see that the software development phases can be overlapped.

Exhibit 8.4 shows the main operations for our school program. It follows the well-known Command processor pattern and involves initialization and a main loop which receives commands and processes them. Receiving a command itself consists of showing a menu and reading a command value, while processing the command includes various other modules, each corresponding to one task the user can perform.

Complicated tasks are divided into more detailed modules. For example, registering a student involves asking a course number, checking if it has space, adding the student to the course roster, and adding the course to the student's list of courses. Following the principle of information hiding, we don't need to know how the student data type keeps its list of courses unless we are dealing with that particular module. A change in Student data type will not affect modules that are not directly using it. Also, if we decide to change the registration process, the low-level modules don't need to change, and we can simply arrange them differently.

Sample Code #7-c shows the implementation of this simple database in C/C++. It starts by defining constant numbers such as arrays sizes that are used later to limit the number of elements we can create (such as adding new students and courses). Then, we define the data type in lines 13–39. The program uses global variables to share data among all the functions. Remember that we have many functions, and they all need to share the same data and work with the same students and courses. So, the data cannot be local unless it is defined in main() and then passed to all functions. We will see and discuss this alternative method in the next example.

Also, note that all arrays have an associated integer variable that shows the number of elements that are actually used. This number is different from the array size because the student and course arrays are empty at the start, and we gradually add items. This means that while these arrays have members in them, these members have not been initialized and hold no data.

Function definitions start on line 48. GetText() and GetNumber() are two **helper functions**. They are not really related to the logic of this program but help with some common operations, in this case showing a prompt text and getting data from the user. The Initialize() function (line 61) gives initial value to our global data. Only rooms are initialized with full data, as students and courses will be created by the user. Note that

- ID values are the same as the array index, so the first room (or student or course) has ID equal to zero. We can also start from a given number, so ID is always, for example, 1000 plus the index value.

- Each room can be used from 9 am to noon and from 1 to 4 pm, for 1-hour classes, 5 days a week. So, the total number of classes in a room is 30 per week (MAX _ CLASSES). Each room has an array called schedule with 30 members. A value of −1 for a schedule item means that the related timeslot is not used (scheduled) yet and is available. If the timeslot is used, it will have the ID of the course using it.

AddStudent() and AddCourse() functions (lines 94 and 111, respectively) create new items. They do so only if the number of items is less than the maximum (array size) and increase the number once the item is created. Note that numStudents and numCourses always show the first available array member. For example, when they are zero, the array is empty, and the first data we create goes to the member at index zero (first member). Then, the array counter (numStudents and numCourses) will be increased, so the following data will go to index one.

RegisterStudent() and ScheduleCourse(), on the other hand, ask for the item ID to process (must exist) and then check another array to see if there is space.

The other functions are pretty straight-forward:

- Report functions print the data for an item.

- GetCommand() shows a menu and asks the user for a command.

- ProcessCommand() calls other functions based on the command.

- MainLoop() consists of the two main operations that are being repeated as long as the command is not Exit.

- Finally, main() (line 215) initializes the data and starts the main loop.

Sample Code #7-c

```
1. #include "stdio.h"
2.
3. #define MAX_STUDENTS          100
4. #define MAX_COURSES           50
5. #define MAX_CHARACTERS        20
6. #define MAX_CLASSES           30
7. #define MAX_ROOMS             3
8. #define MAX_COMMANDS          7
9.
10. #define RESULT_OK            0
11. #define RESULT_TOO_MANY      1
12.
13. // Data Types
14. struct Student
15. {
16.     char   name[MAX_CHARACTERS];
17.     int    ID;
18.     int    courses[MAX_COURSES];
19.     int    numCourses;
20.     float  grades[MAX_COURSES];
21. };
22. struct Course
23. {
24.     char   title[MAX_CHARACTERS];
25.     int    ID;
26.     int    students[MAX_STUDENTS];
27.     int    numStudents;
28.     int    room;
```

```
29.    int   schedule;      //0 to MAX_CLASSES, ID for a
       timeslot
30. };
31. struct Room
32. {
33.    char  address[MAX_CHARACTERS];
34.    int   ID;
35.    int   capacity;
36.    int   schedule[MAX_CLASSES]; //-1 if available, other-
       wise course ID
37. };
38.
39. // Global Variables
40. Student students[MAX_STUDENTS];
41. int      numStudents;
42. Course  courses[MAX_COURSES];
43. int      numCourses;
44. Room    rooms[MAX_ROOMS];
45. int      numRooms;
46.
47.
48. // Functions
49. void GetText(const char* prompt, char* data)
50. {
51.    printf("%s ", prompt);
52.    scanf_s("%s", data, MAX_CHARACTERS);
53. }
54. void GetNumber(const char* prompt, int* data)
55. {
```

```
56.      printf("%s ", prompt);

57.      scanf_s("%d", data);

58. }

59.

60.

61. void Initialize()

62. {

63.      //students

64.      numStudents = 0;

65.

66.      //courses

67.      numCourses = 0;

68.

69.      //rooms

70.      numRooms = MAX_ROOMS;

71.      for (int i = 0; i < MAX_ROOMS; i++)

72.      {

73.          rooms[i].ID = i;

74.          rooms[i].capacity = 20;

75.          GetText("Please enter building name and room #:",
    rooms[i].address);

76.          for (int j = 0; j < MAX_CLASSES; j++)

77.              rooms[i].schedule[j] = -1; //all available

78.      }

79. }

80. void ShowMenu()

81. {

82.      printf("\nChoose one of the following commands:\n");

83.      printf("  1-Add Student\n");
```

```
84.      printf("  2-Add Course\n");

85.      printf("  3-Register Student\n");

86.      printf("  4-Schedule Course\n");

87.      printf("  5-Report Student\n");

88.      printf("  6-Report Course\n");

89.      printf("  7-Report Room\n");

90.      printf("  0-Exit\n");

91. }

92.

93.

94. int AddStudent()

95. {

96.      int result = RESULT _ OK;

97.      int n = numStudents; //so we don't have to repeat this

98.      if (n < MAX _ STUDENTS - 1)

99.      {

100.          students[n].ID = n;

101.          students[n].numCourses = 0;

102.          GetText("Please enter student name:", students[n].
     name);

103.          numStudents ++;

104.      }

105.      else

106.          result = RESULT _ TOO _ MANY;

107.

108.      return result;

109. }

110.

111. int AddCourse() //same as AddStudent
```

```
112. {
113.     int result = RESULT _ OK;
114.     int n = numCourses; //so we don't have to repeat this
115.     if (n < MAX _ COURSES - 1)
116.     {
117.        courses[n].ID = n;
118.        courses[n].numStudents = 0;
119.        GetText("Please enter course title:", courses[n].title);
120.     numCourses ++;
121.     }
122.      else
123.         result = RESULT _ TOO _ MANY;
124.
125.      return result;
126. }
127.
128. int RegisterStudent()
129. {
130.     int result = RESULT _ OK;
131.     return result;
132. }
133.
134.
135. int ScheduleCourse()
136. {
137.     int result = RESULT _ OK;
138.     return result;
139. }
140.
```

```
141. int ReportStudent()
142. {
143.     int result = RESULT _ OK;
144.     return result;
145. }
146.
147. int ReportCourse()
148. {
149.     int result = RESULT _ OK;
150.     return result;
151. }
152.
153. int ReportRoom()
154. {
155.     int result = RESULT _ OK;
156.      return result;
157. }
158.
159. int GetCommand()
160. {
161.     int command;
162.     ShowMenu();
163.     GetNumber("Command:", &command);
164.     return command;
165. }
166.
167. int ProcessCommand(int command)
168. {
169.     int result = RESULT _ OK;
```

```
170.
171.    switch (command)
172.    {
173.    case 0:
174.        printf("Good bye!\n");
175.        break;
176.    case 1:
177.        AddStudent();
178.        break;
179.    case 2:
180.        AddCourse();
181.        break;
182.    case 3:
183.        RegisterStudent();
184.        break;
185.    case 4:
186.        ScheduleCourse();
187.        break;
188.    case 5:
189.        ReportStudent();
190.        break;
191.    case 6:
192.        ReportCourse();
193.        break;
194.    case 7:
195.        ReportRoom();
196.        break;
197.    default:
198.        printf("Invalid command!\n");
```

```
199.          break;
200.      }
201.
202.      return result;
203. }
204.
205. void MainLoop()
206. {
207.      int c = -1;
208.      while (c != 0) //exit command
209.      {
210.          c = GetCommand();
211.          ProcessCommand(c);
212.      }
213. }
214.
215. void main()
216. {
217.      Initialize();
218.      MainLoop();
219. }
```

☑ **Practice Task:** The functions GetText() and GetNumber() provide reusability as they can be part of any other program. They also help with information hiding as the programmer doesn't need to know how they are implemented. Identify other parts of the program that can turn into reusable modules.

Hint: How can we reuse MainLoop() and ProcessCommand()?

8.3.2 Library Database

Another example that follows the same command processor pattern is a database program for a small library. The main data items for a library can be grouped under User and Book. In the previous example, we assumed that the program was being used by a school administrator. To add some complexity and experience with some new features, let's imagine that our program is an interface to the library system, which can be used by both administrators and borrowers. The main difference between these two user types is that the first group can add new books and users to the system.

The following list shows the high-level data types for our program. As in the school example, the User type includes an array referencing the Book items borrowed by this user. We will have two arrays for these two main types.

- User

  - Identifying data (name, ID number, password)

  - List of borrowed books

  - Type (administrator or borrower).

- Book

  - Identifying data

  - Borrower

The User and Book arrays can be empty at the start or be initialized from a data file. The main operations for a borrower user will be

- Authentication that uses the ID number and password

- Borrowing a book that uses and changes the borrower ID of the book user wants to borrow and the list of borrowed items (See Exhibit 8.5 for a process model)

- Returning a book that again uses and changes the same items as borrowing

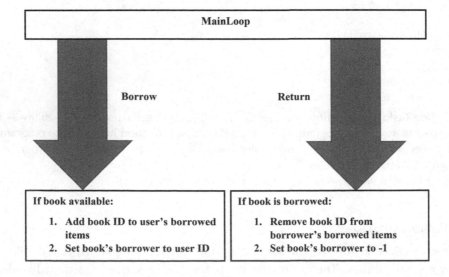

EXHIBIT 8.5 Borrowing and returning.

Note that we can borrow a book if it is not borrowed by another user, which means its Borrower data is set to −1 (invalid user ID). Once it is borrowed, that data will change to the ID for the borrower, and after returning, it is changed back to −1.

An administrator user can perform the following extra operations:

- Adding a new user that uses and changes the User list
- Adding a new book that uses and changes the Book list

Just like the previous example, we will need to have variables that keep track of how many items (user and book) we have. When discussing the school program, you probably noticed that we had a few variables that were essential (main arrays and integer numbers showing their sizes) that were not part of any user-defined types. The concept of modularization suggests that we put together these data into a new **high-level module**, i.e., information about the whole program. In this example, we do this by defining a new type called Library:

- Library
  - List of Users (array of User items)
  - List of Books (array of Book items)
  - Number of users
  - Number of books
  - Any other data that is related to the whole system

This results in a hierarchical structure for our program, as shown in Exhibit 8.6. Such a structure is very common in programs and sometimes is referred to as **Object Aggregation**, as it includes complex items (such as Library) that use simpler ones (such as User and Book).

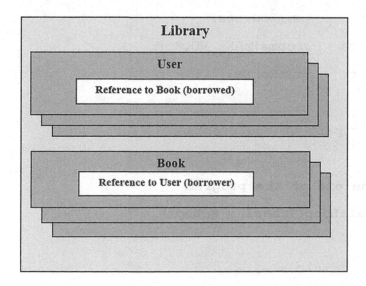

EXHIBIT 8.6   Hierarchical data items in the library program.

> ☞ **Key Point:** Object Aggregation refers to user-defined types that include members of other user-defined types. For example, Library is an aggregated module that has User and Book members.

Sample Code #8-c shows changes to Sample Code #7-c to create such a high-level module for the School program. It also replaces the use of global data with a single high-level data that is local to main() but passed to all functions by reference. This method has multiple advantages:

- There are no global data and external dependency for functions.

- Functions can modify the original data without the need to pass and return copies.

- All data are packaged into a single manageable item.

Note how the data is created in main() and then passed to other functions as parameter.

Sample Code #8-c (partial)

```
1. //defined after other data types
2. struct School
3. {
4.      Student students[MAX_STUDENTS];
5.      int     numStudents;
6.      Course  courses[MAX_COURSES];
7.      int     numCourses;
8.      Room    rooms[MAX_ROOMS];
9.      int     numRooms;
10. };
11. // other part of the program
12. // …
13. //at the end of the program
14. void MainLoop(School * school)
15. {
16.         int c = -1;
17.         while (c != 0) //exit command
```

```
18.             {
19.                     GetCommand(school, &c);
20.                     ProcessCommand(school, c);
21.             }
22. }
23. void main()
24. {
25.             School school;   //all essential program data
26.
27.             Initialize(&school);
28.             MainLoop(&school);
29. }
```

Sample Code 9-p follows a similar structure and implements the library program in Python. The code is quite straight-forward with two low-level modules User and Book and a high-level one, Library (lines 2, 10, and 15, respectively).

All functions receive the Library variable (lib) as a parameter similar to the #8-c. In Python, variables that are based on a class are always passed by reference, and there is no need for pointers. Simple variables such as integer numbers, on the other hand, are automatically passed by value.

The program tracks the current user, and the function Login() performs authentication (line 23). It stays in a loop and tries checking user ID and password until a successful login has happened. Once we have a valid current user, the other functions use that value for all operations.

AddUser() and AddBook() are very similar to Add functions in #7-c. Note that members can be added to Python arrays using append(). The length of an array can also be determined by len() function, so technically, we don't need to have variables like numUsers and numBooks. I have added them to keep things similar to the C/C++ code.

Borrow() and Return() manages the books. To borrow a book, it has to be available (no existing borrower). Line 53 shows how to check if the book iD is valid, and there is no existing borrower. If the book is available, we need to make sure that the book module has the current user as a borrower (line 54), and the user has added the book to their list (line 55). To return a book, we need to set the borrower value back to -1 (available) and remove the book id from the list of books for the current user. Python has a remove() function that removes the first occurrence of a value from the array.

The remaining functions work in a very similar way to C/C++ code to report, initialize, get commands, and process commands.

Sample Code #9-p

```
 1. # Data Types
 2. class User :
 3.      name = ""
 4.      ID = 0
 5.      password = ""
 6.      books = []
 7.      numBooks = 0
 8.      type = 1   #0 admin, 1 patron
 9.
10. class Book :
11.        name = ""
12.        ID = 0
13.        borrower = -1
14.
15. class Library :
16.        users = []
17.        books = []
18.        numUsers = 0
19.        numBooks = 0
20.        currentUser = -1 #no login yet
21.
22. # Functions
23. def Login(lib) :
24.          lib.currentUser = -1
25.          while lib.currentUser == -1 :
26.              uid = int(input("enter user ID: "))
27.              if uid < lib.numUsers :
28.                  password = input("enter password: ")
```

```
29.                    if password == lib.users[uid].password :
30.                        lib.currentUser = uid
31.                    else :
32.                        print("incorrect password")
33.             else :
34.                 print("invalid ID")
35.
36. def AddUser(lib) :
37.         newUser = User()
38.         newUser.ID = lib.numUsers
39.         newUser.name = input("enter new name: ")
40.         newUser.password = input("enter new password: ")
41.         lib.users.append(newUser)
42.         lib.numUsers += 1
43.
44. def AddBook(lib) :
45.         newBook = Book()
46.         newBook.ID = lib.numBooks
47.         newBook.name = input("enter book name: ")
48.         lib.books.append(newBook)
49.         lib.numBooks += 1
50.
51. def Borrow(lib) :
52.         bid = int(input("enter book ID: "))
53.         if bid < lib.numBooks and lib.books[bid].borrower ==
    -1 : #existing and available
54.             lib.books[bid].borrower = lib.currentUser
55.             lib.users[lib.currentUser].books.append(bid)
56.             lib.users[lib.currentUser].numBooks += 1
57.
```

```
58. def Return(lib) :
59.          bid = int(input("enter book ID: "))
60.          uid = lib.books[bid].borrower
61.          if bid < lib.numBooks and uid != -1 : #existing and
     borrowed
62.               lib.books[bid].borrower = -1
63.               lib.users[uid].books.remove(bid)
64.               lib.users[uid].numBooks -= 1
65.

66. def Report(user) :
67.          print(user.ID)
68.          print(user.name)
69.          for b in range(user.numBooks) :   #instead of user.
     numBooks we can use len(user.books)
70.               bid = user.books[b]
71.               print(bid)
72.               print(lib.books[bid].name)
73.

74. def Initialize(lib) :
75.          newUser = User()
76.          newUser.name = "Admin"
77.          newUser.password = "1234"
78.          newUser.type = 0
79.          lib.users.append(newUser)
80.          lib.numUsers += 1  # or = 1 (started with zero)
81.          lib.numBooks = 0
82.          Login(lib)
83.

84. def ShowMenu() :
85.          print("Choose one of the following commands:")
```

```
86.          print("  1-Borrow Book")
87.          print("  2-Return Book")
88.          print("  3-Report")
89.          if lib.users[lib.currentUser].type == 0 :
90.              print("  4-Add User")
91.              print("  5-Add Book")
92.          print("  6-Switch User")
93.          print("  0-Exit")
94.
95. def GetCommand() :
96.          ShowMenu()
97.          c = int(input("enter a command: "))
98.          return c
99.
100. def ProcessCommand(lib, c) :
101.          if c == 0 :
102.              print("bye")
103.          elif c == 1 :
104.              Borrow(lib)
105.          elif c == 2 :
106.              Return(lib)
107.          elif c == 3 :
108.              Report(lib.users[lib.currentUser])
109.          elif c == 4 :
110.              AddUser(lib)
111.          elif c == 5 :
112.              AddBook(lib)
113.          elif c == 6 :
114.              Login(lib)
```

```
115.        else:
116.            print("invalid command")
117.
118. def MainLoop(lib) :
119.        c = -1
120.        while c != 0 :
121.            c = GetCommand()
122.            ProcessCommand(lib, c)
123.
124. lib = Library()
125. Initialize(lib)
126. MainLoop(lib)
```

8.4 A MODULAR GAME PROGRAM

In the early years of computer games, game production was mainly a software activity as art assets, and the gameplays were quite simple. With the advances in computer hardware and software, computer games turn into large scale productions with much expertise involved.[5] The design of game software also evolved, and depending on genre, gameplay, and various technical details, the game programs involve many features and design variations. In this section, I go back to the simple 2D game we have worked on previously and demonstrate how its design can be improved using a modular approach.

SIDEBAR: GAME PRODUCTION CYCLE AND TEAM

The process (lifecycle) model for game production resembles that of a general software combined with more artistic productions such as movies. While variations exist, the following phases generally exist for game projects:

- Concept Development (equivalent to requirement analysis and early design in software projects)
  - Results in a summary (high concept) and a pitch document.
- Pre-production (proof of concept, equivalent to design and early implementation in software projects)
  - Results in a design document ("the Project Bible") and some prototypes.

---

[5] See the Sidebar on Game Production.

- Production (the equivalent of Implementation for software projects. Includes both code and art assets)
  - Results in testable game.
- Test (alpha and beta)
  - Results in a game that is ready to be released.
- Release
  - Results in the first official version.
- Maintenance (including updates)
  - Results in new versions and upgrades.

The game production is one of the most multidisciplinary efforts. Going through the above phases requires a team of experts who get involved in various stages of production. They include

- Designers, such as game designers, level designers, and writers
- Programmers, such as AI and Logic, UI, Networking, Physic, Effects, and Tools
- Artists, such as concept, character, background, modeling, texture, animation, music, and sound
- Testers and test leads
- Internal and external producers and assistants.

8.4.1 2D Side-Scroller Game

In Chapters 6 and 7, we reviewed the basic structure of a 2D game program. It included an initialization and a main loop, which itself consists of updating and drawing. We also talked about how the game is made of data related to the **Game World** (a.k.a. levels or rooms) and **Game Objects**, which can be player-controlled, game-controlled, and stationary. In many simple games, game objects hold similar information, and the basic operations performed on them are changing shape, changing position, and drawing on the screen.

Based on the above general information about games, I demonstrate the process of modular design for a simple 2D side-scroller (platformer) game. Here is the basic idea of our game:

> The character moves horizontally along a flat surface where he can face stationary and moving enemies attacking and different types of score points to pick up. The player can either escape from the enemies, jump over them (will discuss that in the next section), or pass through them by going to defense mode. This mode is activated by pressing the Up key and deactivates automatically after a short time. The player shape changes during this mode for a visual feedback. The player picks up the score items simply by hitting them. The game world is bigger than the screen size, and the player character is always shown in the middle of the screen. Using Right and Left arrow keys, the player can scroll horizontally to see the game world. You can imagine the screen is from a camera that always follows the player. The player wins by reaching the end and loses/dies if losing all health before that.

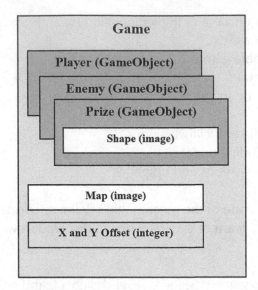

EXHIBIT 8.7   Game program visualization.

Exhibit 8.7 visualizes our game program. We have an all-encompassing entity called Game that includes a series of GameObject-type entities and some other information such as background image, screen position within the game world, and player's current health.

> ✋ **Reflective Question:** Some of the information in the Game type belong to the player. Would it make sense to have different types for game objects, so, for example, the Player type has health as a member, and the Enemy type has damage value? Such information is not shared among different game objects and as such, they cannot be a member of a common GameObject type.

Sample Code #10-j shows the game program in Javascript. It starts with the definition of GameObject module and its functions plus the higher-level Game that includes GameObject items. Compared to the game programs in Chapter 7, this game has two new features: scrolling and the defense mode.

8.4.2 Scrolling

Exhibit 8.8 demonstrates the basic idea of scrolling: the image we show is bigger than the screen, so it doesn't fit (unless we zoom out which we will discuss in Chapter 9). We can scroll in any dimension where the image is bigger than the screen. Think of this action as having a **virtual camera** and moving it to see different views. The display screen represents the camera view, while the image represents the world. In Section 7.3, we talked about linear transformation, such as translation that shifts a point's coordinates. Translation commonly happens with scrolling as the coordinate of a point (for example, the location of an object) in the camera view (screen) is not the same as its location in the image (world).

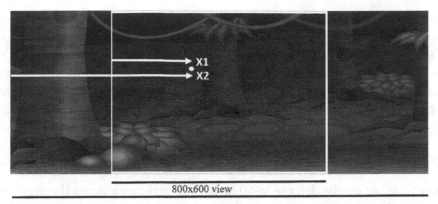

EXHIBIT 8.8 Scrolling and screen vs. world coordinate systems. X1 and X2 are the X coordinates of the selected point with respect to the Screen (camera view) and World (actual image) coordinate systems.

Scrolling is controlled with offset values in X and Y directions, i.e., how much we shift the image or how much we move the camera in either direction. Remember that it doesn't matter if we move the image or camera. It is the relative position that determines what we see. In this book, I define the offset values as the coordinates of the virtual camera relative to the game world. So, if we move the player to the right side of the game world, the offset values will also increase. On the other hand, all drawing functions should subtract the offset values to transform the world coordinates to screen coordinates.

According to our first rule of programming, for every piece of information, we need a variable. So, our first step is to define two variables for offset in X and Y directions (lines 43 and 44, but Y is not used in this game). After that, our standard **3HQ** (3 how questions) method can help with modifying the code:

- Offset values are initialized to zero when created, so even without any scrolling, we can display the game correctly.

- Offset values are changed when the player moves left or right (lines 102 and 119). Imagine that we have a virtual camera that is attached to the player, so its location changes with the player.

- Offset values are used in drawing functions (lines 84–87).

8.4.3 Defense Mode

Similar to scrolling, we need to add a new variable to represent the defense mode. The only difference is that this mode has to automatically change back after a certain amount when the player moves to the right or left. We can define a variable called `playerDefense` that can be `true` or `false`; set it to `true` when we press the Up key and count movements and set it back to `false` when we have counted enough (moved a certain number of pixels). This solution requires two variables: one for the mode and another to count.

Note that counting happens when the mode variable is `true`, so if we know we are counting, then there is no need for a second mode variable.

If we set the defense variable to zero at the start and then set it to a value N when we press the Up key, we can decrease it (count down) every time we move, and as long as the value is more than zero, we are in the defense mode. In the Sample Code #10-j, we define the `playerDefense` in the Game module (line 45), change it to 50 (line 112), decrease it when player moves (lines 104 and 121), and check it when we hit the enemy (line 74). Note that

1. We chose the value 50 to give enough time to pass the enemy

2. Every time we set the defense to 50 or back to zero, we change the player shape as visual feedback.

> ✋ **Reflective Question:** Should the `defense` variable be in the `Game` module or `GameObject`?

Sample Code #10-j

```
1. <!DOCTYPE html>
2. <html>
3. <head>
4. <script>
5. //Data Types
6. function GameObject()
7. {
8.   this.shape = new Image();
9.   this.x = 0;
10.  this.y = 0;
11.  this.visibility = 100;
12. }
13. function DrawObject(obj, ox, oy) // offset values for X/Y
14. {
15.  if(obj.visibility != 0)
16.    ctx.drawImage(obj.shape, obj.x-ox, obj.y-oy);
17. }
```

```
18. function Distance(obj1, obj2)

19. {

20.  var dx = obj1.x - obj2.x;

21.  var dy = obj1.y - obj2.y;

22.  var d = Math.sqrt(dx*dx + dy*dy);

23.  return d;

24. }

25. function Game()

26. {

27.  this.player = new GameObject();

28.  this.player.shape.src = "Player.png";

29.  this.player.x = 320;

30.  this.player.y = 380;

31.  this.enemy = new GameObject();

32.  this.enemy = new GameObject();

33.  this.enemy.shape.src = "Enemy.png";

34.  this.enemy.x = 500;

35.  this.enemy.y = 380

36.  this.prize = new GameObject();

37.  this.prize = new GameObject();

38.  this.prize.shape.src = "Prize.png";

39.  this.prize.x = 700;

40.  this.prize.y = 380

41.  this.map = new Image();

42.  this.map.src = "map.jpg";

43.  this.offsetX = 0;

44.  this.offsetY = 0; //not used

45.  this.playerDefense = 0;

46. }
```

```
47. //Global Variables

48. var c;

49. var ctx;

50. var game;

51. //Regular Functions

52. function Init()

53. {

54.   // initialize graphics screen

55.   c = document.getElementById("myCanvas");

56.   ctx = c.getContext("2d");

57.   //initialize objects

58.   game = new Game();

59.   //set main loop

60.   setInterval(MainLoop, 100);

61.   //set event handler for keyboard input

62.   window.addEventListener('keydown', KeyInput);

63. }

64. function MainLoop()

65. {

66.   Update();

67.   Draw();

68. }

69. function Update()

70. {

71.   game.enemy.x += Math.floor(Math.random() * 7)-3;

72.   //game.enemy.y += Math.floor(Math.random() * 7)-3;

73.

74.   if(Distance(game.player,game.enemy) < 5 && game.playerDe-
      fense == 0)
```

```
75.    Reset();
76.  if(game.prize.visibility != 0)
77.  {
78.    if(Distance(game.player,game.prize) < 5)
79.       game.prize.visibility = 0;
80.  }
81. }
82. function Draw()
83. {
84.  ctx.drawImage(game.map,-game.offsetX,0);  //subtract offset
85.  DrawObject(game.enemy, game.offsetX, 0);
86.  DrawObject(game.prize, game.offsetX, 0);
87.  DrawObject(game.player, game.offsetX, 0);
88. }
89. function Reset()
90. {
91.  game.player.x = 320;
92.  game.player.y = 380;
93.  game.offsetX = 0;
94. }
95. function KeyInput()
96. {
97.  switch (event.keyCode)
98.  {
99.   case 37: // Left
100.  game.player.x --;
101.  game.offsetX --;
102.  if(game.playerDefense > 0)
103.   {
```

```
104.    game.playerDefense --;
105.    if(game.playerDefense == 0)
106.      game.player.shape.src = "Player.png";
107.  }
108.  break;
109.  case 38: // Up
110.  if(game.playerDefense == 0)
111.  {
112.    game.playerDefense = 50;
113.    game.player.shape.src = "PlayerD.png";
114.  }
115.  break;
116.  case 39: // Right
117.  game.player.x++;
118.  game.offsetX ++;
119.  if(game.playerDefense > 0)
120.  {
121.    game.playerDefense --;
122.     if(game.playerDefense == 0)
123.      game.player.shape.src = "Player.png";
124.  }
125.    break;
126. }
127. }
128. </script>
129. </head>
130. <body onload="Init()">
131. <p>Canvas to fill:</p>
132. <canvas id="myCanvas" width="640" height="480"
```

133. `style="border:1px solid #d3d3d3;">`

134. `Your browser does not support the HTML5 canvas tag.</canvas>`

135. `<p><button onclick="Reset()">Reset</button></p>`

136. `</body>`

137. `</html>`

8.4.4 Physics: Platforms, Jumping, and Falling

The example we discussed in the previous section is a typical side-scrolling game but has two missing features:

- Jumping over things

- Different ground levels and platforms.

Both of these features require one essential ability in our code: dealing with gravity. When our game characters jump, we not only give them initial speeds but also need to slow them down to zero speed and then reverse the movement, so they come back down. Similarly, when characters get off platforms, they start falling down at an increasing speed. Jumping and falling are examples of a physical phenomenon called acceleration, or change in speed. While acceleration can be caused by any force, the particular type that happens in jumping and falling is the result of gravity, the force that earth applies on any object to pull them down.

Without gravity acceleration, a jumping object/character will not "come back down," and if you decide to make it go up a certain height and reverse movement, it will look very odd and unrealistic. Acceleration causes the speed to gradually change until it gets to zero and then gradually increases in the opposite direction.

Implementing physics is a good example of information hiding, abstraction, and hierarchical design. If you are programming the gameplay, you may not want to get involved in how the jumping and falling work. You simply need a module that takes an object and make it have an accelerated movement, or even decide whether accelerated movement or any other movement happens at all. Other programmers, or you at another time, create modules that take an object and its surroundings and decide if it has to move and how. Such a hierarchy consists of three possible levels:

1. Gameplay level where we only deal with player activities, for example, we press the Right arrow to move to that direction.

2. Intermediate Physics level where we check all the forces and surrounding objects to decide what type of movement should happen.

3. Detailed Physics level where we control the movement using particular methods of choice such as a realistic accelerated movement vs. simplified constant speed jump and fall.

If you are at levels 1 or 2, you don't need to know how the lower level(s) work, and you may have different options to choose from to perform them. For example, a level 2 module can have two options for level 3 module based on the method of choice, and the program can change at any time to switch between modules or use a new one. Also, if you are at levels 2 or 3, you don't need what the higher level does and why. A physics module is independent of gameplay and can be used (and reused) in any game.

Sample Code #11-c shows our simple platformer game in C/C++. It has all the features of Sample Code #10-j plus jumping and falling. But before I explain how we achieve that feature, let's take a quick look at the organization of this program.

8.4.5 Data Types

Similar to the Javascript version, the program uses two types for GameObject and Game (lines 8 and 27). As you notice, the `GameObject` has much information that is only relevant to the player. We may choose to create separate types for different game objects, or have one with all the required members. Note that if you have separate types, in explicitly typed languages like C/C++, functions like `DrawObject()` need to have different versions. Javascript and Python don't identify a data type, so their functions can receive any data as a parameter as long as they have the members necessary. `DrawObject()` only needs `shape`, `x`, and `y`, which are common among all game objects.

8.4.6 Game Functions

The main() function (line 261) creates the Game variables and passes it to the `RunGame()` function (line 254). It is not really necessary to have the second function as `main()` could do the job, but it is common to have specific functions for your application and not do much in `main()`. Next, `RunGame()` calls `Init()` and `MainLoop()` as two important parts of the game code (lines 101 and 244), and `MainLoop()` itself consists of `Update()` and `Draw()`, on lines 148 ad 232, respectively.

The `Init()` function is very straight-forward and initializes all the game objects, including the background, player, enemy, and prize. Note that we have an array of prizes, and arrays are almost always processed using a loop. Also, note how the code deals with transparent background for shapes. If you are using a graphics library that supports GIF and PNG image file formats, you can have transparency embedded in the images. If not (as is the case for SDL), you have multiple options:

- If the graphics library supports transparency through alpha values, you can go through all pixels after you load the image and set the alpha channel to 0 for all pixels that are supposed to be transparent.

- If the graphics library supports transparency through a mask color for each image, you can define that color after you load the image.

Standard SDL (without extensions) supports only BMP files but both of these options. The `SDLX _ LoadBitmap()` function has parameters for using a mask (see the code in SDLX.cpp for implementation).

- If no parameter is passed in addition to the filename (as in line 127), then a white mask is assumed, so all white pixels are considered transparent.

  - The default mask color is available through a global variable named color _ mask. It can be changed at any time affecting all bitmaps loaded afterward.

- If the second parameter after the filename is false, then there is no transparency (the third parameter will be ignored).

- If the second parameter after the filename is true, then the third parameter must be a variable of type SDLX _ Color* that is the transparent color (usually the background for that image).

```
game->map = SDLX_LoadBitmap("map.bmp", false, NULL);
game->player.shape1 = SDLX_LoadBitmap("Player.bmp", true, &c);
game->player.shape2 = SDLX_LoadBitmap("PlayerD.bmp", true, &c);
game->enemy.shape = SDLX_LoadBitmap("Enemy.bmp");
```

Update() and Draw() work in a way similar to previous versions. The Up key starts the defense mode, and movements decrease it. The collision detection with the enemy uses the defense mode. The Right and Left keys no longer affect the position directly. Instead, they set a direction value that is used in the new Move() function. The Space key starts the jump, which is part of the new Physics-based movement control.

8.4.7 Movement

The new Move() function is called after the keyboard input is processed. It is in charge of determining if the player can move and how much. Move() is part of a three-level modular design for movement control:

1. *Gameplay Level* happens when the player determines the direction of the desired movement by pressing Left, Right, and Space keys. The program simply calls Move(), which can be provided by a library or reused, so the gameplay programmer doesn't need to know how it is implemented.

2. *Intermediate Physics Level* is implemented in Move(), which determines if the object can move. It is done for both horizontal and vertical movements:

   a. For the horizontal movement, we detect where the object wants to go (direction gives us the newX), and then, the program compares the ground level at current X and new X. In this simple game, the ground level changes only once at x=800. So, lines 67–69 compare X to 800 to find out the ground level.[6] As we saw in Section 7.3, more complicated methods can be used for more complicated ground levels. I will show some of these methods in Chapter 12. If the new ground level

---

[6] Note that the ground level information is hard-coded in this sample, for code simplicity. If you choose a different background image with different ground level changes, then this code and the values have to be revised.

will be higher (lower Y value), then we are facing a bump and can't move forward without a jump. The direction value is set to zero to show that movement was not possible. The `Update()` function that calls `Move()` uses this to see if scrolling needs to happen. Note that scrolling has nothing to do with physical movement and is a game-level activity. One game designer may want to have scrolling, and one may not, but the physics-based movement will be the same.

b. Vertical movement is performed if the `jumping` value is `true`. The trick is that falling off a cliff or bump is basically a negative jump and follows the same physical rules (accelerated movement with gravity). So, we again detect if the object is at the ground level. If it is "above ground," then jumping starts but without an initial vertical speed, `obj->sy=0` (free falling).

3. *Detailed Physics Level* of movement is implemented in yet another function, `JumpFall()`, which determines how exactly the vertical movement happens. It follows the formula from mechanics describing any accelerated movement:

$$Y = Y0 + V0 * t + 0.5 * G * t^2$$

where Y0 and V0 are initial Y and initial speed, t is the time from the start of the movement, and G is gravity acceleration. We calculate time using the number of frames. The variable t is set to zero at the start of the movement and incremented at each frame. Play with the values of V0 (`sy`) and G to see which ones give a better and more appropriate jump and fall.

☑ **Practice Task:** Try modifying the code to use multiple bumps in the ground level.

✋ **Reflective Questions**

- Do you see the value of three physics levels? These are the abstraction layers we discussed earlier and provide information hiding and reusability.
- Do you feel comfortable with the level of Physics knowledge required in this example? In games and animation, a minimum understanding of some physics concepts is very helpful, if not necessary. Mechanics and optics are good examples.

Sample Code #11-c

```
1. #include "SDLX.h"

2. #include "stdlib.h"

3. #include "time.h"

4.
```

```
 5. #define NUM _ LIVES   3

 6. #define NUM _ PRIZES 2

 7.

 8. struct GameObject

 9. {

10.      SDLX _ Bitmap* shape;

11.      SDLX _ Bitmap* shape1;

12.      SDLX _ Bitmap* shape2;

13.      int x; //location

14.      int y;

15.      int sx; //speed

16.      int sy;

17.      int dir;

18.      int visibility;

19.      int defense;

20.      int score;

21.      int life;

22.      //gravity

23.      int t; //time (num of frames) for acceleration

24.      bool jumping;

25.      int y0;

26. };

27. struct Game

28. {

29.      GameObject player;

30.      GameObject enemy;

31.      GameObject prize[NUM _ PRIZES];

32.      SDLX _ Bitmap* prize _ shape; //load only once

33.      SDLX _ Bitmap* map;
```

```
34.     int   offsetX;

35.     int   offsetY;  //not used

36.     bool quit;

37. };

38.

39. void DrawObject(GameObject* obj, int offsetx=0, int offsety=0)

40. {

41.     if(obj->visibility != 0)

42.         SDLX _ DrawBitmap(obj->shape, obj->x-offsetx,obj->
    y-offsety);

43. }

44.

45. float Distance(GameObject* obj1, GameObject* obj2)

46. {

47.     int dx = obj1->x - obj2->x;

48.     int dy = obj1->y - obj2->y;

49.     float d = sqrt(dx*dx + dy * dy);

50.     return d;

51. }

52.

53. int JumpFall(GameObject* obj)

54. {

55.     int gravity = 1;

56.     int y = obj->y0 + obj->sy*obj->t + gravity * obj->t*obj->
    t / 2; //in jumping sy is the "initial" velocity

57.     return y;

58. }

59. void Move(GameObject* obj)

60. {
```

```
61.     //X movement
62.     int newx = obj->x;
63.     if (obj->dir == 1)
64.         newx++;
65.     if (obj->dir == 2)
66.         newx--;
67.     int ground = 380;
68.     if (obj->x > 800)
69.         ground = 170;
70.     int newground = 380;
71.     if (newx > 800)
72.         newground = 170;
73.     //no move if there is a bump
74.     if (ground > newground && obj->jumping == false)
75.         obj->dir = 0;
76.     else
77.         player.x = newX;
78.
79.     //Y movement
80.     if (obj->y < ground && obj->jumping == false)
81.     {
82.         obj->jumping = true;
83.         obj->y0 = obj->y;
84.         obj->t = 0;
85.         obj->sy = 0;
86.     }
87.     if (obj->jumping)
88.     {
89.         obj->t++;
```

```
90.          int newy = JumpFall(obj);

91.          if (newy >= ground)

92.          {

93.                  obj->y = ground;

94.                  obj->jumping = false;

95.          }

96.          else

97.          {

98.                  obj->y = newy;

99.          }

100.     }

101. }

102.

103. void Init(Game* game)

104. {

105.     //init SDL library and graphics window

106.     SDLX _ Init("Simple 2D Game", 640, 480);

107.     srand(time(NULL));

108.

109.     //game map (background)

110.     game->map = SDLX _ LoadBitmap("map.bmp", false, NULL);

111.     //player

112.     SDLX _ Color c; //transparent color

113.     c.r = c.b = 0;

114.     c.g = 128;

115.     c.a = 255;

116.     game->player.shape1 = SDLX _ LoadBitmap("Player.bmp",
         true, &c);

117.     game->player.shape = game->player.shape1;
```

```
118.    //shape2 is when in defense mode (UP key)
119.    game->player.shape2 = SDLX _ LoadBitmap("PlayerD.bmp",
   true, &c);
120.    game->player.x = 320;
121.    game->player.y = 380;
122.    game->player.visibility = 100;
123.    game->player.score = 0;
124.    game->player.life = NUM _ LIVES;
125.    game->player.defense = 0;
126.    game->player.jumping = false;
127.    game->player.dir = 0;
128.
129.    game->enemy.shape = SDLX _ LoadBitmap("Enemy.bmp");
130.    game->enemy.x = 500;
131.    game->enemy.y = 380;
132.    game->enemy.visibility = 100;
133.
134.    game->prize _ shape = SDLX _ LoadBitmap("Prize.bmp");
135.    for (int i = 0; i < NUM _ PRIZES; i++)
136.    {
137.        game->prize[i].shape = game->prize _ shape;
138.        game->prize[i].visibility = 100;
139.        game->prize[i].y = 380;
140.    }
141.    game->prize[0].x = 400;
142.    game->prize[1].x = 600;
143.
144.    game->quit = false;
145.    game->offsetX = 0;
```

```
146. }

147.

148. void Update(Game* game)

149. {

150.     SDLX_Event e;

151.     //keyboard events

152.     if (SDLX_PollEvent(&e))

153.     {

154.         if (e.type == SDL_KEYDOWN)

155.         {

156.             //press escape to end the game

157.             if (e.keycode == SDLK_ESCAPE)

158.                 game->quit = true;

159.             //update player

160.             if (e.keycode == SDLK_RIGHT)

161.                 game->player.dir = 1;

162.             if (e.keycode == SDLK_LEFT)

163.                 game->player.dir = 2;

164.             if (e.keycode == SDLK_UP)

165.             {

166.                 if (game->player.defense == 0)

167.                 {

168.                     game->player.defense = 50;

169.                     game->player.shape = game->player.shape2;

170.                 }

171.             }

172.             if (e.keycode == SDLK_SPACE && game->player.jumping == false)
```

```
173.                {
174.                        game->player.jumping = true;
175.                        game->player.y0 = game->player.y;
176.                        game->player.t = 0;
177.                        game->player.sy = -20;
178.                    }
179.            }
180.            if (e.type == SDL_KEYUP)
181.            {
182.                if (e.keycode == SDLK_RIGHT)
183.                        game->player.dir = 0;
184.                if (e.keycode == SDLK_LEFT)
185.                        game->player.dir = 0;
186.        }
187.    }
188.    //process player move
189.    Move(&game->player);
190.    if (game->player.dir == 1) //right
191.    {
192.        game->offset++;
193.        if (game->player.defense > 0)
194.        {
195.                game->player.defense--;
196.                if (game->player.defense == 0)
197.                    game->player.shape = game->player.
    shape1;
198.        }
199.    }
200.    if (game->player.dir == 2) //left
```

```
201.        {
202.                    game->offset--;
203.                 if (game->player.defense > 0)
204.                   {
205.                      game->player.defense--;
206.                    if (game->player.defense == 0)
207.                         game->player.shape = game->player.
     shape1;
208.                   }
209.        }
210.
211.     //update enemy
212.     game->enemy.x += rand() % 7 - 3;
213.     //collision detection
214.      if (Distance(&game->player, &game->enemy) < 5 && game-
     >player.defense == 0)
215.      {
216.                    game->player.life--;
217.                 if (game->player.life == 0)
218.                     game->quit = true;
219.      }
220.      for (int i = 0; i < NUM_PRIZES; i++)
221.      {
222.           if ((game->prize[i].visibility != 0) && Distance
     (&game->player, &game->prize[i]) < 5)
223.           {
224.                       game->prize[i].visibility = 0;
225.                       game->player.score++;
226.                    if (game->player.score == NUM_PRIZES)
227.                          game->quit = true;
```

228.   }

229.   }

230.

231. }

232. void Draw(Game* game)

233. {

234.   SDLX _ DrawBitmap(game->map, game->offsetX, 0);

235.   //Draw the objects (in order: background to foreground)

236.   for (int i = 0; i< NUM _ PRIZES; i++)

237.   DrawObject(&game->prize[i], game->offsetX,0);

238.   DrawObject(&game->enemy, game->offsetX, 0);

239.   DrawObject(&game->player, game->offsetX, 0);

240.

241.   //Update the screen

242.   SDLX _ Render();

243. }

244. void MainLoop(Game* game)

245. {

246.   // Main Loop

247.   while (!game->quit)

248.   {

249.   Update(game);

250.   Draw(game);

251.   SDLX _ Delay(1);

252.   }

253. }

254. void RunGame(Game* game)

255. {

```
256.        Init(game);

257.        MainLoop(game);

258.        SDLX _ End();

259. }

260.

261. int main(int argc, char *argv[])

262. {

263.        Game game;

264.        RunGame(&game);

265.        return 0;

266. }
```

8.4.8 Timers and Callback Functions

If you look at the `MainLoop()` function, you will notice that it sets up the timing of the program using a call to `SDLX _ Delay()`, which causes the program to pause for a certain amount of time (in this case 1 millisecond only). This pause is necessary to make sure each frame takes 1 second. The problem with this method is that the actual time between the start of one frame and the start of the next frame is 1 second plus the execution time of `Update()` and `Draw()`, the two parts of the loop. An alternative method is to use timers.

Timers are signals that computers can send to programs to let them know a certain timing interval has passed. They use the computer's real-time clock and provide precise timing events. Using timers requires two important steps as we saw in Section 7.1 for Javascript:

- Defining a callback function to handle the timer event
- Calling a system function to set up the interval and link the callback function to it

☞ **Key Point:** Timers generate timing-related events. To set up a timer, you need a callback function and a time interval for system to call it.

We can modify the game in Sample Code #11-c to use timers with a few changes. In SDL, a callback has to be a global function with a standard format: it has to return an integer value that is the interval for the next timer; it receives the current interval; and it receives a pointer parameter that can be any type. `Uint32` is defined by SDL as 32-bit unsigned integer. Since the global function has no access to the `Game` class, we use its parameter to pass a pointer to our `Game` object. Our callback function can simply call `Update()` and

Draw() functions of this object to perform all the required actions for creating a new frame. This is not a good idea, though, as timer callback functions run at a high priority in the system and can prevent the program from reading the mouse and keyboard events. It is not recommended to perform long tasks in callback functions. Instead, we only let the program know that we have received a timer signal. According to our first rule of programming, this new information (timer signal to create a new frame) needs a new variable, so we add a new member to the Game class and initialize it to false.

```
//in Game.h
    Bool frameTimer;
//in Game:Game()
    frameTimer = false;
//new function in Game.cpp
Uint32 GameCallback(Uint32 interval, void *param)
{
    //g->Update();
    //g->Draw();

    Game* g = (Game*)param; //tell program this is a Game pointer
    g->frameTimer = true;

    return interval;
}
```

Once we have the callback function set, we need to set the interval. We can do that at Game::Run(), which is called at the start of the game. Note how we pass a pointer to the Game object using the keyword this that always shows the current object in C++.

```
//set timer at the start
void Game::Run()
{
    SDL_TimerID my_timer_id = SDL_AddTimer(30, GameCallback,
this);
    MainLoop();
    delete ps;
    SDLX_End();
}
```

Finally, we modify our main loop to update and draw only when a timer signal has been received. We set the signal back to false to prepare for the next event.

```
//update and draw when timer signal has been received
void Game::MainLoop()
{
    // Main Loop
    while (!quit)
```

```
    {
        if (frameTimer)
        {
            Update();
            Draw();
            frameTimer = false;
        }
        SDLX_Delay(1);
    }
}
```

HIGHLIGHTS

- Manageability and reusability are two important features of a good software design.

- Interface of a module is what the users need to know about it. It does not include what is "inside" the module.

- Information hiding is a basic principle of software design that deals with exposing programmers to only what is necessary. Abstraction can be one way to achieve this.

- Object Aggregation refers to user-defined types that include members of other user-defined types. For example, Library is an aggregated module that has User and Book members.

- Timers generate timing-related events. To set up a timer, you need a callback function and a time interval for system to call it.

- Physics module can control all the movements in a game.

- Three HOW Questions (3HQ) can help design the algorithm. They define how the program should behave and are related to both requirement analysis and design.

END-OF-CHAPTER NOTES

A. Things I Should Mention

- The Unix Philosophy and Ousterhout's notion of deep functions are not the same but share the notion of simplicity. Reusability, on the other hand, usually results in more complicated interfaces as it demands more parameters for customization. The trade-off is really hard to prescribe. Your experience will be the final judge.

- To learn programming and to be a programmer, you don't need to know a lot of math or physics. But it sure helps; many programming problems have mathematical and scientific solutions, and a mathematical mind can strengthen your logical creativity.

- Timers, callback functions, and event handlers are closely related to the topic of **multi-threading** (the ability of a running program to have multiple sequences of execution in parallel) and system priorities. These are beyond the scope of this book, but definitely important as you become more experienced programmers.

B. Self-Test Questions

- What is a Command Processor software pattern?
- What arc deep and shallow functions?
- What is a module interface?
- How do you control the jumping and falling objects using physics formulas?
- What is a timer?

C. Things You Should Do

- Check out these web resources to learn about multi-threading:
  - https://www.tutorialspoint.com/operating_system/os_multi_threading.htm
- Take a look at these books to learn about timers, threads, and processes:
  - *Mathematics & Physics for Programmers (Programming Series)* by Danny Kodicek
  - *Modern Operating Systems* by Tanenbaum and Bos.

D. Reflect on the Experience of Reading This Chapter

- What did you expect from this chapter before reading it?
- What was it about, and what did you learn?
- What tasks did you perform, and what difficulties did you face?
- How did you feel about the material and tasks presented in this chapter?
- How can you improve your learning experience?
- How do you see this topic in relation to the goal of learning to develop programs?
- How do you feel about the notion of good software design? Did the text help you understand how to design better modules?
- What will you do to manage the trade-off between reusability and simplicity? Do you think making a module more reusable will result in being more complex?
- Do you feel comfortable with physics content? Do you need to improve your knowledge of math and physics?

PART 4

Object-Oriented Programming

Code never lies.

Comments sometimes do.

GOAL

In the previous parts of this book, we saw how a program is made of code and data, and how modularization is the key concept that allows managing these two elements in our programs.

Now is the time to realize that the distinction and separation of code and data are not as rigid as you may have thought. In the following chapters, I introduce Abstract Data Types and Objects to see how data and code are inseparable and how we can define modules that integrate them efficiently.

Modularization of Data and Code

OVERVIEW

I concluded the previous part of this book with examples of well-structured and modularized programs with multiple new data types and various functions that processed them. I organized both data and code modules into hierarchies and showed how **aggregation** (classes/structures including other classes/structures) and **abstraction layers** (high-level modules on top of low-level ones) provide efficient ways of modularization where the programmers need to deal with details only when necessary.

Our design process was a data-centric one that started with identifying main data items (and types) and then defining the operations performed on them (functions).

Our approach clearly shows how data and code are generally related: **functions are usually written to process certain data**. While such a clear connection exists between code and data, the Structured Programming (SP) paradigm did not make it explicit. Looking at any of the sample programs, it is not quite clear which functions are related to which data types. It is up to the programmer to put them together and add comments to show the connections, and to remember which ones to use together. Such a reliance on the programmer to properly organize the code is not very efficient.

This chapter extends the notion of modularization and introduces objects and classes as a means of creating a new module that explicitly integrates (**encapsulates**) all the related code and data. This integration is the basis of a popular programming paradigm called **Object-Oriented Programming (OOP)**. In the following sections, I will show how OOP increases the manageability and reusability of the program modules by making them more self-sufficient as all elements are integrated together. You will see that OOP provides a more appropriate program structure for interactive applications like games compared to more linear SP.

The familiar Library and Game programs are used in this chapter again to demonstrate the OOP approach.

9.1 OBJECTS AND CLASSES

In previous chapters, we saw how modules of data are related to modules of code. Even if they are not explicitly combined, a programmer who uses the type `GameObject` needs to be familiar with the function `DrawObject()`. **Objects** are a type of data that makes this connection more explicit and formal. In this chapter, we will see how objects change the way we design and implement our programs, but before doing that, let's understand the relationship between code and data better.

9.1.1 Abstract Data Types

Different programming languages and programmers implement various concepts differently. For example, the standard data type integer consists of a single value, but it also has four standard mathematical operations[1] that can be performed on that value. While the behavior of an integer may seem similar in all programming languages, the way that is implemented "behind the scene" may not be the same in C and Python, for example. Similarly, a GameObject may have common properties in many games, but we can be sure that it is not implemented the same way in different games.

To help programmers define and use frequently used data types, we commonly define them as a mathematical model with a core value plus operations performed on them. The model is called the **Abstract Data Type (ADT)**. It is abstract because it does not define how the data and its operations are implemented. Let's consider an example.

Lists are frequently used elements in many programs. While they can be implemented in different ways and some programming languages have them as a built-in part of the

---

[1] Addition, subtraction, multiplication, and division.

language, they have a basic definition. A list is a series of data elements with operations such as

- Create a new list
- Add a new data member
- Access an existing data member
- Get the number of data members.

Note that conceptually a list is more than just an array. It may be implemented using an array, but it includes more information (length) and requires operations. Sample Code #1-c shows one possible implementation of such a list in C/C++. Variations in this implementation can include different algorithms for functions or different members for the structure (or no structure at all).

Sample Code #1-c

```
1.    #include "stdio.h"
2.    #include "stdlib.h"
3.
4.    #define MAX _ SIZE 10
5.    #define FAIL 0
6.    #define SUCCESS 1
7.
8.    struct List
9.    {
10.           int data[MAX _ SIZE];
11.           int size;
12.    };
13.    //list functions
14.    void InitList(List* list)
15.    {
16.           list->size = 0;
17.    }
18.    int GetListSize(List* list)
```

```
19.    {
20.        return list->size;
21.    }
22.    int GetListData(List* list, int n)
23.    {
24.        if (n >= 0 && n < list->size)
25.            return list->data[n];
26.        else
27.            return -1; //any number that is "out of range"
28.    }
29.    void SetListData(List* list, int d)
30.    {
31.        if (list->size < MAX_SIZE)
32.        {
33.            list->data[list->size] = d;
34.            list->size++;
35.        }
36.    }
37.    //main function
38.    void main()
39.    {
40.        List myList;
41.        InitList(&myList);
42.        for (int i = 0; i < 2 * MAX_SIZE; i++)
43.        {
44.            SetListData(&myList, i);
45.            printf("%d\n", GetListData(&myList, i)); }
46.
47.    }
```

☞ **Key Point:** ADT is a description of a group of related data and the operations that are performed on them. ADT can be implemented in different forms.

☑ **Practice Task:** Write a different implementation of a list.
☑ **Practice Task:** Sample Code #1-c returns an error code for `GetListData()`. This is a common and very helpful practice. Use more #define statements and different error codes for things that can go wrong in the functions and have them all return a "result code" (different errors or success).

✍ **Reflective Question:** If a function returns result codes instead of the actual result value, you may use by-ref parameters to receive the results. What do you think the advantages and disadvantages of this method are?

Other examples of ADTs are

- Stacks are lists with access only to the last member. They are accessed using a first-in-last-out method, so we can add (write) new members to the end of the list and access (read) the last one we have added.

- Queues are lists, but we add new members to one side and access the members from the other size, so we read the first one written.

- Linked Lists are lists where each member points to the next member in a linear form.

- Graphs are an extension of lists where moving from one member to another does not follow a single line (order).

☑ **Practice Task:** Try writing programs using some of the above ADTs. For example, Stack is an array with two related functions Push() and Pop(), for writing a new element to the end of the array and reading one from the same end, respectively.
Hint: You need to keep track of where the end of array is. The "end" here refers to the end of used portion of the array, not the maximum size.

9.1.2 Encapsulation

While a detailed discussion of ADTs is beyond the scope of this book, their introduction is helpful to demonstrate the notion of data–code relationship. ADTs explicitly integrate a value with a set of operations. In other words, they suggest the idea of **Encapsulation**, a term that in the context of computer programming is defined as one or a combination of two things:

- A module that hides some content (data or code) inside and accessible only to the internal code. This definition is sometimes referred to as **Information Hiding**, which we saw in Chapter 8.

- A module that includes both data and the code that operates on that data. This definition is what I use in this book, as it conceptually separates encapsulation and information hiding and allows more flexibility (for example, having a module with both data and code but no information hiding).

In C, encapsulation is not officially possible as C has no language construct that can include both code and data. C structures are limited to data, and an attempt to create a single encapsulated module would look like Sample Code #2-c. In this code, the module functions are implemented in `MyModule.c` while `MyModule.h` shows the interface for the module, anything that other parts of the code need to know in order to use `MyModule`.

Sample Code #2-c

```
1. // MyModule.h
2. Struct MyModule
3. {
4.     //data members here
5. }
6. //functions that process MyModule data
7. // they are implemented in MyModule
8. void ProcessData1(MyModule* m);
9. MyModule*  ProcessData2();
10. int ProcessData3(MyModule* m);
11.
12. // MyModule.c
13. #include "MyModule.h"
14. void ProcessData1(MyModule* m)
15. {
16. //function code here
```

```
17. }
18. MyModule*   ProcessData2()
19. {
20.     //function code here
21. }
22. int ProcessData3(MyModule* m)
23. {
24.     //function code here
25. }
```

While it seems that the above solution allows us to create a module of related data and code, it has a few problems:

- All related parts are organized together by the programmer. There is nothing in the language that clearly identifies these parts as related and dependent on each other. As such, the programmer may mistakenly change this organization later on.

- Reusing this module requires the organization to be kept by any programmer who changes or uses this code, and that all programmers are aware of the code organization and what it means. Otherwise, they may copy only part of it to another application and not understand the dependencies.

- Even though the data is related to a set of functions, we still need to pass the data to them as parameters, which makes it harder to share data among those functions.

What we expect to have as an efficient way of defining modules is something like

Sample Code #2-x   (not in any real language but similar to C/C++)

```
Module MyModule
{
    //data
    int data1;
    int data2;
    //other data here

    //code
    void ProcessData1()
    {
        //function code here
    }
```

```
    void ProcessData2()
    {
            //function code here
    }
    int  ProcessData3()
    {
            //function code here
    }
};
```

The important features of the above code are

- It explicitly shows which functions are members of the module.

- All related data and code are grouped clearly.

- The member functions no longer need to be given a parameter for data. The data members of the module act like global variables for the member functions.

The above features not only group data and code together but also provide a clear integration where the function can easily access the data. In programming, we refer to such modules as **Objects** which can explicitly combine data and code. The data variables that are part of an object are usually referred to as **Data Member** or **Property**, while the functions are called **Member Function** or **Method**.

> ☞ **Key Point:** Encapsulation combines code and data into a single module. The result is an object. It can be implemented without specific language support, but many programming languages provide specific facilities for this purpose.

C++ was introduced as an extension to C with support for objects. Other languages such as Python and Javascript, and also Java and C#, also have the ability to define objects. **OOP** is a programming paradigm commonly used in these languages.

9.1.3 C++ and Classes

The C++ code for creating objects resembles Sample Code #2-x very closely (Sample Code #3-c[2]). Similar to Python and many other OOP languages, C++ uses the keyword `class` to define a new type (similar to keyword `struct`), and **objects** are variables created using a **class** type (sometimes called **instances** of that class). Apart from the use of this keyword, the structure of Sample Code #3-c is pretty much the same as #2-x with two data items and three functions that process them.

Note how the member functions of the class can use the data members directly without being passed as a parameter. If you define these functions outside the class, you will receive

---

[2] Remember that we use -c for both C and C++ code.

a compile error saying that the variables grades and gpa are not defined. But inside the class, these variables act like global ones.

The keywords class and struct both define new types. In order to use this type, another part of the code has to create variables of that type. In Sample Code #3-c, this happens in main(). The variable s is of type Student (a class), and so it is called an object. The main() function creates two objects (s1 and s2, lines 31 and 36). Just like structures, the members of a class type are accessed using the name of the object plus a period followed by the name of the member. This applies to both data and functions.

Sample Code #3-c

```
1. #include "stdio.h" //for printf()

2. #include "stdlib.h" //for rand()

3.

4. class Student

5. {

6. public:

7.     int grades[10];

8.     float gpa;

9.

10.         void Init()

11.         {

12.             for (int i = 0; i < 10; i++)

13.                 grades[i] = rand() % 100;

14.         }

15.         void CalculateGPA()

16.         {

17.             gpa = 0;

18.             for (int i = 0; i < 10; i++)

19.                 gpa += grades[i];

20.             gpa = gpa / 10;

21.         }

22.         void Report()
```

```
23.              {
24.                  for (int i = 0; i < 10; i++)
25.                      printf("%d\n",grades[i]);
26.                  printf("GPA=%f\n", gpa);
27.              }
28. };
29. void main()
30. {
31.              Student s1;
32.              s1.Init();
33.              s1.CalculateGPA();
34.              s1.Report();
35.
36.              Student s2;
37. }
```

☞ **Key Point:** A class is a user-defined type with encapsulation. An object is a data item (variable) based on a class type.

☑ **Practice Task:** Rewrite Sample Code #1-c using a class.
  Hint: There are no parameters for the functions if they need access to the data members.

☑ **Practice Task:** Look at the structures in previous chapters, and try defining classes instead. Which members will they have?

Some OOP languages do not require the concept of a class. For example, in Javascript, we can create an object using a function, although Javascript has the keyword class as well. The user-defined types we saw in Javascript and Python are, in fact, classes, and the variables we made based on those types were objects. Even though we did not define any member functions for them, they could have functions in addition to data. Similarly, a C++ class may have only functions, only data, or both.

The only other new thing in Sample Code #3-c was the keyword public. Let's look at another example in Sample Code #4-c and also explain that keyword. In Chapter 8, we worked on a library example made of a few data types such as Book. This data type included three data members: name, ID, and user. Two main operations were performed on a book: `Borrow()` and `Return()`. We also had `AddBook()`, which involved creating a new book and also adding it to the main `Library` data.

The keyword `class` defines the new type `Book` (line 4). Note how this type includes both data (lines 7-9) and code (lines 11–29). The code is defined as two functions that directly work on the data members. Since these functions are now part of the same module as the data, they treat the data members just like global variables; there is no need to pass the data as parameters, and once a function changes a data (as in line 16), the value is changed for any other part of the program.

The `main()` function (line 31) uses the new type to create a variable called b. A variable made based on a class is an object. This is a key point to remember: classes are the types, and objects are the variables. A class is a template (blueprint) for creating objects. Once the object is created, we can access its members the way we saw in previous chapters for user-defined types (lines 34–39).

Sample Code #4-c

```
1. #include "stdio.h"

2. #define MAX _ CHARACTERS 30

3. #define MAX _ BOOKS 30

4. class Book

5. {

6. public:

7.       char name[MAX _ CHARACTERS];

8.       int  ID;

9.       int  user;

10.

11.          void Init(int id)

12.          {

13.                  printf("enter book name: ");

14.                  scanf _ s("%s",name, MAX _ CHARACTERS);

15.                  user = -1;

16.                  ID = id;

17.          }

18.          void Borrow()
```

```
19.                {
20.                    if (user == -1)
21.                    {
22.                        printf("enter user ID: ");
23.                        scanf_s("%d", &user);
24.                    }
25.                }
26.        void Return()
27.        {
28.                user = -1;
29.        }
30. };
31. void main()
32. {
33.        Book b;
34.        b.Init(1000);
35.        printf("Created. user=%d\n", b.user);
36.        b.Borrow();
37.        printf("Borrowed. user=%d\n", b.user);
38.        b.Return();
39.        printf("Returned. user=%d\n", b.user);
40.        Book anotherBook;
41.        anotherBook.user = 10;
42.        anotherBook.ID = 1000;
43. }
```

A couple of important notes about using a class in C++ are worth mentioning here:

- Member functions of a class can access its data members directly and similar to a global variable. But other functions (outside the class) need to access class members through the name of an object. This is necessary because we may have multiple objects of the same type in our program. Outside functions need to identify which data (belonging to which object) they intend to access. For example, in line 41 and 42, we are accessing the data for `anotherBook`.

- In C++, members of a class are accessible by member functions, but access to them from other parts of the program can be controlled. The keyword `public:` (line 6) defines all the members as accessible in all parts of the program. If you delete or comment out line 6, you will get compile errors for lines 34–42 saying: `Cannot access private member declared in class 'Book'`. These lines are part of `main()`, which is not a member of `Book`. As such, they cannot access the members of Book, which are non-public by default. Later, we will see how defining members as **public** or **private** can help design better objects.

- In C++, the keywords `struct` and `class` both define a new class, meaning that even a structure can have member functions. This is one of the main differences between C and C++. The only distinction between `struct` and `class` is that if using `struct`, then all members are public by default, so structures don't need to use the keyword `public`. This makes C++ code compatible with old C code using struct.

- Members of an object are not initialized by default. The part of the program that creates the object (in this case, lines 33 and 40 in `main()`) has to make sure that the data members have received proper initial value (lines 34, 41, and 42). If the class has an initialization function, the program can use it to give a default initial value to data. But the program may also directly access the data to set other values, as shown in line 41.

Since the programmer may not define an initialization function or not call it before using the object, C++ and many other OOP languages streamline the process of initializing objects through a special function, called a **Constructor**.

☞ **Key Point:** C structures included only data. C++ classes can have both data and code. C++ also uses structures, but they are similar to classes with the exception that their members are public by default, so the keyword public is not necessary in C++ structures. This is for compatibility with C.
Note that Javascript and Python don't have the notion of structure.

☑ **Practice Task:** Think about some software programs you use. Which classes do you think they have? Identify the members of those classes. There is no need to write the code.

🖐 **Reflective Question:** Do you feel comfortable with the notions of class and structure, and how structures have evolved from C to C++? This evolution is one of the differences between two languages.

9.1.4 Constructor

In most cases, when we create a new object, we need to initialize its data members. Imagine buying a new electronic device; you may need to charge the batteries for a certain amount of time, set up the clock, or even remove some wrapping before you can use it. These actions are the standard initialization for that device, and without them, the device cannot be used because certain elements (members) are not properly set.

In Sample Code #4-c, we have given initial value to the `user` member because it has to be clear that the book is available; otherwise, the function borrow won't work. To stream-line the process of object creation, most OOP languages have a specific member function for classes that is automatically called when an object is created. This function is referred to as **Constructor** and is in charge of initializing the data. In C++, constructors are identified with two important features, as shown in Sample Code #5-c:

- They have the same name as the class.

- They have no type.

Sample Code #5-c (part 1)

```
1. class Book
2. {
3. public:
4.     char name[MAX _ CHARACTERS];
5.     int  ID;
6.     int  user;
7.     //constructor
8.     Book()
9.     {
10.             ID = user = -1;
11.             name[0] = 0; //null terminated string
12.     }
13.     //other functions
14.     void Init(int id)
15.     {
16.             printf("enter book name: ");
17.             scanf _ s("%s",name, MAX _ CHARACTERS);
```

```
18.                    user = -1;
19.                    ID = id;
20.            }
21.        void Borrow()
22.        {
23.                if (user == -1)
24.                {
25.                    printf("enter user ID: ");
26.                    scanf_s("%d", &user);
27.                }
28.        }
29.        void Return()
30.        {
31.                user = -1;
32.        }
33. };
34. void main()
35. {
36.        Book b;
37.        b.Init();
38.        printf("Created. user=%d\n", b.user);
39.        //the rest of the code here
40. }
```

The constructor (line 8) can easily be identified as it has the name Book and no type. It makes sure that all data members have an initial value. Note that both ID and user are set to -1, which in our code is used for the invalid value. In the case of the user, this means there is no user associated with the book (it is available). In the case of the variable ID, -1 means we have not assigned an ID yet. If we initialize the ID with 0, for example, it could be confused with a valid ID equal to 0. Assigning an ID is usually an action that the program performs based on a more global system, and that is the responsibility of another (probably higher-level) object such as the library.

The Init() function is still called by the main() which is responsible for creating objects and managing them. The main() code passes an ID value to Init(), which in turn initializes both ID and name. This two-step initialization is very common, although not always necessary. It is helpful in cases when we want to create an object but not use it right away. To expand our book library example and demonstrate this initialization, let look at the following code:

Sample Code #5-c (part 2 with new class and modified main())

```
1. class Library
2. {
3. public:
4.      Book books[MAX _ BOOKS];
5.      int  numBooks;
6.      void AddBook()
7.      {
8.           if (numBooks < MAX _ BOOKS-1)
9.           {
10.               books[numBooks].Init(1000+numBooks);
11.               numBooks++;
12.           }
13.      }
14.   };
15.
16. void main()
17. {
18.        Library lib;
19.        lib.numBooks = 0;
20.        lib.AddBook();
21.        lib.books[0].Borrow();
22.        lib.books[0].Return();
23. }
```

Once the `main()` function creates an object of type `Library`, all members of that class will be created for the new object. This includes an array of `Book` objects. But as we say in the previous chapter, we would like to have a user command for adding new books, so the array at the start of the program is "empty," even though the `Book` objects are created. These objects need to have a minimum initialization, so they don't have any undefined values. This will be done by the constructor automatically as soon as the objects are created. Then, and when we want to add a new book, we call the `Init()` function to properly initialize a new book. Note that the ID values start from 1000.

> ☞ **Key Point:** Constructor is a special function in a class that is automatically called when an object is created. Constructors usually initialize the object. In C++, constructors can easily be identified as they have the same name as the class and no type/return.

> ☑ **Practice Task:** Write a constructor function for the `Library` class.
> ☑ **Practice Task:** Write the UI code for Sample #5-c.

> ✋ **Reflective Questions:** Do you understand the reason for having a constructor and a separate `Init()` function? In what cases, such a two-step initialization can help?

The constructor can also have parameters, as shown below. When using such a constructor, the code that creates an object needs to provide the parameter, for example, `Book b(1000)`.

```
Book(int id)
{
     ID = id;
     user = -1;
     name[0] = 0; //null terminated string
}
```

9.1.5 Python and Javascript Classes
Sample Codes #6-p and #6-j demonstrate defining classes with data and code in Python and Javascript.

Sample Code #6-p

```
1. #class

2. class Book :
```

```
3.    name = ""
4.    ID = 0
5.    borrower = -1
6.    def Borrow(self) :
7.        if self.borrower == -1 :
8.            uid = int(input("enter user ID: "))
9.            self.borrower = uid
10.    def Return(self) :
11.        self.borrower = -1
12.    def Available(self) :
13.        if self.borrower == -1 :
14.            print("available")
15.        else :
16.            print("unavailable")
17. #main code
18. book = Book()
19. book.Available()
20. book.Borrow()
21. book.Available()
22. book.Return()
23. book.Available()
```

The Python code is fairly straight-forward and similar to C++ in concept, although the syntax is a little different.

- Data members are defined without any keyword or types (as in other Python programs), but any data defined inside the class is a member.

- Member functions are defined with keyword `def` and have a standard parameter called `self` that refers to the object for which the function is called.

- Member functions need to use the keyword `self` when referring to a member of the class. Otherwise, the program assumes a local variable with that name is being used.

Python classes can also have a constructor with the standard name _ _ init _ _ (). It has to have the parameter self. If using a constructor, you no longer need to separately define the data members because Python defines new variables the first time they are used. The only trick is to use the keyword self.

```
def __init__(self):
    self.name = ""
    self.ID = 0
    self.borrower = -1
```

In Javascript, we can use a function to create a user-defined type (object), but including functions (object methods) in another function may look confusing. Javascript programs can also use the keyword class to define classes, as shown in Sample Code #6-j.

Note that while the concept stays the same, Javascript has some syntax differences:

- The constructor is identified with the special name constructor.

- Similar to Python, data members (properties) can be defined inside the constructor but have to be referred to (in all class functions) with the keyword this (similar to self in Python).

- Creating an object based on a class required the keyword new (line 23) instead of var.

Sample Code #6-j

```
1.     <html>

2.     <head>

3.     <script>

4.     class Book

5.     {

6.             constructor()

7.             {

8.                     this.name = "";

9.                     this.ID = 0;

10.                    this.borrower = -1;

11.             }

12.             Borrow()

13.             {
```

```
14.                    this.borrower = userid.value;
15.                    document.getElementById("text").innerHTML =
    "book unavailable";
16.                }
17.            Return()
18.                {
19.                    this.borrower = -1;
20.                    document.getElementById("text").innerHTML =
    "book available";
21.                }
22.        }
23.    myBook = new Book();
24.    </script>
25.    </head>
26.    <body >
27.    <p id="text">book available</p>
28.    <p><input id="userid"></p>
29.    <p><button onclick="myBook.Borrow()">Borrow</button> </p>
30.    <p><button onclick="myBook.Return()">Return</button></p>
31.    </body>
32.    </html>
```

Just like C++, the constructor for Python and Javascript classes can have parameters to give initial values to data members. Such parameters have to be passed when creating objects.

☞ **Key Point:** C++, Python, and Javascript have the keyword class for defining new classes. There are, however, some syntax differences.

☞ **Key Point:** In Python, constructors are identified with the name _ _ init _ _ (). In Javascript, they are called constructor.

SIDEBAR: PROGRAMMING PARADIGMS

Programmers design and code their programs in different styles and approaches. Through learning from each other and evolving these styles, certain patterns of programming have emerged that are commonly used by many programmers. These common patterns are called programming paradigms. For example, Structure Programming is a paradigm that focuses on organizing code using iterations, selections, and functions. OOP is another paradigm that is based on encapsulation, integration of code, and data into modules. Two important notes to remember about programming paradigms are

- They overlap. A program may use features of more than one paradigm.
- They identify programming style, not the language. While a particular language may facilitate a certain paradigm, languages are not directly attached to a single paradigm.

It is common though to see phrases such as "C is an SP language," or "C++ is an OOP language," as they refer to main characteristics of the languages that facilitate those paradigms. It is possible though to write Structured Programs in C++ or even Object-Oriented Programs in C (without using the keyword `class` and by manually organizing the code).

Some popular programming paradigms are

- Imperative vs. Declarative (Section 3.1): Languages that describe the operations vs. those that describe the output
- Structural: Using selection and iteration vs. `GoTo` statement
- Procedural: Organizing the code into functions and using function calls
- Structured: Commonly used for structural and procedural
- Functional: Avoiding variables and relying mostly on functions
- Object-Oriented: Using modules of code and data
- Aspect-Oriented: Using common "concerns" that cross-cut class boundaries and are shared among the objects of different types
- Event-based: Organizing the code as a series of responses (handlers) for events
- Reflective: Making programs that can inspect and manipulate themselves.

9.2 OBJECT-ORIENTED PROGRAMMING

OOP has enjoyed increasing popularity as it can potentially make many programs more manageable and reusable. Some computer scientists and programmers do not support OOP as a better option, and indeed most agree that it may not be a preferred option in many cases. When the program is not dealing with significant data and is mainly a series of operations, particularly when they are defined in a linear way, OOP may add overhead by defining new language constructs.[3] But most modern applications have two important features:

---

[3] Linus Torvalds, the creator of Linux Operating System, I quoted to say "limiting your project to C means that people don't screw things up with any idiotic 'object model'."

- They have different data items that can be grouped into separate entities with their own operations.

- They have a non-linear behavior that consists of entities interacting with each other at non-pre-defined times instead of a predefined flow of actions.

Organizing programs as a series of potentially interactive entities, each capable of performing operations, is the key advantage of OOP, which is made possible by encapsulation, the first property of Object-Oriented languages. Encapsulation provides multiple benefits to programmers:

- More manageable programs as the related code and data are clearly grouped

- More reusable programs as the modules are more self-sufficient and easier to copy to another application (source code or compiled binary)

- Information hiding is better supported through defining objects and their interfaces[4]

- Interactive applications are easier to design with Object-Oriented Programs compared to procedure-oriented ones.

I will discuss and illustrate these advantages in this and the next section. OOP has two other basic properties called **Inheritance** and **Polymorphism** that I will introduce in Part 5.

The starting point of any object-oriented program, which is the key to creating a manageable and reusable code, is to select and define the right classes. Recall the data-centered approach for designing programs that I introduced earlier in the book:

- Visualize the program.

- Identify main data items, and define variables for them.

- Identify the operations performed on main data items, and write the algorithm and code for them.

- Write the remaining code to connect all parts.

The process for writing an object-oriented program is very similar:

- Visualize the program.

- Identify main classes (types of objects) and their instances (objects of that type).

- For each class, identify data items, and define variables (data members or properties) for them.

---

[4] Recall that interface is what a module "exposes" to other part of the program, such as the API that is the interface for a library.

- For each class, identify the operations performed on data (member functions or methods), and then, write the algorithm and code for them.

- Write the remaining code to create, and connect all objects (a.k.a. the "**glue code**").

In a typical object-oriented program, we rarely have major data items or functions that are not in any class. Common exceptions are

- Initialization functions to create objects

- Helper functions to perform common tasks

- Global variables that are necessary but not directly related to any of classes.

☞ **Key Point:** The basic steps in writing an object-oriented program are (1) visualizing the program, (2) identifying the objects and classes, (3) identifying data members for each class, (4) identifying the functions (class member or global), and (5) implementing the functions.

☞ **Key Point:** Object-oriented programs may still need data and cod that are not part of any class. However, we try to avoid them.

9.2.1 Identifying Classes and Objects

As in any other design process, and just like variables and functions, there is no unique solution for the design of an object-oriented program. As such, there is no standard process to come up with the classes and objects used in the program. Different programmers may organize the program differently while achieving the same result in terms of functionality. This may sound scary to beginner OO programmers. But there are general guidelines that help us take the first step, which is identifying the classes and objects.

In addition to common notes such as integrating related data and code, information hiding, and reusability, the guidelines for identifying classes include considering three types of information:

- Real-world entities that are represented in the program. Examples are given as follows:

  - A program that controls a library can have a class called User and one called Book to encapsulate all the data and code about users and books.

  - A program that manages a school can have classes such as Student, Course, and Classroom.

  - A program that processes banking information can have classes such as Customer, ATM, and Branch.

- Important groups of related data (variables). Examples are given as follows:

  - A banking program can have classes such as transaction, account, and statement. Note that some of these correspond to real-world physical entities, and some are "virtual" and purely information.

  - A health and fitness app can have classes for Food, activity, and demographic information.

  - A game can have classes for Puzzles, Abilities, Game Objects, and User Profiles.

- Important groups of related code (functions). Examples are given as follows:

  - A program like Photoshop that performs various operations on input images can have classes such as Filter, Transformation, and Brush.

  - An autonomous vehicle's software can include classes such as Vision, Navigation, and Cruise Control.

  - An Integrated Development Environment (IDE) software can include classes such as User Interface (UI), Compiler, Debugger, and Optimizer.

☞ **Key Point:** To identify required classes, look for real-world entities, collections of related data, and collections of related operations.

Note that the above three types of classes overlap in many cases. For example, a typical class like User, which is commonly seen in many programs, is both representing a real-world entity and a collection of data. You can also notice that classes such as User and UI are shared by many programs and, as such, are used frequently. Reviewing and defining more examples provide you with a set of frequently used classes that can be a good starting point for the design of many object-oriented programs.

Another point that is worth mentioning is that many programs include objects that consist of other objects. This is called **Object Aggregation** and means there will be classes that have members of other class types. One of the most common examples is an all-encompassing class that represents the whole software system and includes other classes as parts. Such a class is commonly called Application, System, App, or Game, and we saw it as a user-defined type in SP examples of previous chapters.

Exhibit 9.1 visualizes an object-oriented library program I discussed earlier. It includes the User and Book classes plus an all-encompassing Library class that has arrays of the other two classes, and a single User who is admin. Assuming that the array sizes are N and M, for User and Book, respectively, this program has

- Three classes: Library, User, and Book

- Up to M + N + 2 objects: One instance of Library, one instance of User (for admin), up to M User objects for patrons, and up to N Book objects for library items.

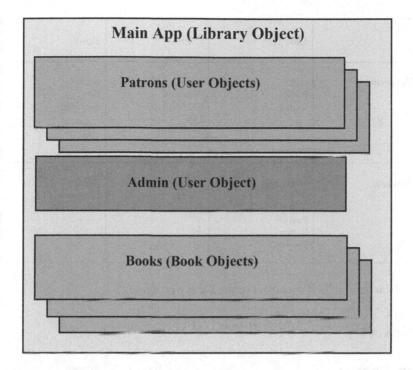

EXHIBIT 9.1   OOP program. Object-oriented library.

☑ **Practice Task:** Identify the required classes, and their members, for the following systems:

- Word Processor such as MS-Word
- Spreadsheet such as MS-Excel
- Drawing Program such as Windows Paint.

Hint: You don't need to have a complete list. Think about major items.

9.2.2 Class Development

Once we have identified the classes and their instances (objects) in a program, the next step is to identify data and functions in each class. Exhibit 9.2 shows the classes of Exhibit 9.1 with their members, as described and used in Chapter 8.

Sample Code #7-p shows the object-oriented implementation of this library system in Python. It is based on Sample Code #9-p in Chapter 8. All Python classes have been extended to include their related functions. They have constructors for initialization, which in the case of User and Book receive parameters. Note how the parameters can have default values. Such values allow the program to call the constructor (create an object) without providing any parameter.[5]

---

[5] C++ offers similar feature in addition to multiple versions of any function with different parameters. I will illustrate such functions later in this chapter.

	Library	User	Book
Properties	Users Books Current User	Name Password ID Type Books List Number of Books	Name ID Borrower
Methods	Login Add User Add Book Borrow Return Report	Report Borrow Return	Borrow Return Is Available

EXHIBIT 9.2 Classes and their members in the library program.

Sample Code #7-p

```
1. class User :
2.    def _ _init_ _ (self, _name="", _password="", _id=-1,
      _type=1):
3.          self.name = _name
4.          self.password = _password
5.          self.ID = _id
6.          self.type = _type    #0 admin, 1 patron
7.          self.books = []
8.          self.numBooks = 0
9.    def Report(self) :
10.         print(self.ID)
11.         print(self.name)
12.         print("borrowed books:")
13.         for b in range(self.numBooks) :
14.             bid = self.books[b]
15.             print(bid)
16.    def Borrow(self,bid) :
```

```
17.                  self.books.append(bid)
18.                  self.numBooks += 1
19.     def Return(self,bid) :
20.                  self.books.remove(bid)
21.                  self.numBooks -= 1
22.

23. class Book :
24.     def __ init __ (self, name="", id=-1):
25.                  self.name = name
26.                  self.ID = id
27.                  self.borrower = -1
28.     def Borrow(self, uid) :
29.                  if self.borrower == -1 :
30.                      self.borrower = uid
31.     def Return(self) :
32.                  self.borrower = -1
33.     def Available(self) :
34.                  if self.borrower == -1 :
35.                      print("available")
36.                  else :
37.                      print("unavailable")
38.

39. class Library :
40.     users = []   #properties can be defined here too
41.     books = []
42.     currentUser = -1 #no login yet
43.     def __ init __ (self):
44.                  newUser = User("Admin", "1234", 0, 0)
45.                  self.users.append(newUser)
```

```
46.              self.numUsers = 1
47.              self.numBooks = 0
48.              self.Login()
49.     def Login(self) :
50.              self.currentUser = -1
51.              while self.currentUser == -1 :
52.                  uid = int(input("enter user ID: "))
53.                  if uid < self.numUsers :
54.                      password = input("enter password: ")
55.                      if password == self.users[uid].password :
56.                          self.currentUser = uid
57.                      else :
58.                          print("incorrect password")
59.                  else :
60.                      print("invalid ID")
61.     def AddUser(self) :
62.              newUser = User()
63.              newUser.ID = self.numUsers
64.              newUser.name = input("enter new name: ")
65.              newUser.password = input("enter new password: ")
66.              self.users.append(newUser)
67.              self.numUsers += 1
68.     def AddBook(self) :
69.              newBook = Book()
70.              newBook.ID = lib.numBooks
71.              newBook.name = input("enter book name: ")
72.              self.books.append(newBook)
73.              self.numBooks += 1
74.     def Borrow(self) :
```

```
75.            bid = int(input("enter book ID: "))
76.            if bid < lib.numBooks :
77.                self.books[bid].Borrow(self.currentUser)
78.                self.users[self.currentUser].Borrow(bid)
79.    def Return(lib) :
80.            bid = int(input("enter book ID: "))
81.            uid = lib.books[bid].borrower
82.            if bid < lib.numBooks and uid != -1 :
83.                lib.books[bid].Return()
84.                lib.users[uid].Return(bid)
85.    def Report(self) :
86.            self.users[self.currentUser].Report()
87.
88. def ShowMenu() :
89.        print("Choose one of the following commands:")
90.        print("  1-Borrow Book")
91.        print("  2-Return Book")
92.        print("  3-Report")
93.        if lib.users[lib.currentUser].type == 0 :
94.            print("  4-Add User")
95.            print("  5-Add Book")
96.        print("  6-Switch User")
97.        print("  0-Exit")
98.
99. def GetCommand() :
100.        ShowMenu()
101.        c = int(input("enter a command: "))
102.        return c
103.
```

```
104. def ProcessCommand(lib, c) :
105.        if c == 0 :
106.            print("bye")
107.        elif c == 1 :
108.            lib.Borrow()
109.        elif c == 2 :
110.            lib.Return()
111.        elif c == 3 :
112.            lib.Report()
113.        elif c == 4 :
114.            lib.AddUser()
115.        elif c == 5 :
116.            lib.AddBook()
117.        elif c == 6 :
118.            lib.Login()
119.        else:
120.            print("invalid command")
121.
122. def MainLoop(lib) :
123.        c = -1
124.        while c != 0 :
125.            c = GetCommand()
126.            ProcessCommand(lib, c)
127.
128. lib = Library()
129. MainLoop(lib)
```

As I mentioned earlier, UI is an important part of many programs. Still, it may be overlooked easily as programmers may not realize its role at the early stages. For example, Exhibits 9.1 and 9.2 only show the main classes that represent the core functionality of a library. But the Sample Code #7-p illustrates that there are other tasks to be done that are not directly part of the core classes, and UI is one of them.

In our library program, the UI tasks are handled through four functions that are currently outside all classes:

- `MainLoop()`

- `ProcessCommand()`

- `GetCommand()`

- `ShowMenu()`

These functions are not part of the `Library` because the same exact library system can be designed and implemented using a completely different UI. So, for reusability, it may not make sense to include UI functions in the `Library` class. That leaves us with two options:

- Keeping the UI functions outside all classes, which works but is not quite object-oriented and loses the advantage of encapsulation and modularization.

- Creating a new class for UI, which is the better design as it creates a module for UI which can be managed and reused more efficiently

Apart from the UI, Sample Code #7-p has two lines of code that start the program (lines 128 and 129). These are necessary in Python as the program cannot start from a function or class, and at least one stand-alone line of code is needed. Similarly, in C++, you need to have a `main()` function that is not part of any class. In Java and C#, on the other hand, the program can start from a member function of a class. So, programs in these languages can be fully object-oriented, where everything is a member of a class.

☑ **Practice Task:** Revise Sample Code #7-p and add a UI class.
Hint: Library can be a member of UI or not. The program starts by creating UI and calling `MainLoop()` which is a member of UI.

9.2.3 Constructors and the Order of Execution

Computer programs are executed in a sequential way by default. The sequential execution is fairly clear in SP as the program starts from a particular point and moves instruction by instruction. Iterations, selections, and function calls, as we have seen in many of our examples, affect this sequential execution. Events such as keyboard input, mouse click, and timers also affect the sequential execution of the program. Object-Oriented programs, while made of various active objects, still follow similar sequential execution, but the presence of objects allows more possibility of actions initiated by different parts of the program. That is what makes OOP more suitable for programs consisting of interacting elements. In the remaining examples of the book, I will demonstrate many cases where objects initiate actions in the program in response to other objects or external events. But for now, it is important to note the effect of constructors in program execution.

Constructors are member functions of a class that are called automatically every time an instance of that class (an object) is created. In Sample Code #7-p, the program starts on line 128 where we create an object of type Library. Every time an object is created, the following actions are performed:

- All data members of that object are created. In this case, the members are two empty arrays and an integer number.

- The constructor of the class is called. This will create one User object, which is then added to the users array.

  - Note that creating the first User will cause the constructor o User class to be called.

The timeline of actions for the program is as follows:

- Line 128 defines a new object.

  - Library data is created for the new instance of Library.

  - Library constructor is called for the new instance.

    - A new User is defined.

      - User data is created.

      - User constructor is called.

    - The new User is added to the array.

    - The rest of Library constructor is executed.

- Line 129 calls the MainLoop() function.

It is important to pay attention to this order as there are many "hidden" constructors, which may be called multiple times (for example, when we create an array of objects). The key to understanding the order of execution is that every time we create an object, its data will be created first, and then its constructor will be called.

☞ **Key Point:** When we create an object, the data members are created and then the constructor is called. If any data member is itself an object, its constructor will be called when that member is being created. The order of constructor calls is important.

☑ **Practice Task:** If a Library class has 20 members of type User and 100 Books, determine the order of constructor calls.

✋ **Reflective Questions:** If an object includes a member which itself is another object, why should the constructor of the member object be called first? Does the order mentioned in "Key Point" above make sense to you?

9.2.4 Class View in Visual Studio

Visual Studio allows you to see your project in different ways. The default way is to use Solution Explorer, which shows all the files. The main advantage of using Solution Explorer is that it provides a view to your project that corresponds to the file and folder structure you have. When we are working with multiple files and want to switch between them, this is very useful. But there are times that we want to focus on modules (such as functions and classes) and don't particularly care (or know) in which file they are located. When you prefer to browse through your project's modules (functions and classes), then you can use the **Class View** in Visual Studio, as shown in Exhibit 9.3.

Class View list all classes and global functions. Double-clicking on a class will take you to the class definition (usually the header file), while double-clicking on a member function will show the implementation of that function. All your global functions that don't belong to any class are listed together, including `main()`.

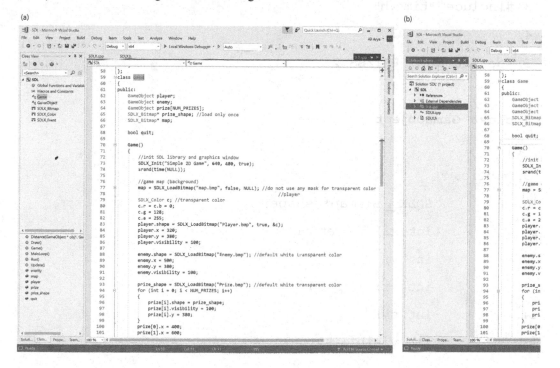

EXHIBIT 9.3   Class View vs. Solution Explorer.

9.3 OBJECT-ORIENTED GAMES

In previous sections, I introduced OOP as a way of designing programs that is most suitable for cases with groups of related data and code, and also interacting entities as opposed to sequential operations. Games are good examples of such interactive programs where multiple modules can perform independent actions and also interact with each other.[6]Examples of these interacting entities in games are visible items such as player characters, enemies, environments, and also behind-the-scene modules such as gameplay manager, renderer, physics engine, and networking. Using event handlers such as those we saw in Chapters 6 and 7 (for example, timer), objects can initiate their own actions at certain times or in response to various external events, independent of each other.

Sample Code #8-c demonstrates a simple OO game based on our previous examples. For the sake of simplicity, it does not include features such as scrolling or gravity. The player moves in two directions and picks up prizes. Hitting an enemy will end the game.

I will discuss two main classes of this program (GameObject and Game) in the next two sections, and then, we wrap up this chapter with a brief discussion on how to structure C++ files when creating multiple classes.

Sample Code #8-c

```
1. #include "SDLX.h"

2. #include "stdlib.h"

3. #include "time.h"

4.

5. #define NUM _ PRIZES 2

6.

7. class GameObject

8. {

9. public:

10.        SDLX _ Bitmap* shape;

11.        int x; //location

12.        int y;

13.        int sx; //speed

14.        int sy;

15.        int visibility;
```

---

[6] This can include interacting with user which is what we commonly consider as an "interactive" program or system.

```
16.
17.          GameObject()
18.          {
19.              x = y = sx = sy = 0;
20.              visibility = 100;
21.          }
22.          void Draw()
23.          {
24.              if (visibility != 0)
25.                  SDLX _ DrawBitmap(shape, x , y);
26.          }
27.          void SetSpeedX()
28.          {
29.              sx = rand() % 7 - 3;
30.          }
31.          void SetSpeedX(int s)
32.          {
33.              if (s > 5)
34.                  sx = 5;
35.              else if (s < -5)
36.                  sx = -5;
37.              else
38.                  sx = s;
39.          }
40.          void SetSpeedY()
41.          {
42.              sy = rand() % 7 - 3;
43.          }
44.          void SetSpeedY(int s)
```

```
45.              {
46.                  if (s > 5)
47.                      sy = 5;
48.                  else if (s < -5)
49.                      sy = -5;
50.                  else
51.                      sy = s;
52.              }
53.          void Move()
54.          {
55.              x += sx;
56.              y += sy;
57.          }
58.      };
59. class Game
60.  {
61.  public:
62.          GameObject player;
63.          GameObject enemy;
64.          GameObject prize[NUM _ PRIZES];
65.          SDLX _ Bitmap* prize _ shape; //load only once
66.          SDLX _ Bitmap* map;
67.
68.          bool quit;
69.
70.          Game()
71.          {
72.              //init SDL library and graphics window
73.              SDLX _ Init("Simple 2D Game", 640, 480);
```

```
74.              srand(time(NULL));
75.
76.              //game map (background)
77.              map = SDLX _ LoadBitmap("map.bmp", false, NULL);
78.              //player
79.              SDLX _ Color c; //transparent color
80.              c.r = c.b = 0;
81.              c.g = 128;
82.              c.a = 255;
83.              player.shape = SDLX _ LoadBitmap("Player.bmp",
     true, &c);
84.              player.x = 320;
85.              player.y = 380;
86.              player.visibility = 100;
87.              //enemy
88.              enemy.shape = SDLX _ LoadBitmap("Enemy.bmp");
89.              enemy.x = 500;
90.              enemy.y = 380;
91.              enemy.visibility = 100;
92.              //prize
93.              prize _ shape = SDLX _ LoadBitmap("Prize.bmp");
94.              for (int i = 0; i < NUM _ PRIZES; i++)
95.              {
96.                  prize[i].shape = prize _ shape;
97.                  prize[i].visibility = 100;
98.                  prize[i].y = 380;
99.               }
100.             prize[0].x = 400;
101.             prize[1].x = 600;
```

```
102.
103.            quit = false;
104.    }
105.
106.    void Update()
107.    {
108.            SDLX_Event e;
109.            //keyboard events
110.            if (SDLX_PollEvent(&e))
111.            {
112.                if (e.type == SDL_KEYDOWN)
113.                {
114.                    //press escape to end the game
115.                    if (e.keycode == SDLK_ESCAPE)
116.                        quit = true;
117.                    //update player
118.                    if (e.keycode == SDLK_RIGHT)
119.                        player.SetSpeedX(1);
120.                    if (e.keycode == SDLK_LEFT)
121.                        player.SetSpeedX(-1);
122.                }
123.                if (e.type == SDL_KEYUP)
124.                {
125.                    if (e.keycode == SDLK_RIGHT ||
    e.keycode == SDLK_LEFT)
126.                        player.SetSpeedX(0);
127.                }
128.            }
129.            //process player move
```

```
130.              player.Move();

131.

132.              //update enemy

133.              enemy.SetSpeedX();

134.              enemy.Move();

135.

136.              //collision detection

137.              if (Distance(&player, &enemy) < 5)

138.              {

139.                   quit = true;

140.              }

141.              for (int i = 0; i < NUM_PRIZES; i++)

142.              {

143.                   if  ((prize[i].visibility   !=   0)   &&
       Distance(&player, &prize[i]) < 5)

144.                   {

145.                        prize[i].visibility = 0;

146.                   }

147.              }

148.

149.     }

150.     void Draw()

151.     {

152.              SDLX_DrawBitmap(map, 0, 0);

153.              //Draw the objects (in order: background to
       foreground)

154.              for (int i = 0; i< NUM_PRIZES; i++)

155.                   prize[i].Draw();

156.              enemy.Draw();

157.              player.Draw();
```

```
158.
159.            //Update the screen
160.            SDLX _ Render();
161.      }
162.      void MainLoop()
163.      {
164.            // Main Loop
165.            while (!quit)
166.            {
167.                  Update();
168.                  Draw();
169.                  SDLX _ Delay(1);
170.            }
171.      }
172.      void RunGame()
173.      {
174.            Init();
175.            MainLoop();
176.            SDLX _ End();
177.      }
178.
179.      float Distance(GameObject* obj1, GameObject* obj2)
180.      {
181.            int dx = obj1->x - obj2->x;
182.            int dy = obj1->y - obj2->y;
183.            float d = sqrt(dx*dx + dy * dy);
184.            return d;
185.      }
186.  };
```

187.

188.   int main(int argc, char *argv[])

189.   {

190.   Game game;

191.   game.RunGame();

192.   return 0;

193.   }

9.3.1 GameObject Class

C++ classes have to define data members (properties) explicitly as variables. For the simple version of GameObject, these variables include shape, location, speed, and visibility. These are the only data we need for the simple game. Once the data members are defined, we identify and implement the functions (methods).

GameObject() is simply initializing all data to default values.

Draw() is very similar to DrawObject() function we had in non-OOP version of the game. The main difference is that it no longer needs to receive a GameObject as a parameter because the function is a class member and has access to all the data belonging to the object.

Move() changes the location of the object based on its speed. The advantage of having this function and letting the object control its movement is that we can later change the way the object moves without changing other parts of the code. This program design follows the principle of Information Hiding and modular design since the module is taking care of its internal tasks.

SetSpeedX() and SetSpeedY() are used to give value to speed data. Each of these has two versions: one with no parameter that randomizes the speed and one with a value that is used directly. As mentioned before, C++ allows multiple versions of a function with different parameters. This feature is called **Function Overloading**. Since the speed values can be negative, we no longer need a direction variable. Movement along two axes (X and Y) can be performed independently. This is an alternative solution to having a direction variable which may be more suitable in some cases.

☞ **Key Point:** Function Overloading in C++ allows multiple functions with the same name but different parameters. This feature is helpful to have the same operation performed ion different data or with different input types.

Note that any part of the program that called these SetSpeed functions could easily access the speed data and set it directly. For example, in line 119, when the user presses the Right key, we have player.SetSpeedX(1). Alternatively, we could say player.sx=1, which would seem simpler and wouldn't need a new member function. The advantages of having a Set function for data members of a class are

- The class function can check if the given value is acceptable. In this case, we are limiting the speed range.

- The class function can perform related tasks if the value of a data member has changed; for example, we can change a dependent data or report to the user.

Later, when I discuss private data members, we will see that for such members, a public **Set** function is required to change their values from any part of the program outside the class. Also, such private data will need a **Get** function to read its value from outside the class.

☞ **Key Point:** Get/Set functions can be used to read and write the value of a class member. They are required if the data is not public.

✋ **Reflective Question:** Do you see the value of Get/Set functions?
✋ **Reflective Question:** There are two possible ways to define the speed and direction of movement. We can have a direction variable, which can have different values for different directions (such as be top, bottom, left, and right), and a speed variable which is a positive value in the specified direction. Alternatively, we can have a speed in X direction (positive or negative) and another in Y direction. Do you see any advantage in either method? How many different directions can we have in each case?

9.3.2 Game Class

The Game class is an all-encompassing entity that holds all game objects and implements the game logic. In this example, it has the player, enemy, prizes, background (game map), and a variable that identifies if the game has to end (bool quit). Similar to GameObject, the functions for the Game class are the same as functions in the non-OO version of the game program.

Game() is the constructor and replaces the Init() function. The code is almost identical to what we had in the non-OO version except that there is no reference to a game, as all data is now owned by the object itself. Note that we could have the constructor in addition to an Init() function. The initialization tasks that have to happen as soon as the object is created will be in the constructor, and some tasks that need to happen later can be in another member function to be called when needed.

Update(), Draw(), MainLoop(), and Run() are also identical to the global functions we had in the non-OO version.

Distance() can be a member of the Game class or remain a global function. If it is a global function, then it has to be defined or declared before being used in Game.

Remember that function declarations (or prototypes) are the description of a function that includes return type, name, and parameters but without the body (implementation). We used them in header files when we had multiple source files or were using libraries. To achieve better manageability and reusability, it is recommended to implement each class in a different file. This is particularly important when the program grows in size or is being developed by multiple programmers. In the next section, I review the most common and convenient way to structure your files when doing OOP.

9.3.3 Multiple Source Files

It is common and recommended practice in OOP to have separate source files for each class. This practice makes the classes more manageable and reusable. When programming in Python and Javascript, having multiple source files is fairly easy. If you have a class defined in Class1.py, you can simply add the following line in any other source file that needs to use that class. This is what we have been doing in our Python samples that use a library or multiple files (See Chapter 6, Section 2).

```
from Class1 import *
```

When adding Javascript code to an HTML file, we can also use multiple source files using the following line:

```
<script src="Class1.js">
```

In C++, having multiple source files (and libraries) requires the use of header files. Recall that header files usually include #define statements, user-defined types, and declaration (prototype) for functions implanted in other sources. For example, if multiple source files in our program are using the same user-defined type (e.g., struct Student), we can have the following header file:

```
//Student.h

struct Student
{
        int ID,
        float GPA
};
//SomeSoucrceFile.cpp

#include "Student.h"
```

Classes can be used in a similar way, but the only difficulty is that a class definition, as we have done so far, includes both data and code. C++ allows function prototypes to be repeated as many times as we want, but function definitions (implementation) can only be done once in the whole program. That means if we have a full class and include it in multiple source files, we will get a compiler error. To avoid this problem, C++ allows defining classes in an alternative form. Sample Code #9-c shows this form for GameObject.

The alternative form of defining classes has two important features:

- Inside the class, we only have function prototypes (lines 1–18).

- The full implementation of the functions is written outside the class, just like any other function definition. The only difference is that we need to add the class name and the ":::" symbol before the function name. This symbol lets the program know that the function we are defining is not a global function but a member of a class.

☑ **Practice Task:** Remove GameObject:: from the lines 19, 24, and others that define a function. See what errors you get. Do they make sense? Is it clear why you need to have the class name?

Sample Code #9-c

```
1. class GameObject
2. {
3. public:
4.     SDLX _ Bitmap * shape;
5.     int x; //location
6.     int y;
7.     int sx; //speed
8.     int sy;
9.     int visibility;
10.
11.         GameObject();
12.         void Draw();
13.         void SetSpeedX();
14.         void SetSpeedX(int s);
15.         void SetSpeedY();
16.         void SetSpeedY(int s);
17.         void Move();
18. };
19. GameObject::GameObject()
```

```
20. {
21.          x = y = sx = sy = 0;
22.          visibility = 100;
23. }
24. void GameObject::Draw()
25. {
26.          if (visibility != 0)
27.                  SDLX _ DrawBitmap(shape, x, y);
28. }
29. void GameObject::SetSpeedX()
30. {
31.          sx = rand() % 7 - 3;
32. }
33. void GameObject::SetSpeedX(int s)
34. {
35.          if (s > 5)
36.                  sx = 5;
37.          else if (s < -5)
38.                  sx = -5;
39.          else
40.                  sx = s;
41. }
42. void GameObject::SetSpeedY()
43. {
44.          sy = rand() % 7 - 3;
45. }
46. void GameObject::SetSpeedY(int s)
47. {
48.          if (s > 5)
```

```
49.                    sy = 5;
50.            else if (s < -5)
51.                    sy = -5;
52.            else
53.                    sy = s;
54. }
55. void GameObject::Move()
56. {
57.            x += sx;
58.            y += sy;
59. }
```

Separating function implementation from class definition does not mean they have to be in separate files. But having multiple source files is the main advantage of this ability. Organizing C++ programs with separate source files for each class can be achieved through the following steps:

1. For each class, create a Header and a CPP file. Use Add New Item command in Visual Studio to add these empty files to the project. Name them the same as the class.

   • Alternatively, you can Add New Class command, which automatically creates two files for you. This is the default Add command when using Class View.

2. For each class, write the class definition with function prototypes in the Header file.

3. For each class, write the function implementations in the CPP file.

4. If you have global functions used by your classes (such as helper functions), put all of them in a new file (I call it Globals.cpp), then create a header file with prototypes of all global functions and call it Globals.h.

5. Create a new header file with any name (I call it Main.h), and add the following items to it (in this order):

   • #include for all your standard C/C++ libraries such as stdio and stdlib

   • #include for all non-standard libraries such as SDL

   • All the #define statements you have in your program

   • #include Globals.h

   • #include for all the class headers in the order of dependency. For example, Game class uses GameObject, so it has to be included later.

6. Create a Main.cpp where you have your `main()` function and any other global function used by `main()` and not used by classes.

   - If you have an existing file, you can rename it to Main.cpp or any other reasonable name such as ProjectX.cpp

7. Include Main.h file in all your CPP files.

8. Every time you create a new class, make sure you have the H/CPP pair and add the header to Main.h

9. Every time you add a new global function that classes need, make sure Globals.cpp and Globals.h are updated.

This process is not the only way to organize your C++ programs. It may not be the most efficient either because it includes everything in all source files, but in my experience, it is the most convenient and least confusing, so the best option for beginners. A few quick notes about this process:

- C/C++ compiler gives error when something is defined multiple times. When including header files, it is possible to include them more than once. For example, you may have GameObject.h included in GameObject.cpp and Main.h, and then include Main.h in GameObject.cpp. This will result in GameObject.h being included twice. To avoid this situation, you can use a combination of `#define` and `#ifdef`, but an easier way is to use `#pragma once` at the start of any header file. This command will make sure the header file is included only once.

- The names Main.h, Main.cpp, Globals.h, and Globals.cpp are arbitrary. Feel free to name your files as you find fit, but use meaningful names.

- You may divide your global functions into multiple source files too. Group them based on related functionality and have a separate header file for each of these source files.

☞ **Key Point:** A common and straight-forward method for having separate files for your classes is to define the class in a header file and the function implementations in a cpp file. Include all your header files in a single new header file, and include that in all your cpp files. Pay attention to the order of header files when you include them.

HIGHLIGHTS

- ADT is a description of a group of related data and the operations that are performed on them. ADT can be implemented in different forms.

- Encapsulation combines code and data into a single module. The result is an Object.

- A class is a user-defined type with encapsulation. An object is a data item (variable) based on a class type.

- C structures included only data. C++ classes can have both data and code. C++ also uses structures, but they are similar to classes with the exception that their members are public by default, so the keyword public is not necessary in C++ structures. This is for compatibility with C.

- Constructor is a special function in a class that is automatically called when an object is created. Constructors usually initialize the object. In C++, constructors can be easily identified as they have the same name as the class and no type/return.

- In Python, constructors are identified with the name _ _ init _ _ (). In Javascript, they are called constructor.

- The basic steps in writing an object-oriented program are (1) visualizing the program, (2) identifying the objects and classes, (3) identifying data members for each class, (4) identifying the functions (class member or global), and (5) implementing the functions.

- To identify required classes, look for real-world entities, collections of related data, and collections of related operations.

- When we create an object, the data members are created, and then, the constructor is called. If any data member is itself an object, its constructor will be called when that member is being created. The order of constructor calls is important.

- Function Overloading in C++ allows multiple functions with the same name but different parameters. This feature is helpful to have the same operation performed on different data or with different input types.

- Get/Set functions can be used to read and write the value of a class member. They are required if the data is not public.

- A common and straightforward method for having separate files for your classes is to define the class in a header file and the function implementations in a cpp file.

END-OF-CHAPTER NOTES

A. Things I Should Mention

- I discussed OOP after SP, but historically OOP was introduced earlier.

- There are programming languages that have objects but no explicit concept of a class. For example, Javascript allows us to create objects using a constructor function and no class. There is a group of languages called prototype-based that support "classless" OOP.

- The C++ language was developed by Bjarne Stroustrup in 1979. It was initially called "C with Classes."

B. Self-Test Questions

- What is an object?

- What is a class?

- What is the object-oriented design process?

- What does a constructor do?

- How do we identify the potential classes and objects for a program?

- What are the most common classes in a game?

- What is an easy way to structure files in an object-oriented program?

C. Things You Should Do

- Use a game engine like Unity or Unreal, and identify the objects and classes they are using. See how they build on top of the basic modular OO design in this chapter and introduce many other objects.

- Take a look at these books to learn more about the design of programming languages and different programming paradigms and language concepts:

  - *Programming Language Design Concepts*, by David A. Watt

  - *Programming Language Explorations*, by Ray Toal et al.

  - *C++ Programming Language*, by Bjarne Stroustrup

D. Reflect on the Experience of Reading This Chapter

- What did you expect from this chapter before reading it?

- What was it about, and what did you learn?

- What tasks did you perform, and what difficulties did you face?

- How did you feel about the material and tasks presented in this chapter?

- How can you improve your learning experience?

- How do you see this topic in relation to the goal of learning to develop programs?

- Do you understand the need for a class as a blueprint or template for objects?

- Does the suggested design process make sense?

- Can you think of other ways to identify the classes you need in a program?

- Are you comfortable with using source and header files for classes in C++?

CHAPTER 10

Object-Oriented Design

Topics
- Object-Oriented Design Process

At the end of this chapter, you should be able to:
- Design fairly complicated software programs made of different classes
- Use dynamically created objects for increased program efficiency
- Use Destructors to free object resources

OVERVIEW

In Chapter 9, I introduced the concepts of class and object as the basis for a more efficient modularization that brings together data and code. I mentioned that the main advantages of Object-Oriented Programming (OOP), achieved through encapsulation of related data and code, are (1) increased manageability, (2) increased reusability, (3) better information hiding, and (4) suitability for non-sequential programs. While the examples provided in Chapter 9 did briefly review these advantages, they were too simplified to show the real-world role that OOP plays in providing programs with these features. In this chapter, I focus on two examples and show how OOP achieves those advantages when implementing full functionality programs, a game, and a visual effect tool.

At the start of this book, I defined design as a modular approach that decides on entities and how they are related. When designing smaller programs with very few data items and operations, the design is primarily for algorithms, that is operations and how they are ordered. Once the programs grow in size and complexity, the design process becomes more involved with choosing the right modules and organizing them in different ways. We start this chapter with the selection and organization of modules (classes and objects) and show how this is the key to the advantages of OOP. We refer to this modular design as **Object-Oriented Design (OOD)**.

OOD is the primary step in OOP, but it can be used for other purposes. OOD can be implemented without the actual use of classes and objects. For example, many older operating systems or software libraries have an OOD but are implemented in C using structures and related functions, as we saw in Section 9.1. Component Object Model (COM), which is a standard interface system by Microsoft for software components, is implemented for Windows in C and without the use of any C++ classes to maintain compatibility with C programs. OOD can also be used for the design of any modular system and not just software.

The process for designing object-oriented programs was demonstrated briefly in the last chapter. It included identifying the classes and objects and then following the data-centered approach by identifying main data items and operations for each class. The process ended with adding any extra code that is needed. In this chapter, I follow this process and explain it in more detail. While there is no standard way to identify classes, objects, properties, and methods, it is crucial to use guidelines and review examples to develop the skills for designing efficient object-oriented programs.

I will also discuss two important subjects when working with classes: access control for class members and dynamic objects. We saw that in C++, we need to define if the class members are public or not. Non-public members are essential in achieving information hiding, and we will see how to use them in this chapter. Earlier in the book, when I introduced pointers in C/C++, I stated their advantages for passing function parameters by reference and also dynamic memory allocations. In this chapter, we are going to revisit that as well, see how we can allocate blocks of data dynamically, and then extend that idea to creating objects dynamically, which is a feature frequently used in OOD.

10.1 SOFTWARE DESIGN WITH CLASSES

10.1.1 OOD Process for a Game

A typical computer game is made of multiple modules in charge of various tasks, such as rendering the implementing the main logic of the game (gameplay), advancing the story, controlling the mechanics, interacting with the user, rendering the scenes, managing physics-based movements, and connecting to a network. While some of these tasks are specific to a given game, many of them are generic. For example, the same rendering module can be used for many games regardless of what is the gameplay as they all involve drawing similar objects and maps.

The modular and reusable nature of computer game software has resulted in the notion of **Game Engines**, which are a collection of such reusable modules that programmers can assemble and modify to create new games. These engines commonly have tools to simplify the process, such as level editors to easily create new game environments and APIs to access code modules. Unity3D, Unreal, CryEngine, and GameMaker are some of the most popular examples of these game engines. They are all made of various modules that combine related data and code.

The clear focus on manageability and reusability, independent and interacting modules, and combination of data and code in each module makes games and game engines perfect candidates for OOP and OOD. While the design and development of a complete game engine are beyond the scope of this book, here I review the process for a simple game made of the most common modules. This game and its modules can be easily reused to develop new games.

As discussed in the Introduction, the software development process starts by identifying the requirements of the program; what it needs to do. In the case of games, this is usually described through what is called the **Game Design Document (GDD)**. This document is the main source of information for any game development project and involves the following information:

- Basic gameplay (goals, rules, challenges, and actions)

- Story, characters, and setting

- Art style and visual look

- User interactions

- Platform specific features such as compatibility with certain consoles

- Software-specific requirements may also be included to require certain software design features. Technically, this is not part of GDD as game design does not have to specify how the software is organized, but it can be included.

☞ **Key Point:** GDD is the description of how a game should work and look.

GDD provides input to both artists and software developers in the game production team. For the software group, the GDD is equivalent to the **Requirement Document** in other software projects. With the information about what the software needs to do available, the software team can go through the design process:

- Identify classes and objects

- Identify the properties and methods for each class

- Design the algorithms (starting with the 3HQ method) and implement the methods

- Identify the extra data and code (globals) and implement them.

10.1.2 Requirements

Our main goals in this example are

- Design and development of a series of reusable modules for 2D platformer games

- Design and development of a sample game using the above modules.

Software requirements are generally defined as **functional** (what to do) and **nonfunctional** (how to do it). Functional requirements cover items such as the response to particular inputs, the type of movement, and what will be seen on screen. Non-functional requirements include items such as timing (how fast actions are performed) and quality. In this book, I focus mainly on functional requirements. For a detailed discussion of requirement management and game design, please refer to great books such as *Software Engineering* by Ian Sommerville and *Fundamentals of Game Design* by Ernest Adams.

The main idea of our sample game is a character that moves left and right along different platforms (surfaces at different altitude), jumps and falls between platforms, avoids or destroys enemies/obstacles, and collects prizes. Many variations can be imagined for this basic idea that will lead to different games, but they all have the following common requirements:

10.1.3 Game-Related Requirements

1. Rendering:

    1. All scenes shall be in 2D.

    2. The game shall have side-scrolling with player character always in the middle horizontally.

2. Game Objects:

    1. Each object shall have location and shape.

    2. Each object shall have visibility levels (fully transparent to fully opaque)

    3. Each object shall have different speeds in X and Y direction that can be randomized or be set within an acceptable range.

3. Player Movement:

    1. The Player Character shall be controlled by Left and Right arrow keys.

    2. The Player Character shall not be able to scroll beyond the game world limits.

    3. The Player Character shall be stopped when ground level suddenly changes.

    4. The Player Character shall fall when moving past the edge of a platform or ground level.

    5. The Player Character shall be able to jump by pressing the Space bar.

    6. Jumping and Falling shall follow the laws of physics, such as gravity.

    7. The ground shall be a continuous terrain with multiple step-like levels or independent platforms, as shown in Exhibit 10.1.

EXHIBIT 10.1   Ground levels.

10.1.4 Software-Related Requirements (for Reusability and Manageability)

4. Rendering:

1. A Render object shall be shared among objects to perform all the drawing tasks.

2. A Render object shall allow buffered batch rendering for every frame using a Begin command that draws the background, drawing commands for each object, and End command that draws the render buffer to screen.

5. Physics:

1. A Physics object shall control all movements.

6. Game:

1. A Game object shall manage the main gameplay.

2. The Game object shall implement a main loop with Update and Draw parts.

10.1.5 Specific Requirements for Sample Game

In addition to the above requirements that are common and define the game engine (reusable part), the software will also have the following requirements that are specific to the sample game:

7. The game shall include a player character, three randomly moving enemies, and three prize objects.

8. The player shall have a normal mode and a defense mode.

1. The defense mode shall be activated by pressing the D key.

2. The defense mode shall end automatically after a few seconds.

3. The defense mode shall be visually distinguishable.

9. The game shall end in one of the following situations:

1. The player is in contact with enemies while not in defense mode (Losing case).

2. The player collects all prizes (Winning case).

10. The enemies shall have a random vibration movement.

11. The prizes shall be picked up upon contact and shall disappear.

SIDEBAR: GAME DESIGN DOCUMENT

GDD is the bible of a game project. While just like any other design document, GDD is a dynamic item that can evolve throughout the project, its first and basic version is prepared at the Pre-production phase, after Concept Development, and before Production.

GDD includes everything that is needed to develop the game. The idea is that you can pass it to another team, and they can make the game you had in mind. That may sound a little too much of an expectation but gives you the general idea. It is important to note that GDD is not a software design document; it does not say "how" to make the software for the game. GDD is more of a Requirement Document for the software team (a.k.a. programming team) that describes "what" the software has to do. Based on GDD, the programming team will make a software design and then develop the software.

Different game projects and studios don't follow the same format for GDD, but the most common parts are

- Concept
- Gameplay
  - Goals
  - Rules
  - Challenges (obstacles)
  - Actions (interaction and mechanics).
- Setting
- Story and characters
- Levels and progression
- Visual Design
  - Art style
  - Characters and environment.
- User Interface
  - Menus
  - Views
  - Controls
  - Accessibility.
- Sound and music.
- Special features such as monetization and networking.
- Platform-specific notes.

A good discussion of GDDs can be found here https://www.gamasutra.com/view/feature/3384/the_anatomy_of_a_design_document_.php

10.1.6 Game Software Design

Considering the above requirements, Exhibit 10.2 shows a software design visualization that includes a series of classes for common modules (Engine) and a single class for the specific sample (Game). The Game module is the only one that includes non-reusable parts. The other modules are independent of the specific design of the sample game and are generic.

To identify the classes, we followed the guidelines mentioned in Chapter 9:

- `GameObject` class corresponds to actual entities.

- `Game` class corresponds to the collection of data as it holds all the important data (objects and their information) and also has the functions related to them.

- `Render`, and `Physics` classes correspond to collections of related functions as they don't hold much data on their own and are more operation-related.

While rendering and physics operation could be done in Game class or within each GameObject, the addition of two specific classes for these operations results in a more reusable and manageable design. It implements information hiding because the knowledge of "how to render" or "how to move" is put into a single module that needs to have it. For example, if we intend to draw all objects with half visibility (semi-transparent), the actual objects don't change. So, the process of showing an object as semi-transparent should not affect the objects themselves but the rendering. Having a separate rendering module allows us to make such decisions without changing the rest of our game.

Once the classes have been identified, we need to choose their members, which determine the actual objects created in the program.

10.1.6.1 Render

Sample Code #1-c shows the header file (the list of members) for `Render` class. In charge of all drawing in the game, this class only has two data members that define the amount of scrolling in X and Y directions. In Chapter 8, I discussed scrolling as a transformation

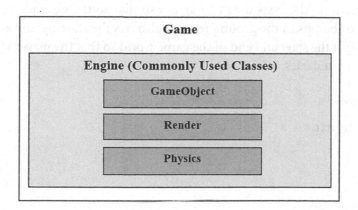

EXHIBIT 10.2   OOD for sample game and game engine.

from word to screen coordinate system. You can think of a virtual camera that follows the player and, as such, sees part of the game world. The scrollX and scrollY values are basically the movements of this camera.

In addition to the data members, Render class has three member functions as described in the requirements.

Sample Code #1-c

```
1. class Render
2. {
3. public:
4.     Render();
5.
6.     int  scrollX; //amount of scrolling on x direction
7.     int  scrollY; //if used
8.
9.     void Begin(SDLX _ Bitmap* map);
10.    void Draw(GameObject* a);
11.    void End();
12.    void Init();
13. };
```

10.1.6.2 Physics

Sample Code #2-c shows the header file for the Physics class. In charge of controlling all movements, this class assumes that the game world has multiple ground levels, as illustrated in Exhibit 10.1. The class uses two arrays similar to the example in Section 7.3 to represent steps or bumps in the ground level. It also has the starting and ending X values to limit scrolling at the start and end of the game world so that the player won't scroll past the game world boundaries.

Sample Code #2-c

```
1. class Physics
2. {
3. public:
4.     Physics();
5.
```

```
6.      void   Move(GameObject* obj);
7.      int    JumpFall(GameObject* obj);
8.      bool   Collision(GameObject* obj1, GameObject* obj2);
9.      float Distance(GameObject* obj1, GameObject* obj2);
10. };
```

10.1.6.3 Game

Sample Code #3-c shows the header file for the Game class. There is nothing new here compared to the previous version as the class collects all the main objects used in the game.

Sample Code #3-c

```
1. class Game
2. {
3. public:
4.      GameObject player;
5.      GameObject enemy;
6.      GameObject prize[NUM _ PRIZES];
7.      SDLX _ Bitmap* prize _ shape; //load only once
8.      SDLX _ Bitmap* map;
9.
10.     bool quit;
11.
12.     Game();
13.     void  Update();
14.     void  Draw();
15.     void  MainLoop();
16.     void  Run();
17. };
```

10.1.7 GameObject, Information Hiding, and Class Access Control

Sample Code #4-c shows the header file for GameObject class. In addition to data related to location and shape, the class has members for speed and also acceleration-related data, including time t and initial position y0 as used in formula $y = y0 + v0 * t + 0.5 * g * t^2$.

The initial speed for acceleration is also needed but is stored in sy. The class includes Draw() and Move() functions, but as we will see later, they can perform the actions themselves or pass the job to the new Physics and Render classes. New variables are added to the GameObject class to determine this choice (lines 15 and 16).

This class has a restriction on the amount of speed the object can have. Similarly, the visibility value has to be between 0 and 100 (%). We can assume any part of the code outside the class that changes these variables will assign valid data. But this assumption is not safe as the other parts of the program may be written by someone else who doesn't know about these restrictions (or by us later after we forget, which happens many times). The principle of Information Hiding states that any module should take care of its own internal working, and as such, it doesn't need to expose its internal information to outside. If the way that speed and visibility are implemented is "internal affairs" of the object, then any code outside the class should not have access to them to avoid any incorrect changes. Such information hiding also allows the implementation of this internal information to change without affecting the rest of the program.

In C++, access control to class members can be defined using the keywords public and private (and protected as we will see in Chapter 11). Members of a class are by default private, which means they cannot be accessed from any code outside the class. Note that if there is no access defined, the members are assumed to be private. This default access is the only difference between C++ classes and C++ structures. If we define a module using the keyword struct, then all members are public by default. In both cases of class and struct, the **Access Modifiers** (the keywords public and private) can be used to change the default.

These modifiers apply to all members from the point they have been used to the next modifier, if any. So, they can be used as many times as we want in a class. In languages such as Java and C#, the same keywords are used to define access to class members, but they have to be used for each member separately.

Private data members cannot be read or written from outside, which means the Game, Physics, and Render classes cannot directly access the GameObject speed and visibility. To provide access, **getter** and **setter** functions are provided. Using public Get() and Set() functions for private data is a very common OO practice to implement information hiding. These functions ensure that

- Outside code does not depend on how private data is implemented.

- Outside code cannot set the private data to an invalid value.

- If something has to be done after the value of a private data is changed (such as recording the changes in a log file), it will be done.

☞ **Key Points**
- Members of a C++ class are by default private, which means no code outside the class can access them.

- The keywords `private` and `public` are access modifiers and control.
- Get/Set member functions can provide access to class private data for outside code.

Sample Code #4-c

```
1. class GameObject
2. {
3. private: // default - no need to write this
4.     int sx;          //speed
5.     int sy;
6.     int visibility;
7.
8. public:
9.     SDLX _ Bitmap * shape;
10.    int x;           //location
11.    int y;
12.    int t;           //time for acceleration.
13.    bool jumping;
14.    int y0;
15.    bool usePhysics;
16.    bool useRender;
17.
18.    GameObject();
19.    void Draw();
20.    void Move();
21.
22.    int  GetSpeedX();
23.    void SetSpeedX();
24.    void SetSpeedX(int s);
25.    int  GetSpeedY();
```

```
26.      void SetSpeedY();

27.      void SetSpeedY(int s);

28.      int  GetVisibility();

29.      void SetVisibility(int v);

30. };
```

10.1.8 Implementing Class Methods

Once the members of each class are identified, the implementation of their methods is fairly straight-forward. Most of these functions have been discussed earlier. Following, I will review a few of the important functions. For full access to the code, please refer to the online companion.

10.1.8.1 GameObject

Sample Code #5-c shows the GameObject functions that have been revised compared to the previous example. Move() and Draw() can perform the actions themselves or pass the job to the new Physics and Render classes. New variables are added to the class to determine this choice, which is, by default, true (lines 6 and 7).

Sample Code #5-c

```
1. //GameObject.cpp

2. GameObject::GameObject()

3. {

4.      x = y = sx = sy = 0;

5.      visibility = 100;

6.      usePhysics = true;

7.      useRender = true;

8.      shape = NULL;

9. }

10. void GameObject::Draw()

11. {

12.      if (useRender)

13.       SDLX _ DrawBitmap(shape, x, y, visibility);

14.      else

15.          render.Draw(this);
```

```
16. }
17. void GameObject::Move()
18. {
19.     if(usePhysics)
20.         physics.Move(this);
21.     else
22.     {
23.         x += sx;
24.         y += sy;
25.     }
26. }
27. void GameObject::SetVisibility(int v)
28. {
29.     if (v >= 0 && v <= 100)
30.     {
31.         visibility = v;
32.     }
33. }
```

10.1.8.2 Game

The Game class is almost identical to the previous example. The function Distance() is moved to Physics class, so there is no implementation for it in Game.cpp.

10.1.8.3 Render

The Sample Code #6-c shows the functions for Render class in Render.cpp. Both Begin(), which draws the background image (map), and Draw(), which is for objects, use scrollX and scrollY values to move the virtual camera. Note that the global library function SDLX _ DrawBitmap() receives the bitmap to draw, the location, and also visibility percentage. If no value is provided for visibility, it is assumed 100.

Sample Code #6-c

```
1. Render::Render()
2. {
3. }
```

```
 4. void Render::Init()
 5. {
 6.     scrollX = 0;
 7.     scrollY = 0;
 8. }
 9. void Render::Begin(SDLX_ Bitmap* map)
10. {
11.     SDLX_ DrawBitmap(map, -scrollX, -scrollY);
12. }
13. void Render::Draw(GameObject* obj)
14. {
15.     if (obj->GetVisibility() != 0)
16.     {
17.         int x = obj->x - scrollX;
18.         int y = obj->y - scrollY;
19.         SDLX_ DrawBitmap(obj->shape, x, y, obj->
    GetVisibility());
20.     }
21. }
22. void Render::End()
23. {
24.     SDLX_ Render();
25. }
```

☑ **Practice Task:** Allow a general visibility variable to control the visibility of all drawn objects regardless of their own visibility setting.

☑ **Practice Task:** Try other effects in Render class.

10.1.8.4 Physics

The Physics class functions are again almost exactly the same as `JumpFall()`, `Move()`, and `Distance()` global functions used in Section 8.4. Note that similar to the code in 8.4, we assume the game has a flat ground with only one bump (level change) at $x = 800$ (lines 14-19).

Sample Code #7-c

```
1. #include "Main.h"

2. Physics::Physics()

3. {

4. }

5. int Physics::JumpFall(GameObject* obj)

6. {

7.      float gravity = 0.2;

8.      int y = obj->y0 + obj->GetSpeedY()*obj->t + gravity * obj-
   >t*obj->t / 2; //in jumping sy is the "initial" velocity

9.      return y;

10. }

11. void Physics::Move(GameObject* obj)

12. {

13.     int newx = obj->x + obj->GetSpeedX();

14.     int ground = 380;

15.     if (obj->x > 800)

16.         ground = 170;

17.     int newground = 380;

18.     if (newx > 800)

19.         newground = 170;

20.     //no move if there is a bump

21.     if (ground > newground && obj->jumping == false)

22.         obj->SetSpeedX(0);

23.     else

24.         {
```

```
25.        obj->x = newx;
26.        ground = newground;
27.    }
28.
29.    //Y movement
30.    if (obj->y < ground && obj->jumping == false)
31.    {
32.        obj->jumping = true;
33.        obj->y0 = obj->y;
34.        obj->t = 0;
35.        obj->SetSpeedY(0);
36.    }
37.    if (obj->jumping)
38.    {
39.        obj->t++;
40.        int newy = JumpFall(obj);
41.        if (newy >= ground)
42.        {
43.            obj->y = ground;
44.            obj->SetSpeedY(0);
45.            obj->jumping = false;
46.        }
47.        else
48.        {
49.            obj->y = newy;
50.        }
51.    }
52. }
53. bool Physics::Collision(GameObject* obj1, GameObject* obj2)
```

```
54. {
55.     float d = Distance(obj1 , obj2);
56.
57.     if (d < COLL _ DISTANCE)
58.         return true;
59.     else
60.         return false;
61. }
62. float Physics::Distance(GameObject* obj1, GameObject* obj2)
63. {
64.     int dx = obj1->x - obj2->x;
65.     int dy = obj1->y - obj2->y;
66.     float d = sqrt(dx*dx + dy * dy);
67.     return d;
68. }
```

☑ **Practice Task:** Use the two-array method of Section 7.3, and modify the Move() function in class Physics to support different numbers and sizes of platforms.

☑ **Practice Task:** Add a GroundLevel() function that returns the ground level at any given X. Use it instead of the code in lines 14-19.

10.2 DYNAMIC OBJECTS

10.2.1 Dynamic Arrays and Objects

In Chapter 7, when I introduced arrays, I showed how C/C++ is more restricted compared to Javascript and Python when it comes to array sizes. While in Javascript and Python, and some other languages, the size of an array can be defined and changed when the program is running, C/C++ arrays have a fixed size that the programmer needs to know and use as a constant number:

```
int data1[6]; or:

#define MAX _ SIZE 10

int data2[MAX _ SIZE];
```

As I mentioned earlier, any information that is constant, and as such known and set at the time of writing the program (**compile-time**), is generally referred to as **Static**. All things that can be defined differently every time that program is executed (**run-time**) are called **Dynamic**. Every time a program creates a variable or loads content from storage to memory (for example, loading an image file), a block of memory will be allocated to hold the information. In cases like creating a single variable, the size of this block of memory is known. For example, `int  x;` will result in allocating a memory block with the size of an integer number, which is usually 4 bytes. Similarly, creating an object of a class type requires a known amount of memory as the class is clearly defined. On the other hand, loading an image file requires a block of memory with an unknown size because we don't always know what the file size is. Creating a block of memory with a size that is only known at run-time is called **Dynamic Memory Allocation**.

All programming languages allow programmers to allocate memory dynamically. For example, the arrays in Javascript and Python are created dynamically. In C/C++, dynamic memory allocation (for arrays and other purposes) is possible through standard functions, particularly `malloc()` defined in stdlib.h:

```
int size;

//some code here to get the size

void* data = malloc(size);
```

The `malloc()` function receives a size in bytes and returns a pointer to the start of a memory block allocated with that size. It is the responsibility of the program to calculate the required size, and the data type is not known to the function, so the pointer is returned a `void*` (unknown type). It is also the responsibility of the program to change the data type before it is used. So, to use the `malloc()` function to create an array with a size given by a variable, the program needs to do two things:

1. Calculate the memory size

2. Change the data type of the pointer.

For example, to create an integer array with a length given by user, we can use the following code:

```
int size;

printf("enter array length:");

scanf_s("%d", &size);

//int info[size]       //we can't do this in C/C++

int* info = (int*)malloc(size * 4); //size in bytes
```

Note that, ideally, once we have the array size as a variable, we would say `int info[size];` something that is possible in languages such as Java and C# with a slightly different syntax. Since in C/C++ using a variable for array size is not possible, we dynamically allocate memory for an array with the alternative line `int* info = (int*)malloc(size * 4);` This line has two important tasks that I mentioned earlier:

1. It calculates the size in terms of bytes, not the number of elements. If the array has 10 elements, then the memory is 10 * 4 bytes because each integer data is 4 bytes. So, in general, for an integer array with a given `size`, the memory is `size*4` bytes.

   a. The above point is based on the assumption that the size of an integer is 4 bytes. In general, we can use the standard `sizeof()` function to get the size of any variable or type.

   b. Technically, the `sizeof()` function returns the size relative to the size of `char` type, which is usually 1 byte but defined by `CHAR _ BIT` (number of bits for a `char`). So, the safe method of using malloc() is

   ```
   int* info = (int*)malloc(size * sizeof(int) * CHAR _ BIT / 8);¹
   ```

2. The return value from `malloc()` is `void*`. The array we are creating (data) is int, so we create an `int*` to hold the address of memory and change the return value to `int*` (**typecasting**).

☞ **Key Points**
- Dynamic memory allocation allows programmers to create blocks of memory of any size.
- Arrays with dynamic sizes can be created using dynamic memory allocation.

☑ **Practice Task:** Write a simple console application with a user-defined type. Then, ask the user to enter a number, and create an array of that user-defined type with the given size.

Hint: Similar to the above example, but instead of `int` you have the other data type.

🖐 **Reflective Questions**
- Do you understand the concept and role of algorithms?
- Can you think of algorithms that you follow in non-programming activities?

---

¹ Assuming that the char type has a size of 1 byte is fairly safe with modern computers, except special devices. Assuming that integer and float have 4 bytes is also reasonably safe.

Once the `int*` info is defined and has proper value, it can be used just like any array we saw before. Recall that if info is a pointer (memory address), then `*info` is the actual data. So, we can access the actual integer numbers using `*info`, `*(info+1)`, `*(info+2)`, etc. This is usually referred to as **pointer notation**.

```
//set all numbers to zero
*info = 0 //same as *(info+0) = 0;
*(info+1) = 0;
*(info+2) = 0
```

Alternatively, we can use the more familiar and easy-to-use **array notation**:

```
info[0] = 0;

info[1] = 0;

info[2] = 0;
```

You notice that pointer and array notations are equivalent, and that is because behind the scene, an array is just a block of memory allocated by the program, and its name is simply the pointer to the start of the memory. The two notations can always be used regardless of how the array was created (statically or dynamically).[2]

> ☞ **Key Point:** An array name is a pointer to the start of the array. So, `array_name[0]` is the same as `*array_name`, `array_name[1]` is the same as `*(array_name+1)`, etc. These two are called array and pointer notations, respectively, and can be used interchangeably.

Sample Code #8-c demonstrates the use of dynamic memory allocation for creating arrays with dynamic size. There are two important things in this sample code:

- You can see that the same method for creating dynamic arrays can be used regardless of the array type (in this example, `int` or `Student`). If the size of each array element is not known, we can always use the standard function `sizeof()` to calculate it. So, `sizeof(int)` will return the number of bytes needed for an `int` data, while `sizeof(Student)` will return the number of bytes for one `Student` object.

- Any data that is created dynamically have to be destroyed; otherwise, it stays in memory. Some programming languages such as Java and C# do this automatically through a process called **Garbage Collection** that determines if a block of memory is not used anymore. In C/C++, the programmer has to call the function `free()` to release the memory.

---

[2] I only mention the pointer notation here but avoid using it due to complexity that it creates for new programmers. Using pointer notation for arrays (and any block of data) does have some advantages but is more suitable for experienced programmers as it is easier to make mistakes when using pointers as opposed to standard arrays.

When using dynamic memory allocation in C/C++, it is very important to release dynamically allocated memory to avoid **memory leak** (unused memory increased over time) and other memory issues. A common case of memory leak happens when a function uses a local variable (pointer) to dynamically allocate memory. At the end of this function, all the local variables will be erased from the computer memory, but the memory addressed by the pointer will not be automatically released (unless there is garbage collection). If this function is being called regularly, the pointer will be created again and again, and each time leaves another block of unusable memory behind.

While all the memory used by a program will be released once the program terminates, it is important to use the memory efficiently as long as the program is running. A rule of thumb is that each module that allocates a memory has to release it when done. Many libraries have functions such as `LoadBitmap()` that return a pointer. They are internally using dynamic memory allocation and usually have a counterpart such as `DestroyBitmap()` that releases the memory when that data is no longer needed. The programmers should make sure to call these functions in pairs: if using `malloc()`, call `free()` to release; if using `LoadBitmap()`, then call `DestroyBitmap()`.

Note that even though the array size can be dynamic using malloc(), it cannot be changed after the array is created. Unlike C/C++, Javascript and Python allow arrays to be created with any size and then expanded as much as needed. Some other languages such as Java and C# allow arrays to be created with the same syntax regardless of the size being static or dynamic (unlike C/C++ that has two methods), but have no way of expanding the array size after creation (similar to C/C++). Expanding the array in C/C++, C#, or Java requires creating a new array and copying the old values to it.

☞ **Key Point:** In C/C++, we cannot use a code like the following line to create an array with dynamic size:

```
array_type array_name[array_size];
```

(Replace `array_type`, `array_name`, and `array_size` with proper values.)
Instead we do

```
array_type* array_name = (array_type*)malloc(array_size *
    sizeof(array_type);
```

For example,

```
Student* s = (Student*)malloc(numStudents * sizeof(Student);
```

Once the array is created, we can use the array notation, regardless of how the array was created.

☑ **Practice Task:** Write a C/C++ program that starts with a small and keeps adding data to it. Once the end is reached, ask the user for a new size and create a new array, copy the data to it, and keep adding more data.

✋ **Reflective Question:**
- Do you feel comfortable with the concept of dynamic arrays? The key is to understand how the `malloc()` function can be used to replace a normal array definition code.

Sample Code #8-c

```
1. #include "stdio.h"

2. #include "stdlib.h"

3. struct Student

4. {

5.      int ID;

6.      float GPA;

7. };

8. void main()

9. {

10.     int n;

11.     printf("how many numbers would you like to enter: ");

12.     scanf_s("%d", &n);

13.

14.     int* data = (int*)malloc(n * sizeof(int));

15.     for (int i = 0; i < n; i++)

16.     {

17.         printf("enter data: ");

18.         scanf_s("%d", &data[i]);

19.     }

20.     for (int i = 0; i < n; i++)

21.     {

22.         printf("%d\n", data[i]);
```

```
23.          printf("%d\n", *(data+i));
24.     }
25.     free(data);
26.
27.     Student* s = (Student*)malloc(n * sizeof(Student));
28.     s[0].ID = 1000;
29.     s[0].GPA = 90;
30.     s[1].ID = 1001;
31.     free(s);
32. }
```

10.2.2 Dynamic Object Creation

Dynamic memory allocation not only allows a program to create an array using a dynamic size but also allows it to define a variable for the array at any time but create the array later. For example, the Physics class in our sample games may need to have a member for an array of platform heights. Not knowing how many "bumps" we have in the game world for the ground level, we create two arrays (for X and Y) with an assumed maximum size. This is not an efficient use of memory as we may end up creating an array of size 100 but only have two levels. Instead, we can have two pointer members:

```
int* xLevels;

int* yLevels;
```

Later, when we know how many bumps we have, we will create the arrays:

```
xLevels = (int*)malloc(numBumps * sizeof(int));

yLevels = (int*)malloc(numBumps * sizeof(int));
```

And then, we can use these two as regular arrays:

```
xLevels[0] = 0;

yLevels[0] = 800;

//etc
```

Similarly, we may need to have a member object (single or array), but we may not want to create the actual objects until necessary. An object (even a single one) can take a significant amount of memory. Imagine a shooter game when the player can fire up to 1000 bullets at a time. Creating an array of 1000 GameObjects for the bullets and keeping them in memory for the duration of the game is not a good idea. The better approach is to create an object when needed and destroy it when done.

In C++, we can create an object dynamically using the keyword new and release it using delete.

```
//define a pointer to object
Student* s1;
//create the object right away or later
s1 = new Student();
s1->ID = 1002;
//destroy the object when done
delete s1;
```

Note that defining the variable s1 in the above code does not create an object. It only creates a pointer. The object is created later with s1 = new Student(); which dynamically creates an object and calls its constructor.

> ☑ **Practice Task:** In the above following code, identify where and how many times the constructor of Student will be called:
>
> ```
> Student s[10];
> Student* ps[10];
> for (int i=0; i < 10; i++)
> ps[i] = new Student();
> ```

Pay attention to the fact that when we define a pointer, it has no valid value (it is not initialized). To make sure we don't use a pointer before proper initialization, it is a good practice to set it to NULL when it is not pointing to any valid data. So, the above code can be changed as follows:

```
//define a pointer to object
Student* s1 = NULL;
//create the object right away or later
s1 = new Student();
s1->ID = 1002;
//destroy the object when done
if(s1!= NULL)
{
```

```
    delete s1;

    s1 = NULL; //no longer valid

}
```

Setting an invalid pointer to NULL and making sure it is not NULL prior using it are good practices to avoid memory access run-time errors that can cause a program to crash. This practice is useful not just for the pointer to objects but any other type of pointer.

10.2.3 Object Destructor

Constructor is a member function that is automatically called when the object is created. This can happen in two cases:

- If the object is created statically, then the constructor is called as soon as the variable is defined: Student s1;
- If the object is created dynamically, then the variable is a pointer, and the constructor is called when the new operator is used.
  - Student* s1;   //pointer
  - s1 = new Student();   //object

On the other hand, an object is destroyed in two possible cases:

- If the object is created statically, it will be destroyed when the variable **scope** ends. The scope is the part of the program when the variable is defined:
  - For a variable defined inside the block of code for a loop or if statement, the scope is that block of code.
  - For a local variable defined in a function, the scope is that function.
  - For a global variable, the scope is the whole program.
  - For a class member, the scope is the lifetime of the class instance.
- If the object is created dynamically, it is destroyed when we call delete or when the program ends, whichever happens first.

The delete keyword releases the memory used statically by the object, meaning that all data members will be erased from the computer memory. But if one of the members is a pointer that has been used to dynamically allocate memory, then that allocated memory will not be released. Examples of such pointers are dynamic arrays and objects created with malloc() and new. To allow an object to release all its dynamically allocated memory (and other system resources such as network and files), there is an optional member function in C++ called **Destructor**.

```
Student::~Student()
{
    //any clean-up code
}
```

A destructor is an optional member function of any class that

- Has the same name as the class but starting with a ~

- Has no type

- Is automatically called when the object is destroyed.

If a class uses pointers to dynamically allocate memory (for arrays, objects, loading files, and other purposes), it is a good practice to have a destructor and free all memory that any instance of the class has created. For example, the class Game in Section 10.1 uses pointers to load bitmaps. Now, you can probably guess that SDLX _ LoadBitmap() function uses dynamic memory allocation to create a block of memory large enough for a bitmap to be loaded. SDLX _ DestroyBitmap() safely releases all the allocated memory and should be called when the Game object is destroyed. In our example, this happens at the end of the program when all memory is released anyway. But the proper practice is to define the Game class in a more reusable and safe way.

```
Game::~Game()
{
    SDLX_DestroyBitmap(map);
    SDLX_DestroyBitmap(player.shape);
    SDLX_DestroyBitmap(enemy.shape);
    SDLX_DestroyBitmap(prize_shape);
}
```

Even though we could destroy the bitmaps for GameObject items inside a destructor for that class, it is a safer practice to release memory by the same object that allocates it. For example, in this case, all prizes share one single bitmap and that is why the Game class loads that bitmap (and all the others). If the destructor of any prize releases the bitmap, the other prizes will no longer have a shape, even if they are not being destroyed. The Game class code is a better place to see if and when the bitmaps need to be erased. On the other hand, if each GameObject loads its own bitmap, then they are independent and can be destroyed in GameObject destructor. This is a software design choice that has to be made considering issues such as reusability, manageability, and information hiding.

☞ **Key Point:** Destructor is member function that is automatically called when the object is destroyed. It can happen when the delete is called or when the variable scope ends.

☑ **Practice Task:** Add a destructor for `GameObject` to destroy shapes.

10.2.4 Particle Systems

In computer graphics, many complex substances and phenomena such as water, smoke, and fire are simulated through the use of arrays of simple objects commonly referred to as **particles**. The basic idea is that while these complex subjects have shapes that are very fluid and difficult to model, they can be thought of as a set of very simple and small objects, as shown in Exhibit 10.3. Such simulation reduces the visual complexity and instead relies on a complicated control mechanism to manage the appearance and movement of the particles. Such a combination of particles and their control mechanism is called a **Particle System**.

Due to the dynamic nature of such fluid subjects, the number of particles can be from a few to thousands or more. As such, creating the maximum number of particles, we may need, can be a significant waste of memory. The better alternative is dynamic object creation.

10.2.5 Smoke Particle System Requirements

Consider the task of adding a smoke visual effect to our sample game of Section 10.1. The requirements for this smoke effect can be defined as follows:

1. The smoke system shall have a source point where smoke emission starts, identified by XY coordinates.

2. The smoke source shall have a speed in X and Y directions.

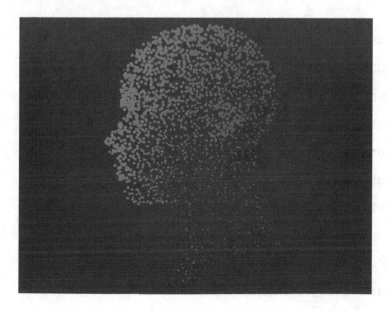

EXHIBIT 10.3   Particle systems. Simple particles can together make a complex shape.

3. The smoke source shall have a smoke particle rate (number of particles created per frame).

4. Smoke particles shall move randomly after creations.

5. Smoke particles shall gradually disappear after creation.

☞ **Key Point:** A particle system is an array of simple objects that together can form complex shapes.

10.2.6 Smoke Particle System Design

Each particle of smoke can be just another GameObject. Following our OOD approach, the ParticleSystem class can be defined and added to the program. Sample Code #9-c shows the header file for this class. The data members are defined directly based on the above requirements. The class has an array for particles, but the array type is not GameObject but GameObject*. So, we are only creating a series of pointers. Since all particles share the same shape, there is no need to load an image file for each particle, and only one image (particle shape, pshape) is used. The class has constructor and destructor, set functions for accessing the private data, and Update and Draw to control and render the particles at every frame.

Sample Code #9-c

```
1. //ParticleSystem.h

2. class ParticleSystem

3. {

4.     GameObject* particles[MAX _ PARTICLES];

5.     SDLX _ Bitmap* pshape;

6.     int sourceX;

7.     int sourceY;

8.     int speedX;

9.     int speedY;

10.     int rate; //particle per frame

11.

12. public:

13.     ParticleSystem();

14.     ParticleSystem(int x, int y, int sx, int sy);
```

```
15.        ~ParticleSystem();
16.        //move source
17.        void Move();
18.        void Move(int x, int y);
19.        //update and draw particles
20.        void Update();
21.        void Draw();
22.        //particle emission rate
23.        void SetRate(float r);   //number of particle per frame
24.        //source speed
25.        void SetSpeedX(int sx);
26.        void SetSpeedY(int sy);
27. };
```

10.2.7 Smoke Particle System Functions

Sample Code #10-c shows the implementation of ParticleSystem functions. The class has two constructors, one with default values and the other with parameters. In both cases, a shared bitmap for all particles is loaded, and the particle pointers are initialized to NULL as no actual particle has been created. The destructor deletes all existing particles (not NULL) and also releases their shared bitmap (line 22).

Particle rate and the source speed and location are private data, so the class has public functions to change them. In case of location (sourceX and sourceY), these two coordinates are changed by one function, Move(), but it has two versions:

- Move() with no parameter moves the particle source using its speed (line 46).
- Move() with parameters moves the particle source to a given position (line 51).

The Draw() function (line 58) is fairly straight-forward as it goes through a loop and draws any particle that is not NULL. The Update() function (line 66) is the most critical part of the particle system that controls how the particles appear, move, and disappear, in order to satisfy the requirements 3, 4, and 5, respectively.

The Update() function is divided into two parts:

- The first part creates new particles. It has two nested loops. The inner loop goes through all particle pointers and finds the first NULL pointer (available particle) and creates a new particle. The outer loop repeats this process for as many as determined by rate. Note that since the particles are constantly generated and destroyed, we cannot be sure that all available particles are at the end of the array or start of it. They can be randomly at any place within the array.

- The second part of the function controls the existing particles. It again loops over all particle pointers in the array, finds the non-NULL ones, and performs the following actions

  - Gives them a random speed

  - Moves them

  - Decreases their visibility (making them gradually disappear)

  - Deletes them once their visibility is zero. When the particle is deleted, the pointer is set to NULL again, so we know the pointer is not valid.

Sample Code #10-c

```
1. #include "Main.h"
2.
3. ParticleSystem::ParticleSystem()
4. {
5.     sourceX = sourceY = speedX = speedY = 0;
6.     for (int i = 0; i < MAX _ PARTICLES; i++)
7.         particles[i] = NULL;
8.     rate = 1;
9.     pshape = SDLX _ LoadBitmap("p.bmp");
10. }
11. ParticleSystem::ParticleSystem(int x, int y, int sx, int sy)
12. {
13.     sourceX = x;
14.     sourceY = y;
15.     speedX = sx;
16.     speedY = sy;
17.     for (int i = 0; i < MAX _ PARTICLES; i++)
```

```
18.            particles[i] = NULL;
19.      rate = 1;
20.      pshape = SDLX _ LoadBitmap("p.bmp");
21. }
22. ParticleSystem::~ParticleSystem()
23. {
24.      for (int i = 0; i < MAX _ PARTICLES; i++)
25.      {
26.           if (particles[i] != NULL)
27.                delete particles[i];
28.      }
29.      SDLX _ DestroyBitmap(pshape);
30. }
31. //particle emission rate (particles per frame)
32. void ParticleSystem::SetRate(float r)
33. {
34.      rate = r;
35. }
36. //source speed
37. void ParticleSystem::SetSpeedX(int sx)
38. {
39.      speedX = sx;
40. }
41. void ParticleSystem::SetSpeedY(int sy)
42. {
43.      speedY = sy;
44. }
45. //move source
```

```
46. void ParticleSystem::Move()
47. {
48.     sourceX += speedX;
49.     sourceY += speedY;
50. }
51. void ParticleSystem::Move(int x, int y)
52. {
53.     sourceX = x;
54.     sourceY = y;
55. }
56.     //update and draw particles
57.     //called once per frame
58. void ParticleSystem::Draw()
59. {
60.     for (int i = 0; i < MAX_PARTICLES; i++)
61.     {
62.         if (particles[i] != NULL)
63.         particles[i]->Draw();
64.     }
65. }
66. void ParticleSystem::Update()
67. {
68.     //create new particles
69.     if (rate >= 1)
70.     {
71.         for (int i = 0; i < rate; i++)
72.         {
73.             for (int i = 0; i < MAX_PARTICLES; i++)
74.             {
```

```
75.              if (particles[i] == NULL)
76.              {
77.                  GameObject* p = new GameObject();
78.                  p->x = sourceX;
79.                  p->y = sourceY;
80.                  p->SetVisibility(100);
81.                  p->shape = pshape;
82.                  p->usePhysics = false;
83.                  particles[i] = p;
84.                  break;
85.              }
86.          }
87.      }
88.  }
89.  for (int i = 0; i < MAX_PARTICLES; i++)
90.  {
91.      if (particles[i] != NULL)
92.      {
93.          particles[i]->SetSpeedX();
94.          particles[i]->SetSpeedY();
95.          particles[i]->Move();
96.          int v = particles[i]->GetVisibility();
97.          v--;
98.          particles[i]->SetVisibility(v);
99.          if (v == 0)
100.         {
101.             delete particles[i];
102.             particles[i] = NULL;
103.         }
```

```
104.                 }
105.       }
106. }
```

> ✋ **Reflective Question:** In the above code, when creating a new particle, we look for the first NULL pointer. Does this make sense to you? Can you think of a simpler way?

10.2.8 Using the Smoke Particle System

The new `ParticleSystem` class can be used in many different forms; we can have one for any `GameObject`, or some of them, if we are designing a game with cars or airplanes and want to simulate the exhaust, or we can have particle sources independent of game objects as some type of environmental effect. To demonstrate, I replace all the game objects with a single particle source that can be moved around as player character, as shown in Exhibit 10.4.

The `Game` class needs to have a new member `ParticleSystem* ps;` which can be public for the sake of simplicity. The constructor of the Game class has to initialize this new member, although this can happen later if we don't want to start with the smoke effect right away.

```
ps = new ParticleSystem(300, 200, 0, 0);
```

Sample Code #11-c shows the new `Update()` and `Draw()` functions for Game class. Pressing the arrow keys set the speed in the appropriate direction, and releasing the keys reset the speed back to zero. After that, the `Move()` and `Update()` for `ParticleSystem`

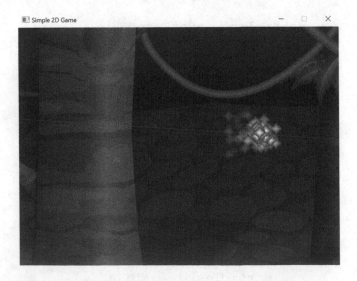

EXHIBIT 10.4   Particle system demo.

have to be called to move the source and control the particles. The Draw() function for Game simply calls the Draw() for ParticleSystem.

Sample Code #11-c

```
1. void Game::Update()
2. {
3.      SDLX _ Event e;
4.      //keyboard events
5.      if (SDLX _ PollEvent(&e))
6.      {
7.          if (e.type == SDL _ KEYDOWN)
8.          {
9.              //press escape to end the game
10.             if (e.keycode == SDLK _ ESCAPE)
11.                 quit = true;
12.             //update particle source
13.             if (e.keycode == SDLK _ RIGHT)
14.                 ps->SetSpeedX(1);
15.             if (e.keycode == SDLK _ LEFT)
16.                 ps->SetSpeedX(-1);
17.             if (e.keycode == SDLK _ UP)
18.                 ps->SetSpeedY(-1);
19.             if (e.keycode == SDLK _ DOWN)
20.                 ps->SetSpeedY(1);
21.         }
22.         if (e.type == SDL _ KEYUP)
23.         {
24.             if (e.keycode == SDLK _ RIGHT || e.keycode == SDLK _ LEFT)
25.                 ps->SetSpeedX(0);
```

```
26.                if (e.keycode == SDLK_UP || e.keycode ==
   SDLK_DOWN)
27.                    ps->SetSpeedY(0);
28.            }
29.        }
30.
31.    ps->Move();
32.    ps->Update();
33.
34. }
35. void Game::Draw()
36. {
37.    render.Begin(map);
38.    ps->Draw();
39.    render.End();
40. }
```

HIGHLIGHTS

- GDD is the description of how a game should work and look.

- Members of a C++ class are by default private, which means no code outside the class can access them.

- The keywords private and public are access modifiers and control.

- Get/Set member functions can provide access to class private data for outside code.

- Dynamic memory allocation allows programmers to create blocks of memory of any size.

- In C/C++, the standard function malloc() performs dynamic memory allocation. It returns a pointe to the start of the allocated memory.

- Arrays with dynamic sizes can be created using dynamic memory allocation.

- An array name is a pointer to the start of the array. So, array_name[0] is the same as *array_name, array_name[1] is the same as *(array_name+1), etc. These two are called array and pointer notations, respectively, and can be used interchangeably.

- Destructor is member function that is automatically called when the object is destroyed. It can happen when the delete is called or when the variable scope ends.

- A particle system is an array of simple objects that together can form complex shapes.

- The key part of a particle system is the logic that moves the particles.

END-OF-CHAPTER NOTES

A. Things I Should Mention

- OOD doesn't need to be used together with OOP. Even if you are using a non-OOP language, you can design your software using OO approach.

- Many visual effects that you see in movies and animations are made using particle systems. Game engines usually have built-in support for various particle systems to create smoke, fire, and water.

B. Self-test Questions

- What is a GDD?

- What are common classes in a game?

- How do we create dynamic objects in C++?

- In the following code, at which line the constructor is called?

  ```
  1. GameObject* player;

  2. …

  3. player = new GameObject();
  ```

C. Things You Should Do

- Design a software using OOD and then implement it in C with no classes.

- Check out these web resources:

  - https://www.gamasutra.com/blogs/LeandroGonzalez/20160726/277928/How_to_Write_a_Game_Design_Document.php

  - http://seriousgamesnet.eu/assets/view/238

  - https://www.khanacademy.org/computing/computer-programming/programming-natural-simulations/programming-particle-systems/a/intro-to-particle-systems

  - https://docs.unity3d.com/Manual/ParticleSystems.html

D. Reflect on the Experience of Reading This Chapter

- What did you expect from this chapter before reading it?

- What was it about, and what did you learn?

- What tasks did you perform, and what difficulties did you face?

- How did you feel about the material and tasks presented in this chapter?

- How can you improve your learning experience?

- How do you see this topic in relation to the goal of learning to develop programs?

- Do you see the relationship between GDD and software design?

- Why and when should we use destructor functions?

- Do you see the need for creating dynamic objects (starting with a NULL pointer and creating the object later)?

PART 5

More about Objects and Classes

Why do programmers always mix up Halloween and Christmas?

Because Oct 31 = Dec 25.

GOAL

In the previous parts of this book, we saw how to define modules of code and data, and also modules that combine both. Object-Oriented Programming (OOP) organizes the program as a series of modules based on encapsulation. It allows increased manageability, reusability, and information hiding, and is more appropriate for software made of interacting elements.

In this part of the book, I introduce the other two aspects of OOP, inheritance, and polymorphism. Inheritance allows a class to be related to another based on sharing and extending the members. Polymorphism follows inheritance and allows objects to behave as any of the related classes. Although I briefly mention how to use inheritance and polymorphism in Python and Javascript, this part of the book is primarily focused on C++, which is more suited for advanced OOP.

More about Objects and Classes

GOAL

CHAPTER 11

Class Hierarchies

> **Topics**
> • Inheritance for Classes

> **At the end of this chapter, you should be able to:**
> • Extend a class using child classes
> • Create a variety of game object classes

OVERVIEW

In the game program of the previous chapter, various types of objects were created based on the same class. Player, Enemy, Prize, Animated Object, and Particle did not exactly have the same role and needed members that were not useful for others. For example, an animated object requires an array of shapes, an enemy may need specific members to chase the player, or a smoke particle may not need to have gravity-related data.

When dealing with objects that have differences, our approach has been to define a `GameObject` class that has all the requirements and then for any object, use only the members that are needed. This approach is good for making reusable classes but is not quite manageable as the classes become increasingly bigger as we define new types of objects and include all their features in a single shared class. An alternative way is to define different classes for each of these object types. While this allows customized and efficient classes, it creates repetition when these classes share many members. Also, it makes it difficult to have classes like `Render` and `Physics` that can deal with all objects in a similar way. The `Render` class, for example, will need to have separate functions to draw any type of object, resulting in loss of reusability. Every time we define a new game object class, the `Render` class has to be modified to support that. This is inconvenient when we are developing our own Render class, and impossible when we are using a game engine without access to its source code.

Object-Oriented Programming (OOP) offers a solution through **Inheritance**, the ability to extend a class by adding new members or modifying existing ones. Through inheritance, we can define new classes that share all members of `GameObject` but have new features. There is no need for repetition or modifying the `GameObject` class, but the new classes can still be treated as `GameObject` if needed. While aggregation was a relationship between classes where one was a member of the other, inheritance defines a new relationship where one (called **derived or child class**) extends the other (called **base or parent class**) and inherits all its members.

The ability to define a class based on an existing one not only provides flexibility in software design, but also allows a new type of programming library called a **Class Library**, a set of existing classes that can be used or extended in a program.

In this chapter, I discuss inheritance and how it works and demonstrate it through the definition of multiple types of GameObject in our sample game. The examples will be discussed using a new graphics library called **OpenFrameworks (OF)**, which is fully object-oriented and designed for C++.

The discussion of inheritance leads us to **Polymorphism**, the ability of an object to act as both parent and child classes. Polymorphism is the subject of our next chapter.

11.1 EXTENDING CLASSES WITH INHERITANCE

11.1.1 C++ Student Class Revisited

In previous chapters, I frequently used a school management program as an example where the fundamental module was a `Student` type defined as structure or class. This module generally had an array of grades plus a GPA as data. Such a representation has quite a few limitations and problems:

1. The number of courses is not fixed for students. A dynamic array can be more helpful in such a case.

2. Grade Point Average (GPA) is not really just a simple average but a weighted average where the credit count of each course is taken into account.

3. Different programs of study apply certain structures to their students, which requires new information and operations. Such programs result in students who share many properties but also differ in some others. For example, an engineering student may have a required internship, while an art student may require to take a studio course with double credit count as regular courses. At the same time, they both share an array of courses with credit counts and a GPA.

The third issue is the key challenge we address in this chapter. It requires defining multiple classes for different programs of study but also dealing with what they share as a base and generic student. Before we work on this challenge and introduce class inheritance as a solution, let's start by addressing the first two issues and restructuring our generic Student class.

A dynamically defined array is what I discussed in the previous chapter. The credits for each course is a new piece of information which, according to our First Rule of Programming require sits own variable. Since we have an array for grades, it makes sense to have an array of similar size for credits. Clearly, since the arrays are defined dynamically, we need another variable to hold the number of courses. We assume the module only needs an ID for personal information. Adding others such as name is pretty straight-forward.

To provide better information hiding, we can also make all the data private and define set/get functions to access them, including setting the ID, calculating the GPA, setting grade and credit for a course, and getting a full report of all data. This results in a new design for the Student class, as shown in Sample Code #1-c.

Sample Code #1-c

```
1. class Student
2. {
3.      int ID;
4.      int GPA;
5.      int* grades;
6.      int* credits;
7.      int numCourses;
8. public:
9.      Student()
10.     {
11.         grades = credits = NULL;
12.         numCOurses = 0;
13.         id = -1;
14.     }
15.     Student(int id)
16.     {
17.         grades = credits = NULL;
18.         numCOurses = 0;
19.         SetID(id);
20.     }
```

```
21.     ~Student()
22.     {
23.       if (grades != NULL)
24.         free(grades);
25.       if (credits != NULL)
26.         free(credits);
27.     }
28.     void SetID(int id)
29.     {
30.       if (id >= 0)
31.         ID = id;
32.     }
33.     int GetID()
34.     {
35.       return ID;
36.     }
37.     void SetNumCourses(int n)
38.     {
39.       if (n > 0)
40.       {
41.         numCourses = n;
42.         grades  = (int*)malloc(n * sizeof(int));
43.         credits = (int*)malloc(n * sizeof(int));
44.         for (int i = 0; i < numCourses; i++)
45.         {
46.           grades[i] = credits[i] = 0;
47.         }
48.       }
49.     }
```

```
50.    int CalculateGPA()

51.    {

52.      GPA = 0;

53.      int totalCredits = 0;

54.      for (int i = 0; i < numCourses; i++)

55.      {

56.          GPA += grades[i] * credits[i];

57.          totalCredits += credits[i];

58.      }

59.      if (totalCredits != 0)

60.          GPA /= totalCredits;

61.      else

62.          GPA = 0;

63.      return GPA;

64.    }

65.    void AddGrade(int i, int g, int c)

66.    {

67.        if (grades!=NULL  &  credits!=NULL  &&  i>=0  &&
       i<numCourses)

68.        {

69.            grades[i] = g;

70.            credits[i] = c;

71.        }

72.    }

73.    void Report()

74.    {

75.      printf("Student %d\n", ID);

76.      for (int i = 0; i < numCourses; i++)

77.      {
```

```
78.              printf("%d %d\n", grades[i], credits[i]);
79.         }
80.      printf("%d\n\n", CalculateGPA());
81.    }
82. };
```

The generic Student class has the following functions:

Student(): Two versions of constructor allow initializing the object with an invalid ID or a given value. They also set the grades and credits array pointers to NULL at the start.

~Student(): The main task for destructor is to free the dynamic arrays if they have been allocated (pointers are not NULL).

SetID(): sets the value of student ID if the given number is above zero. Negative ID values are invalid. That is why we initialize ID with -1 to show it is not set properly yet. Since ID is not public, there is also a GetID() function.

SetNumCourses(): accepts any value greater than zero and allocates memory for the course-related arrays.

CalculateGPA(): performs a weighted average operation using the formula GPA = sum_of (grade * credit) / sum_of (credit). For example, if we have two courses with credit counts of 2 and 3, and grades of 70 and 80, respectively, the average is not (70+80)/2 but (70 * 2+80 * 3)/(2+3). Note that if the total number of credits is zero, then we should not use this formula as it results in a divide-by-zero error. In such cases, we simply set GPA to zero. It is a good practice to always check the divisions for such an error that can cause the program to crash.

AddGrade(): adds both grade and credit count for a course. The course is identified with an index value for the dynamically allocated arrays.

Report(): prints all data for the student.

The main() function or any other part of the program can use this class.

```
Student s1(1001);

s1.SetNumCourses(2);

s1.AddGrade(0, 100, 3);

s1.AddGrade(1, 80, 1);
```

```
s1.AddGrade(2, 70, 1);  //this won't work as we have only two
    courses

s1.Report();
```

Calling `Report()` will show the following lines:

```
Student 1001

100 3

80 1

95
```

🖐 **Reflective Questions:** See how the error checking is done in class methods and how the functions do more than just setting a value.

- Does this demonstrate the significance of private data and set functions? The programmer who uses this class can't just set any value to the data members.
- Does it make sense from information hiding point of view? The programmer who uses this class doesn't need to know how courses are implemented, and the implementation can change in the future without affecting any code that is using the class, as long as they still call the same functions. For example, we may switch to fixed arrays or save the data in a database in future.

11.1.2 Different Yet Similar Classes

Now that we have a new and revised Student class that is suitable for any generic student, let's see how we can deal with the issue of different programs of study. To begin with, imagine that we have two programs that students can enroll in: Science and Art. Science students are fairly generic except that they have a fixed program with 20 courses. Art students similarly have a fixed program size with 15 courses, which usually have higher credit counts. In particular, their first course is a studio with 6 credits (assume typical courses have 3 credits). There are three solutions for including these programs in our software:

Solution #1 does not change anything in the code except adding a program ID (PID) to student data. Everything else will be dealt with manually when entering data. We can initialize PID to -1 in the default constructor, add a new parameter for it in the second constructor, and add a new function `SetPID()` to change the value. This works but is not very reusable and manageable. As we add features to our programs, it will become harder and harder to control them. For example, the `Student` class functions cannot easily limit the number of courses as they are generic.

```
//in Student class

int PID; //set to -1 in constructor

void SetPID(int pid)

{

    if (pid >= 0)

            PID = pid;

}
```

Solution #2 defines multiple classes, one for reach program (ArtStudent and ScienceStudent), instead of a generic class. The classes will look very much like each other (and Student). The differences are

- PID and numCourses are set to fixed values.

- The arrays are created right away based on fixed values.

This solution works but involves a significant repetition of code and also loses the generic class. If we have other modules that don't care about the program and deal with all students the same way, they won't be able to work. As an example, imagine we have another class called Course which has a list of students. Courses are frequently shared between programs or are electives or service courses that any student can take. For such a class, we need to be able to write something like the following code, which is no longer possible.

```
class Course

{

    Student students[50];

    void RegisterStudent(Student* newStudent);

    //rest of the class

};
```

Solution #3 involves extending a class to create a new one, resulting in two classes, a **base** and a new one **derived** from it. In OOP, this feature is referred to as **Inheritance** as it creates a parent–child relationship between two classes where one (**parent**) is the base, and the other (**child**) inherits everything from the base and modifies it if needed. Inheriting the family properties, I am still ARYA, but I am ALI ARYA. This means I have all the generic and common characteristics of any ARYA, but add my own personal flavor to it.

In C++, the inheritance relationship is explicitly added when defining a new class, by adding three elements after the new class name:

- :

- Type of inheritance (almost always public, but can also be protected or private). The keyword `public` shows that all public members will stay public.

- Name of the parent class

☞ **Key Point:** Inheritance is a relationship between a parent (a.k.a. base) class and a child (a.k.a. derived) one where the child receives all members of the parent and can add new members or change existing ones.

So, the following code shows that the new class ScienceStudent is derived from Student and inherits all its members. There is no need to repeat Student members, and the programmer can only deal with added or changed members. As you can see, the constructor calls SetNumCourses() and SetPID() while main() calls SetID() even though they are not defined in the new class. They are being called for the ScienceStudent object because they (and other members) are inherited from the base class.

Sample Code #2-c

```
1. class ScienceStudent : public Student
2. {
3. public:
4.     ScienceStudent()
5.     {
6.         SetPID(0);
7.         SetNumCourses(20);
8.     }
9.     ScienceStudent(int id) : Student(id)
10.    {
11.        SetPID(0);
12.        SetNumCourses(20);
13.    }
14. };
```

```
15. void main()
16. {
17.     ScienceStudent s1;
18.     s1.SetID(5);
19.     ScienceStudent s2(1000);
20. }
```

Note how the constructor of ScienceStudent passes a parameter to Student(). Inheriting all members of the base class means that the new class has to initialize the base class members. This happens by automatically calling the base constructor in the following order:

- Base class members are created.
- Base class constructor is called.
- New members are created.
- New constructor is called.

☞ **Key Point:** When creating instances of a child class, first the base class members will be created and base class constructor will be called.

If the instance of the class (object) is being created with a parameter (line 19, above), then that parameter can be passed to the base class constructor if needed. In the Sample Code #2-c, line 19, an ID is being provided to the new object. Setting ID happens in the base constructor, so the parameter is passed, as shown in line 9. If the child class constructor doesn't pass the parameter, then the default constructor for the base class will be used.

```
//This code calls Student's constructor with parameter
ScienceStudent(int id) : Student(id)

//This code calls Student's default constructor
ScienceStudent(int id)
```

The class ScienceStudent added two new functions (constructors) to the base class. Now, let's consider the art students. Similar to ScienceStudent, the new ArtStudent class has to define new constructors and similarly set PID and numCourses. But it also has another requirement: the first course has to have a credit count of 6. As a programmer, we may design this feature in different ways, but one method is to revise the AddGrade() function and make sure that the credit count for the course at index 0 (the first course) is always 6 (Sample Code #3-c below, line 22).

```
SampleCode #3-c
1. class ArtStudent : public Student
2. {
3.     public:
4.     ArtStudent()
5.     {
6.         SetPID(1);
7.         SetNumCourses(15);
8.     }
9.     ArtStudent(int id) : Student(id)
10.     {
11.         SetPID(1);
12.         SetNumCourses(15);
13.     }
14.     void AddGrade(int i, int g, int c)
15.     {
16.         if (grades != NULL & credits != NULL && i >= 0 &&
            i<numCourses)
17.         {
18.             grades[i] = g;
19.             credits[i] = c;
20.         }
21.
22.             credits[0] = 6;
23.         }
24. };
```

If you add Sample Code #3-c to your program, you will notice that it will have a compile error because the new class is accessing private data of Student class in its revised function AddGrade(). Recall that in C++, class members can be **private** (only accessible by the class itself) and **public** (accessible also by modules in the program). To give child classes access to members but still block access by other modules, C++ has a new type of access called **protected**. Depending on which members need access in the child class, we add a line to the definition of Student class:

```
class Student
{              //start with private (default)
    int ID;
    int GPA;
protected:   //from here, members are protected
    int* grades;
    int* credits;
    int numCourses;
public:        //from here, everything is public
```

> ☞ **Key Point:** The protected access modifier defines a member that can be accessed by the child classes.

The above classes extend the base class by adding new functions and modifying existing functions. Imagine another case, EngineeringStudent, where students can do an internship. There is no credit or grade for this internship, but doing it is an option that has to be recorded. Whether or not the student has done an internship is a piece of information that requires a new variable. So, we need to add a new data member to the class. We also need to have new constructors to initialize this new member, a new function to change it, and also a modified Report() function.

The modified Report() function (line 21) still needs to do everything that the base class function does and then add a report on the new data (internship). To do so, and instead of repeating the code, we can simply call the base class function Student::Report(), as seen on line 23. Not that following C++ general syntax, to show a member of the class, we use the "::" symbol after the class name. To show a member of an object we use . after object name.

Sample Code #4-c

```
1. class EngineeringStudent : public Student
2. {
3.   bool internship;
4. public:
5.   EngineeringStudent()
```

```
6.   {
7.       SetPID(2);
8.       SetNumCourses(25);
9.       internship = false;
10.  }
11.  EngineeringStudent(int id) : Student(id)
12.  {
13.      SetPID(2);
14.      SetNumCourses(25);
15.      internship = false;
16.  }
17.  void SetInternship(bool i)
18.  {
19.      internship = i;
20.  }
21.  void Report()
22.  {
23.      Student::Report();
24.      if (internship)
25.          printf("internship done\n");
26.      else
27.          printf("internship not done yet\n");
28.  }
29.
30. };
```

The main() function or any other part of the program can use these new classes just like Student.

```
ScienceStudent s2(1002);

s2.AddGrade(0, 100, 3);
```

```
s2.AddGrade(1, 80, 1);

s2.AddGrade(2, 70, 1);

s2.Report();


ArtStudent s3(1003);

s3.AddGrade(0, 100, 3);

s3.AddGrade(10, 80, 1);

s3.Report();


EngineeringStudent s4(1004);

s4.Report();
```

Calling `Report()` for s4 (`EngineeringStudent`) will show the following line at the end:

```
internship not done yet
```

Looking at three child classes, we can see that inheritance allows sharing all the base class members and performing three possible modifications:

- Adding new data members

- Adding new functions

- Changing existing functions, also called **Overriding** a base function.

A child class cannot remove an existing member (data or function) or change a data member. Exhibit 11.1 shows our classes in a chart usually referred to as **Class Hierarchy**, a tree-like structure that shows the child–parent relationship between classes. You can draw this with the base at the top or bottom, but either way, nodes represent classes, and the lines are parent–child relationships.

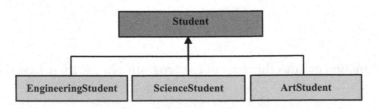

EXHIBIT 11.1   Class hierarchy. Student is the base (parent or superclass) and the other three are derived from it (child or subclass).

☞ **Key Point: A child class can:**
- Add new properties
- Add new methods
- Revise the code for existing methods (override them).

☞ **Key Point:** A class hierarchy is tree structure where base classes expand through child classes.

☑ **Practice Task:** Design a class hierarchy for vehicles (extend base on type and details), university students (extend based on programs and years), and bank accounts (extend based on type and features). Identify new members that child classes add.

🖐 **Reflective Questions**
- Do you think a child class can be defined based on more than one parent?
- If yes, what happens if parent classes have similarly named functions with different code?

SIDEBAR: MULTIPLE INHERITANCE

Inheritance allows classes to be extended through child classes that not only inherit their members but also add/modify them. It should not come as a surprise that classes can inherit members from more than one parent, a concept generally referred to as multiple inheritance. Imagine classes A and B below:

```
class A
{
    int a1;
    int a2;
    void DoA();
};
class B
{
    int b1;
    int b2;
    void DoB();
};
```

Multiple inheritance allows defining a new class, say AB, that extends both A and B.

```
class AB : public A, public B
```

```
{
    //add new data members
    int a3;
    int b3;
    //add new member function
    void DoAB();
    //modify existing functions
    void DoA();
    void DoB();
};
```

Instances of the new class AB have triple identities: A, B, and AB. For example, we can have a class Student and another Employee for a university. Now a new class called TeachingAssistant can inherit from both Student and Employee and represent students who work for the university to assist in some classes.

The concept of multiple inheritance is very powerful when dealing with objects that have overlapping features. We can have base classes that each corresponds to one feature, and allow new classes to inherit from a selection of these "feature" classes to implement those features. The base classes can even be abstract ones (interfaces) with no implementation. For example, we can have banking services such as FinancialAdvising, MoneyExchange, and TellerServices as base features (abstract classes or interfaces), and then, each type of banking service provider, such as branch, ATM, and mobile, can implement a different selection of those features.

The main difficulty with multiple inheritance is conflicting member functions, i.e., two base classes both having the function but with a different implementation. In the example above, imagine both classes A and B have a function called DoX(). In that case, an object of type AB will have two versions of this function, and it is not clear which one needs to be used. Two common solutions for this problem are

- Listing base classes with priority, so the class listed first as the base (A in the above example) will have priority if there is a conflict
- Using Abstract classes (interfaces), so there is no base implementation, and the child class has to have its own implementation for commonly named functions.

11.1.3 Inheritance in Python and Javascript

The syntax for using inheritance in Python and Javascript is a little different from C++, as expected, but the concepts are almost entirely the same. A new class can extend an exsiting one and establish a parent–child relationship where the child can add new data or functions or modify existing functions. In this section, I demonstrate using inheritance in these two languages through another Object-Oriented Design (OOD) example.

In previous chapters, I frequently used a library system as an example. A library includes various item types and is not limited to the book. For example, we can have CD/DVDs and magazines/newspapers (periodicals). All these types of items share some information and operations (properties and methods), but they may also have their own specific ones.

The Book class in our previous examples had the following members:

- Properties
  - Name
  - ID
  - Borrower.
- Methods
  - Init (and/or Constructor)
  - Borrow
  - Return.

Assuming two types of items in the library (Book and Disk, as examples of printed and digital items), we can see that they all share the above members but can have others[1]:

- Book
  - Number of pages.
  - Edition number.
- Disk
  - Format (CD, DVD)
  - Operating system (Mac, Windows, Linux, Universal)
  - Browse (a method to browse and search the electronic content).

We can also notice that once we have different types of items, the information about "what type this item is" also needs a variable, plus the reporting ability with various information we have now. So the following two members should also be added to all item types:

- Type (property)
- Report (method).

---

[1] This added information is just for example. I'm sure other information can be imagined for all these item types.

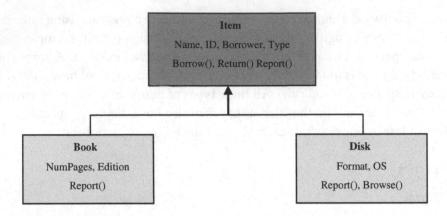

EXHIBIT 11.2   Library class hierarchy.

Using an OOD approach and considering the benefits of inheritance, we can design our system, as shown in Exhibit 11.2, using the class hierarchy.

11.1.4 Python Inheritance Example

Sample Code #5-p shows the base class Item and its child classes Book and Disk. Here are a few things to pay attention to when using inheritance in Python:

- Similar to C++, the base class name comes after the new child class name, but here it is inside brackets ( ).

- You can create a new class without making any changes. In C++, we just leave the new class definition empty. In Python, we write the keyword pass (line 20).

- There are two ways to call the base class function:

  - Using the base class name (line 24), which needs passing self as a parameter.

  - Using the function super() (line 30), which doesn't need self. The keyword super in Python and any other languages refers to the base class.

- The function Browse() in class Disk is new. Since our example doesn't actually implement any content, the code simply prints a message.

```
Sample Code #5-p
1. class Item :
2.     def _ _init _ _ (self, name="", id=-1):
3.         self.name = name
4.         self.ID = id
5.         self.borrower = -1
```

```
6.              self.type = "unknown type"
7.         def Borrow(self, uid) :
8.              if self.borrower == -1 :
9.                   self.borrower = uid
10.        def Return(self) :
11.             self.borrower = -1
12.        def Report(self) :
13.             print("=====")
14.             print(self.name)
15.             print(self.ID)
16.             print(self.type)
17.             print(self.borrower)
18.
19. class Item2(Item) :
20.        pass
21.
22. class Book(Item) :
23.        def _ _init_ _(self, name="", id=-1, np=-1, ed=1):
24.             Item._ _init_ _(self, name, id)
25.             self.type = "book"
26.             self.numPages = np
27.             self.edition = ed
28.
29.        def Report(self) :
30.             super().Report()
31.             print(self.numPages)
32.             print(self.edition)
33.
34. class Disk(Item) :
```

```
35.          def _ _init_ _(self, name="", id=-1, f="dvd",
     o="universal"):
36.          Item._ _init_ _(self, name, id)
37.          self.type = "disk"
38.          self.format = f
39.          self.os = o
40.
41.    def Report(self) :
42.          super().Report()
43.          print(self.format)
44.          print(self.os)
45.
46.    def Browse(self) :
47.          print("search and browse the content")
48.
49. #program execution starts here
50. b = Book("testBook", 1000, 200, 1)
51. b.Report()
52.
53. d = Disk("testDisk",1001)
54. d.Report()
```

Running this program will show an output like the following lines:

```
=====
testBook
1000
book
-1
200
1
```

```
=====

testDisk

1001

disk

-1

dvd

universal
```

☑ **Practice Task:** Write the code for Disk and Periodicals.

11.1.5 Javascript Inheritance Example

If you recall, in Javascript, we can define classes (and objects) using two methods: a constructor function or the class keyword. Both methods allow inheritance, but using the keyword results in a much more straight-forward and less confusing code. It is basically similar to the syntax in C++ and Python.

Sample Code #6-j shows the partial implementation of the library system in Javascript. The main points to pay attention to are

- To define the inheritance relationship, we use the keyword extends (line 15).

- The base class constructor is automatically called with any parameters provided (for example, "testDisk" and 1001 on line 30). If new properties are being added in the child class, then we add a new constructor (line 17) with new parameters. An explicit call to the base constructor is needed (line 19) before defining the new data members (lines 20 and 21). Again, the keyword super is used to refer to the base class.

Sample Code #6-j

```
1. class Item
2. {
3.   constructor(name, id)
4.   {
5.     this.ID = id;
6.     this.name = name;
7.   }
8.   Report()
```

```
9.    {
10.    //some code
11.    }
12.    //more members
13. }
14.
15. class Disk extends Item
16. {
17.    constructor(name, id, f, o)
18.    {
19.       super(name,id);
20.       this.format = f;
21.       this.os = o;
22.    }
23.    Browse()
24.    {
25.    //some code
26.    }
27.    //more members
28. }
29.
30. let d = new Disk("testDisk", 1001);
31. d.Report()
```

11.2 DIFFERENT TYPES OF GAME OBJECTS

In this section, I focus on two specific problems in games and show how we can use inheritance to solve them. These problems include characters using Game Artificial Intelligence (Game AI) and Animation. We will see how the use of inheritance can create new classes that share the basic features of GameObject but extend them to provide specialized abilities.

Before we discuss these cases, though, I'd like to introduce a new graphics library called **OpenFrameworks (OF)**, that is designed specifically for OOP. While everything we see in this (and next chapter) can be done using our familiar SDL/SDLX library, introducing OF has multiple advantages:

- It provides more functionalities for both 2D and 3D games and other applications.

- It is more widely used.

- It is fully object-oriented and provides base classes that can be used directly or extended for customization.

- It provides a basic structure for interactive applications, hiding many details from the programmer.

11.2.1 OpenFrameworks

C and C++ provide programmers with a great set of features that allow highly optimized programs suitable for high-performance applications such as games. For example, pointers and dynamic memory allocation are powerful tools to manage memory exactly the way the program needs without the overhead of behind-the-scene modules such as garbage collection that take memory and time and affect the performance negatively. On the other hand, C and C++ do not have standard support for graphics and multimedia and rely on third-party tools for that purpose. SDL library was one example of these tools. Its main features were a simple interface, support for common 2D operations, and compatibility with C (non-OOP), although it could be used in C++ programs too.

Not using any OOP features is a common feature among C/C++ libraries to keep compatibility with still-popular C language and nor forcing the programmer to use OOP while giving them the option to use C++ and classes. On the other hand, the growing popularity of OOP has motivated some developers of C/C++ libraries to focus only on C++ and offer more facilities for OOP programmers. OF is an example of such cases.

If you look at the sample SDL/SDLX game we have been working on over the last few chapters, then you see that there are basic modules and classes that are independent of the type of our game. SDLX _ Bitmap, GameObject, Physics, and Render are all generic and can be used in any application. Even the Game class includes many standard members that are common but need to be customized. Object-oriented libraries such as OF (a.k.a. **Class Libraries**) offer a collection of such classes that can be used in two ways:

- Directly as is. A program can use classes like GameObject and Render without any customization.

- Through inheritance. A program may use classes such as Game and extend them to implement a particular game logic. In such cases, the base Game class will not have any specifics like object members or details on keyboard handling. Still, it will provide common features such as a main loop, keyboard even handlers, and Draw() and Update() function that needs to be overridden.

In OF, the most commonly used example of the first group is ofImage class that is the counterpart to the SDLX _ Bitmap structure we developed but significantly more advanced. All OF programs use and extend a base class called ofBaseApp, which is equivalent of the Game class in our previous examples. You can see that all OF classes start with prefix of.

As a third-party tool, OF does not come with any IDE pre-installed. Please refer to the Companion Website for information on how to install OF for Visual Studio, and note that all documentation and examples are for OF v.0.9.8. Later versions may change some aspects of the library.

OF projects include multiple special settings that may be difficult to manage manually. As such, creating an empty project in Visual Studio and adding the OF parts and settings is not recommended. To create a new project with OF, you may use two different methods:

- Use the program ProjectGenerator.exe (inside the sub-folder projectGenerator-vs, under the main OF installation folder) to create a new OF project in the location you want. It will automatically add the required settings and starting files to your project.

- You may also install the OF extension for Visual Studio to have a new project type inside the IDE for your convenience. It may have compatibility issues with your version of Visual Studio, though.

Once your first project is created, open the Solution file in Visual Studio. You will notice that there are two projects: the one you created and the OF library. Do NOT play around with the library. You won't need to do anything with it. Also, don't move your project files and folders as they may have specific reference sot the locations. Take these initial actions:

- Build and run the program. You should see a black text window and a gray or white graphics window, both empty. OF allows you to keep your text/console window for textual messages and information. The program ends when you press Escape.

- Using Solution Explorer, inspect the project files. You should have main.h, ofApp.h, and ofApp.cpp.

- Using Class View, inspect the classes and global functions. You should have a class called ofApp and a global function main().

  - The ofApp class is the equivalent of Game class in our previous examples. I will discuss it in detail soon.

  - The main function does two things:

    - It calls ofSetupOpenGL() to start the library, which is OpenGL-based. You can change the dimensions or switch between Windows and full-screen mode.

    - It calls ofRunApp() with a parameter that is a dynamically created ofApp object. No need to do anything with this line or any other part of main.cpp.

11.2.2 ofApp Class

All the OF actions happen inside the `ofApp` class. The automatically generated code has empty functions for all required operations including `setup()`, `update()`, `draw()`, and event handlers (responses to keyboard and mouse events). The main loop and initialization are implemented within the base class `ofBaseApp`, and already call the functions provided to the programmer who can simply add the code within the body of already-created functions that are called automatically; `setup()` will be called at the start; `update()` and `draw()` will be called at each video frame (every iteration of the main loop); the event handlers will be called when the related event happens.

Note that

- The base class is already taking care of ending the main loop when Escape key is pressed.

- The functions `update()` and `draw()` are named in all lower case. Do not rename these or any other functions as they are inherited from `ofApp` and are used with these names.

- OF uses double buffering and hardware acceleration, but the details are hidden from the programmer.

11.2.3 Customizing the Project

To follow the style used in our previous game examples, you can customize the project:

- Open the source files main.cpp, ofApp.cpp, and ofApp.h, and in each of them, do a Replace All to change of App to `Game`.

  - The base application class, as mentioned earlier, is called `ofBaseApp`. The project generator creates a new child class based on `ofBaseApp` and calls it `ofApp`. To make our OF programs similar to our SDL/SDLX ones, and to have more meaningful names, we rename this to `Game`, although this is not necessary.

- In Solution Explorer, locate and rename ofApp.h and ofApp.cpp files to Game.h and Game.cpp.

- In Solution Explorer, right-click on `src` and Add New Item. Choose a header file and name it Main.h. Make sure you right-click on `src` and add `\src` to the location. OF project generator creates all the source files in a folder called `src`.

- Open Main.h and add the following lines. This is to follow our process of having one single header file that includes all classes, as discussed in Section 9.3.

  - `#pragma once`

  - `#include "ofMain.h"`

  - `#include "Game.h"`

- Remove any include command from Game.h, but leave the `#pragma once`.

- Open all main and Game.cpp, and replace all include commands with a single one:

  - `#include "Main.h"`

11.2.4 Adding New Content

Before we can do anything with our game, we need to learn how to add new content. The minimum we need is the ability to have a new class and also adding data files.

- In your project folder, locate the folder `bin` and under that `data`. Create a new folder called `images`, and make sure you save all your image files there.

  - Unlike the SDL projects where the default location was the project folder, in OF the default location for source files is `src` and the default location for data files is `bin\data`.

  - Add a background image and a player character to the new `images` folder.

  - OF can use most common image formats, including GIF and PNG with transparency.

- Add a `GameObject` class. To add a new class,

  - Right-click on `src` and add a new header or cpp file. Make sure the location is the `src` folder.

  - Do not include anything in header files, but have the `#pragma once`.

  - Include only Main.h in cpp files.

  - Include the new header file in Main.h before the class that uses it. Remember that order matters, so include GameObject.h before Game.h.

- Add code for `GameObject`, as explained below.

- Add code for `Game`, as explained below.

11.2.5 GameObject Class

Sample Code #7-c shows the code for OF-based `GameObject` class. Everything is pretty straight-forward. Note that

- The code is for two files: header and cpp.
- The `ofImage` class is similar to SDLX_Bitmap structure. It has a function called `draw()` that `GameObject` can use.
- No `Render` is implemented yet, so drawing is done in the `GameObject` itself.
- To simplify the code, all members are set as public.

Sample Code #7-c

```
1. //GameObject.h
2. #pragma once
3. class GameObject
4. {
5. public:
6.    int x;
7.    int y;
8.    int speedX;
9.    int speedY;
10.   ofImage shape;
11.
12.   GameObject();
13.   ~GameObject();
14.
15.   void Draw();
16.   void Move();
17. };
18. //GameObject.cpp
19. #include "Main.h"
20. GameObject::GameObject()
```

```
21. {
22.        x = y = 0;
23.        speedX = speedY = 0;
24. }
25. GameObject::~GameObject()
26. {
27. }
28. void GameObject::Draw()
29. {
30.        shape.draw(x, y);
31. }
32. void GameObject::Move()
33. {
34.        x += speedX;
35.        y += speedY;
36. }
```

11.2.6 Game Class

Sample Code #8-c shows the OF-based Game class. Three members have been added to the Game class:

- A background image

- A GameObject for player

- An array that has true/false values for each key on the keyboard. This array is necessary because OF has a separate function for checking the keyboard, so it is not done inside the update() function. So, for example, when keyPressed() function is called, we have a piece of information about a particular key (identified with the code parameter) is pressed. This information needs to be stored in a variable if we want to use it later. Since we have many keys, we create an array with one member per key and use the key code as an index to this array. Every time that the keyPressed() is called, we set the corresponding member to true, and once the keyReleased() is called, we set it to false. The array members are initialized to false (lines 39–40 below).

Sample Code #8-c

```cpp
1. //Game.h

2. #pragma once

3. #include "ofMain.h"

4. #include "GameObject.h"

5. class Game : public ofBaseApp

6. {

7.   public:

8.       ofImage bgImage;

9.       GameObject player;

10.

11.      bool keyDown[255]; //all keys

12.

13.      void setup();

14.      void update();

15.      void draw();

16.

17.      void keyPressed(int key);

18.      void keyReleased(int key);

19.      void mouseMoved(int x, int y );

20.      void mouseDragged(int x, int y, int button);

21.      void mousePressed(int x, int y, int button);

22.      void mouseReleased(int x, int y, int button);

23.      void mouseEntered(int x, int y);

24.      void mouseExited(int x, int y);

25.      void windowResized(int w, int h);

26.      void dragEvent(ofDragInfo dragInfo);

27.      void gotMessage(ofMessage msg);

28. };
```

```
29. //Game.cpp
30. include "Main.h"
31. void Game::setup()
32. {
33.    bgImage.load("images/map.bmp");
34.
35.    player.x = 110;
36.    player.y = 250;
37.    player.shape.load("images/r0.gif");
38.
39.    for (int i = 0; i < 255; i++)
40.        keyDown[i] = false;
41.
42.    ofSetFrameRate(40);
43. }
44. void Game::update()
45. {
46.    //assume no speed and change if key pressed
47.    player.speedX = 0;
48.    if (keyDown[OF_KEY_RIGHT] == true)
49.        player.speedX++;
50.    if (keyDown[OF_KEY_LEFT] == true)
51.        player.speedX--;
52.    //now let the player move if speed is not zero
53.    player.Move();
54. }
55. void Game::draw()
56. {
57.    bgImage.draw(0, 0);
```

```
58.
59.     //draw player
60.     //method 1
61.     //player.shape.draw(player.x, player.y);
62.     //method 2
63.     player.Draw();
64. }
65. void Game::keyPressed(int key)
66. {
67.     keyDown[key] = true;
68. }
69. void Game::keyReleased(int key)
70. {
71.     keyDown[key] = false;
72. }
73. //other functions unchanged
```

11.2.7 Enemy vs. Player

Now that we have set up our OF-based game program, it is time to work on the problem I introduced at the beginning of this section. As I discussed and demonstrated in earlier examples, when it comes to movement, computer games usually consist of three types of objects (or characters):

- Player object, controlled and moved by the player
- Moving objects, controlled and moved by the game
- Stationary objects, fixed at a position although may change in shape or occasionally be moved (for example, be picked up).

In previous examples, I showed how to use a standard module, GameObject, to represent all objects in a game using common properties and methods such as

- Location
- Speed

- Shape

- Draw()

- Move()

Operations such as Draw and Move are common among all these objects and can be implemented in a common way. For example, we assign a speed to any object, and the `Move()` function applies that speed to change the location, or the object is passed to a `Physics` module that will consider ground levels in addition to speed and determines the location. There is no need for different objects to have different data or code, as the movement algorithm is the same, and the only difference is the value of speed.

Now, consider a game such as Pac-Man, or any other game in which the enemies perform a certain action that is more complicated than simple linear movement. In Pac-Man, enemies follow different movement patterns depending on situation and character: chase the player, move randomly, scatter around the game world, avoid the player, and so on. An action such as chasing the player gives the player the impression of interacting with an entity that has some form of agency, the ability to decide and act based on the situation. While changing the direction of movement to always follow the player is a very simple action, it demonstrates such an agency.

Artificial Intelligence (AI) is an area of computer science that deals with developing algorithms that allow computers to make decisions and solve problems. In games, the term "AI" is generally used to refer to the ability of a game to make players feel they are dealing with an intelligent agent. The actual implementation of Game AI may not require real intelligence in the computer as everything can be done based on simple pre-determined rules defined by the programmer. Modern games are moving towards having real intelligent components, but that topic is beyond the scope of this book. Regardless of the level of AI used in a Non-Player Character (NPC), the movements are not done by simply changing the location based on speed. A certain new algorithm has to be implemented, which is beyond the `Move()` function in `GameObject`. As we saw earlier, the options are (1) defining a new class for enemies and (2) extending the `GameObject` class to create an `Enemy` class.

Sample Code #9-c shows the new `Enemy` class. It uses Enemy.h and Enemy.cpp. As I explained earlier, to add new files to your project, right-click on src in Solution Explorer, enter your filename, and make sure the location is the src folder. Also, follow our program structure: add a line in Main.h (after GameObject and before Game) to include Enemy.h, and in Enemy.cpp only include Main.h.

The first piece of information that this class needs is a target to follow. To be generic and reusable, this target cannot be hard-coded as a player and can be any `GameObject`. Since the target object is generally an independent object in the game, we should not create a new object but use a reference (pointer) to an existing one. This pointer can be set any time in the program to change the target for chasing. The constructor needs to be updated to initialize this pointer, and `Move()` has to be updated to include the new algorithm. The base (parent) class constructor will still be automatically called before the derived (child) class constructor, but the base class Move() function will not be called unless the derived class calls it explicitly.

Just like any other pointer, target has to be initialized with NULL to show it is not a valid address yet. The algorithm for chasing (moving towards) a target is very simple: if we are on the right side of the target, then we move to the left; if we are on its left side, then we move to the right. If we are close enough (within 2-pixel distance), then we stop moving to avoid bouncing back and forth around the target. In our implementation, we only set the speed and then call the base class Move(). This allows us to rely on the base class if there are other things that need to be done.

☞ **Key Point:** One of the most important roles of AI in games is to control the actions of NPC.

☑ **Practice Task:** In Sample Code #9-c, the enemy moves if the distance to target is less than 2 (lines 23–26). Remove the +2 and –2, and see what happens. Can you explain why?

✍ **Reflective Questions:** Lines 23–26 demonstrate a simple way of following a target. Do you understand how the work? Can you think of more complex ways to do this (for example, going fast if target is far and slowing down when we get closer)?

Sample Code #9-c

```
1. //Enemy.h

2. #pragma once

3. class Enemy : public GameObject

4. {

5. public:

6.     GameObject * target;

7.

8.     Enemy();

9.     void Move();

10. };

11. //Enemy.cpp

12. #include "Main.h"

13.
```

```
14. Enemy::Enemy()
15. {
16.     target = NULL;
17. }
18. void Enemy::Move()
19. {
20.     if (target != NULL)
21.     {
22.         speedX = 0;
23.         if (x > target->x+2)
24.             speedX = -1;
25.         if (x < target->x - 2)
26.             speedX = 1;
27.
28.         GameObject::Move();
29.     }
30. }
```

To use this new class, follow these steps:

- Make sure you have added Enemy to Main.h (after GameObject and before Game).
- Add a line to Game.h for a new member
  - Enemy enemy;
- Initialize the enemy in Game::Game()
  - enemy.x = 50;
  - enemy.y = 250;
  - enemy.shape.load("images/r2.gif");
  - enemy.target = &player;
- Move the enemy in Game::update()
  - enemy.Move();

- Draw the enemy in `Game::draw()`

  - `enemy.Draw();`

Note the order of drawing: anything that is drawn later will show up on top.

> ☑ **Practice Task:** Add another enemy type with a different behavior.

11.2.8 Animation Revisited

In Section 7.3, I explained that animation could be implemented using an array of shapes (frames) with two important notes:

- We need to have a piece of information that tells us which frame to show at any time.

- We need to have an operation that advances this frame number and resets it back to zero when it reaches the end.

Knowing these two requirements, designing a child class, `AnimatedObject`, based on `GameObject`, is fairly straight-forward, as shown in Sample Code #10-c. The class adds two new data members (and no longer uses the existing `ofImage shape`). It overrides the `Draw()` function and adds a new `Advance()` function.

Sample Code #10-c

```
1. //AnimatedObject.h
2. #pragma once
3. class AnimatedObject : public GameObject
4. {
5. public:
6.     ofImage frames[3];
7.     int currentFrame;
8.
9.     AnimatedObject();
10.    void Draw();
11.
12.    void Advance();
13. };
```

```
14. //AnimatedObject.cpp

15. #include "Main.h"

16. AnimatedObject::AnimatedObject()

17. {

18.     currentFrame = 0;

19. }

20. void AnimatedObject::Draw()

21. {

22.     frames[currentFrame].draw(x, y);

23. }

24. void AnimatedObject::Advance()

25. {

26.     currentFrame++;

27.     if (currentFrame == 3)

28.         currentFrame = 0;

29. }
```

To use this new class, follow these steps:

- Make sure you have added AnimatedObject to Main.h (after GameObject and before Game).

- Add a line to Game.h for a new member

    - AnimatedObject prize;

- Initialize the animated prize in Game::Game()

    - prize.x = 300;

    - prize.y = 250;

    - prize.frames[0].load("images/10.gif");

    - prize.frames[1].load("images/11.gif");

    - prize.frames[2].load("images/12.gif");

- Update the prize in `Game::update()`

  - `prize.Advance();`

- Draw the prize in `Game::draw()`

  - `Prize.Draw();`

> ☑ **Practice Task:** Use a different timing for animation. For example, advance the animation every three frames (slower).

HIGHLIGHTS

- Inheritance is a relationship between a parent (a.k.a. base) class and a child (a.k.a. derived) one where the child receives all members of the parent and can add new members or change existing ones.

- When creating instances of a child class, first, the base class members will be created, and the base class constructor will be called.

- The protected access modifier defines a member that can be accessed by the child classes.

- A child class can

  - Add new properties

  - Add new methods

  - Revise the code for existing methods (override them).

- A class hierarchy is a tree structure where base classes expand through child classes.

- Class libraries are class-based (object-oriented) software programming libraries that allow programmers to use existing classes directly or extend them using inheritance.

- OpenFrameworks (OF) is a class library for graphics in C++.

- OF has a standard class that controls the application and provides functions for various aspects of a game such as initialization, draw, and update.

- One of the most important roles of AI in games is to control the actions of a Non-Player Character (NPC).

END-OF-CHAPTER NOTES

A. Things I Should Mention

- The base and derived classes are also called **superclass** and **subclass**. These terms are more common in Java and some other languages.

- Another object-oriented 2D game development library for C++ is **Cocos2D** that allows development for various desktop and mobile platforms.

B. Self-Test Questions

- What is inheritance?

- What can a child class do?

- What is protected access modifier in C++?

- What is the base application class in OF?

C. Things You Should Do

- Try multiple inheritance by defining an AnimatedEnemy class.

- Check out this article on multiple inheritance:

  - https://www.freecodecamp.org/news/multiple-inheritance-in-c-and-the-diamond-problem-7c12a9ddbbec/

D. Reflect on the Experience of Reading This Chapter

- What did you expect from this chapter before reading it?

- What was it about, and what did you learn?

- What tasks did you perform, and what difficulties did you face?

- How did you feel about the material and tasks presented in this chapter?

- How can you improve your learning experience?

- How do you see this topic in relation to the goal of learning to develop programs?

- Do you see the value of inheritance in OOP?

- What would be the main difficulty in having a class library with no inheritance?

Object Identities

At the end of this chapter, you should be able to:
• Use static and dynamic polymorphism to have objects with multiple identities

OVERVIEW

The introduction of inheritance adds a significant improvement for Object-Oriented Programming as it allows classes to be extended. The increased reusability and manageability caused by inheritance is in the form of extending classes, maintaining the existing functionality without the need to repeat them. Creating a child class gives programmers the choice of using the parent or the child(ren) to define new objects, depending on which set of functionality is needed.

A more interesting, and sometimes less noticed, outcome of inheritance is that objects not only can be defined based on either the parent or the child, but also can act as either one once defined as a child. When we have two classes Student and ArtStudent, we have the option of defining an object as either one of these two. If the object is an instance of Student, then it can only have the members of that class and act as a Student. But if we define the object as an ArtStudent, then it has inherited all members of Student, which means it has all it takes to "be a Student." This means the object can have two **Identities** and perform as a generic Student or as an ArtStudent.

The ability of child objects to have multiple identities (since there can be a long lineage of grandparent classes) is crucial in designing reusable objects. For example, our Render class in the basic game example was designed to draw "all" objects. This task was fairly straight-forward when there was only one type of object, but having multiple classes of objects means a function like `Render::Draw()` has to have different versions implemented for different object types. These multiple versions are inconvenient if all object types are known when designing the `Render` class. They are impossible to make if the object types are to be extended at any time to create new classes. Adding a new `Gameobject` class requires changing the code for `Render` class to add new `Draw()` functions, something that is not possible if the `Render` class is provided as a third party tool/library.

Having multiple identities, on the other hand, means the `Render` class needs to have only one `Draw()` function that deals with base `GameObject`. All objects, assuming they are instances of child classes of `GameObject`, can act as `GameObject` and be processed by `Render::Draw()`. The ability to have multiple identities also leads to the ability programmatically choose which identity an object needs to use at any time, a feature usually referred to as **Polymorphism**. It gives programmers the full flexibility to decide which identity is used for any specific access to members of the class. So, if object X is an instance of class B, itself based on class A, and class A has two functions `foo()` and `bar()` that is overridden by class B, the programmer can decide (1) at any moment which identity X has and (2), regardless of current identity, which versions of the functions `foo()` and `bar()` should be used. For example, X can be treated as A (base), so `A::foo()` will be used, but an exception is made and `B::bar()` will be used instead of `A::bar()`.

Polymorphism is very helpful when we want to pass objects to a module to be treated as the base, but we want some specific functions to be always used from the child version. In the case of `Render` class, this can happen when the overall rendering is done by the `Render` class, but some specific aspects such as visual effects need to be done class-specific.

In this chapter, we will look at a couple of examples that demonstrate the use of polymorphism. I start with the default behavior where any child can act as a parent class without the need for any special coding. Then, we move to the concept of **Virtual Functions** that specifically identify the version of class members to be used. Through this discussion, I will also review the notions of abstract classes, interfaces, and static members, and add some features to our sample game. By the end of this chapter, our game program will have all the main features discussed throughout the book and can be used as the base for many other games.

12.1 RENDERING MULTIPLE GAMEOBJECT TYPES

In Chapter 10, I introduced the `Render` and `Physics` classes as a modular way of managing drawing and movement for all objects. These operations can include many variations, such as rendering styles or movement with/without gravity, but they are usually quite generic and independent of any particular game. Instead of each object or a `Game` class carrying the burden of performing these operations, reusable modules such as `Render` and `Physics` classes are good solutions to take care of all these tasks for all objects. Relying on these classes, which can be developed once, or borrowed from libraries and game engines, frees the programmers from dealing with the related tasks so they can develop their own game logic and objects.

One problem that arises, though, is the need for these classes to know what object types they have to work with, so they can provide proper functions. For example, if Render class as to draw the GameObject, then it has to have a function such as

```
void Render::Draw(GameObject* obj);
```

If there is another type of object, say, Enemy, that needs to be drawn, then another function will be needed, which causes a problem when developing programs using a third-party Render module that cannot be changed.

```
void Render::Draw(Enemy* obj);
```

In the previous chapter, I showed how multiple types of objects could be created using inheritance, so they can share the common properties and methods of a base class but extend it to have new and/or modified ones. Inheritance allows better reusability and manageability when defining classes. The ability to extend a class while maintaining the base members offers an added value, which is objects that have more than one **Identity**, which leads to a major property of many systems, including OOP that is generally referred to as **Polymorphism**. In a more general way, polymorphism, which literally means "having multiple forms," is the situation of having entities with different types but the same name or appearance. In programming, this can be functions or objects.

After encapsulation and inheritance, polymorphism is the third major property of OOP. In the next section, I demonstrate the concept and show how it can be used in many situations, such as the above case of Render class.

☞ **Key Point:** Polymorphism is the ability of objects to behave in different forms, as instances of both the parent and the child classes.

12.1.1 OF-Based Render Class

At the end of Chapter 11, I presented a simple game with multiple objects using OpenFrameworks. Now, let's add a Render class to this program, which can serve as a central place to do all the drawings. I demonstrate the value of such a class by introducing the notion of visual effects and the ability to add them when rendering objects. A game engine can come with a standard rendering components and a series of visual effects, so the game developers don't need to worry about such things and focus on their own game-specific issues.

The OF-based Render class is very similar to the SDL-based one that we saw in Chapter 8. Sample Code #1-c shows the initial code for this class. I will improve on this as we move forward in this chapter. See the instructions in Chapter 11 for adding a new class to your OF project.

The class includes the scrolling ability we first saw in Chapter 7 with the platformer/side-scrolling game. The only functions are to draw an image and a GameObject. The image is drawn using the game world coordinates (x and y) and scrolling values that change the position of the virtual camera (screen). Keep in mind that OF or SDL drawing functions use coordinates relative to the top-left of the screen (a.k.a. screen coordinate systems), but the GameObject uses coordinates relative to the game map (world coordinate system). Scrolling values transform these coordinates: ScreenX = WorldX − ScrollX.

Sample Code#1-c

```
1.   //Render.h

2.   #pragma once

3.   class Render

4.   {

5.   public:

6.       Render();

7.

8.       int  scrollX;

9.       int  scrollY;

10.

11.      void Draw(GameObject* obj);

12.   };

13.   //Render.cpp

14.   #include "Main.h"

15.   Render::Render()

16.   {

17.       scrollX = 0;

18.       scrollY = 0;

19.   }

20.   void Render::Draw(ofImage* img,int x, int y)

21.   {

22.       img->draw(x - scrollX, y - scrollY);

23.   }

24.   void Render::Draw(GameObject* obj)
```

```
25.    {
26.        if (obj->visibility != 0)
27.        {
28.         of SetColor(255, 255, 255, obj->visibility*255/100);
29.         Draw(obj->currentShape, obj->x, obj->y);
30.         ofSetColor(255, 255, 255, 255);
31.        }
32.    }
```

To use the new class, we need to add a member of this type to the Game and call its Draw() function when drawing objects. OpenFramworks hides double buffering details from the programmers, so there is no need to transfer the buffer to screen as we did in SDL-based Render class with the Begin() and End() functions. Since no End() function is needed for rendering, I have also removed the Begin() and replaced it with a more generic function Draw() that receives an image, such as the background. We could also define background as a GameObject.

```
void Game::draw()
{
        render.Draw(&bgImage, 0, 0);
        //draw player
        //method 1
        //player.shape.draw(player.x, player.y);
        //method 2
        //player.Draw();
        //method 3 (using Render class)
        render.Draw(&player);
}
```

12.1.2 Function Overloading

In previous chapters, I mentioned that C++ allows creating different versions of functions with the same name but different parameters. This is called **Function Overloading**, or if it a member function in a class, **Method Overloading**. The Render::Draw() has two such versions, one for drawing images and another for GameObject. Function Overloading is an example of **Static Polymorphism**. It creates a polymorphic entity, one that has different implementations under the same name. The static aspect comes from the fact that different versions are identifiable at the time we write the program (compile-time).

If you compare the following two usages of Render::Draw(), it is clear at compile-time which line refers to which function version (based on the parameters).

```
render.Draw(&bgImage, 0, 0);
render.Draw(&player);
```

The compiler can then use the proper code for this function call when translating to machine code. The act of associating a name to proper content is referred to as **Binding** and, when it happens at compile-time, **Early Binding**. Dynamic polymorphism or **Late Binding** happens when at compile-time, it is not clear which version has to be used. I will demonstrate such a case and its advantages later in this chapter, but first, let's focus on the problem of using multiple object types in our Render class.

> ☞ **Key Point:** Binding is the act of associating a name to a content, for example, deciding which version of a function needs to be called. Early Binding happens at compile-time while Late Binding at run-time.

12.1.3 Multiple GameObject Types

In the previous chapter, I showed how to use inheritance to create two specialized classes Enemy and AnimatedObject based on GameObject. Having three types of objects to render, we may imagine three separate functions to perform the task:

```
void Draw(GameObject* obj);

void Draw(AnimatedObject* obj);

void Draw(Enemy* obj);
```

The operation for all three functions will be the same and uses the same information from the objects:

```
Draw(obj->currentShape, obj->x, obj->y); //pass to Draw()
for image
```

Even though the code for functions is the same, we would need three separate functions; that is, GameObject, Enemy, and AnimatedObject were separate classes. As I explained earlier, using separate classes not only causes the inconvenience of creating three functions but also makes it impossible to work with a third-party tool that provides a rendering module. Once we create our new classes, we have no way of adding such functions to it.

Thanks to inheritance, though, the problem can be solved with almost no effort.

12.1.4 Inheritance and Identities

Adding the Render class to our game and combining it with the child GameObject classes we made in the previous chapter, there is no extra code needed to draw the objects.

```
render.Draw(&prize);    //AnimatedObject
```

```
render.Draw(&enemy);    //Enemy
render.Draw(&player);   //GameObject
```

Since child classes inherit all members of the parent, `prize` and enemy include a parent and a child identity, which allows them to act as a `GameObject`. An instance of either one of `AnimatedObject` or `Enemy` classes (or a pointer to it) can be used when a `Gameobject` is needed. Inside the `Render::Draw()`, the pointer will be treated as a `GameObject*`; the program goes to the memory location and accesses the `GameObject` part of the object.

Using base identity with inheritance, as used in our `Render` class, is another example of static polymorphism. It is static because when we draw an object in `Render::Draw()`, even if the original object is `Enemy` or `AnimatedObject`, the pointer type is `GameObject*` and its behavior is also from `GameObject`. So, looking at the code, we can clearly know that the object is being treated as the base class. This is acceptable because the `Render` class has no need of any members of the derived classes (`AnimatedObject` and `Enemy`). There can be cases though that certain behaviors have to come from the child, not the parent, even if the variable/pointer seems to be of the base type. For example, in the Student example of Section 11.1, the `Report()` function needs to show the child class information even if we are using a pointer to the base class. Such a requirement leads to **Dynamic Polymorphism**.

☞ **Key Point:** Instances of the child class can be passed to functions that need a parent object.

12.2 DYNAMIC POLYMORPHISM

In Section 11.1, I revised and extended our `Student` class using inheritance. Imagine the task of adding a UI element in a school management software to get a report. A common way of doing this will be to enter a name or student ID, which the software uses to search for the student and then call the `Student::Report()` function. The code will be something similar to the following:

```
//these two are updated every time we have a new student of
   any type
int numAllStudents;
Student* allStudents[MAX _ STUDENTS];
//search function
Student* SearchStudent(int id)
{
    for (int i = 0; i < numAllStudents; i++)
```

```
        {
            if (allStudents[i]->GetID() == id)
                return allStudents[i];
        }
        return NULL;
    }

    //report function
    void ReportStudent()
    {
        printf("enter student ID: ");
        int id;
        scanf_s("%d", &id);
        Student* s= SearchStudent(id);
        s->Report();
    }
```

Assume that int numAllStudents and the array Student* allStudents[]are updated every time we add a new student of any type (the code not shown here). Having a single array of Student* allows us to deal with all students regardless of their program. Since the Student class has a program ID member (PID), it is possible to detect which program the Student* belongs to and typecase when needed:

```
    ArtStudent* artStudent = (ArtStudent*) student;
```

The main point, for now, is that we can get a pointer to the student with the right ID and then call the Report() function for that student. The problem with this method is that, based on static polymorphism, the program calls the Student::Report(), not the child class function. If the student is from engineering and has an internship, that information will not be reported because the base class function does not include any information specific to child classes.

To allow maximum flexibility when dealing with multiple identities, C++ and many other OOP languages offer **Dynamic Polymorphism** which allows **Late Binding**, meaning that the actual code that is called will be determined in run-time based on the real type of the object and not the pointer type in the ReportStudent() function. Note that in this function, we call Report() for Student* s, so according to static polymorphism, it should be Student::Report(). Using dynamic polymorphism, the program uses the function `EngineeringStudent::Report()` if the original object was `EngineeringStudent`. To identify the need for dynamic polymorphism, C++ programmers have to use the keyword **virtual** before the definition of any function for which the child version should always be used.

```
class Student
{
public:
    virtual void Report()
    {
        //code for Report()
    }
        //the rest of the class
};
```

A **Virtual Function** is one that is expected to be overridden by child classes. So, the version is determined by the object's run-time type instead of the variable/pointer type where the function call is happening. We need to add the keyword `virtual` only in the header file for the base class. The overridden version of this function in all child classes will be assumed virtual.

☞ **Key Points**
- When a child object is passed to a function that needs a parent, the object will be treated as parent within that function. If the function accesses one of the methods, the parent version will be used.
- In the above example, if the method is virtual, then the child version will be used.
- Virtual functions are expected to be overridden, so the child version is always used.

I discuss polymorphism in Python and Javascript shortly after we see more examples of it in C++. Other languages such as C# and Java also support polymorphism but in slightly different ways. While C# uses the keyword `virtual` in a similar way to C++, Java treats all member functions as virtual by default.

☑ **Practice Task:** Add a UI to create dynamic objects of different program types (`ArtStudent`, `EngineeringStudent`, etc.), add the pointers to the `allStudents`, and update `numAllStudents`. Then, add search and report ability to the UI using ID. See the difference when virtual functions are used or not.

```
ArtStudent* a1 = new ArtStudent();
allStudents[numAllStudents] = a1;
numAllStudents++;
```

SIDEBAR: C# AND JAVA

Throughout the book, I mentioned Java and C# frequently, even though I did not discuss them in detail. Both these languages are commonly used. While they have a lot in common with C/C++, their creation was due to certain concerns that existing languages such as C/C++ did not answer (See "Sidebar: Web and Virtual Machines" in Chapter 3).

The syntax in Java and C# is very similar to C++. Some of the main differences are

- Java and C# is fully object-oriented, so there are no global functions.
- In both languages, a function called `Main()` has to exist in one of the classes and will be the starting point of the program.
- There is no explicit pointer in either one of these two languages. Arrays can be created using a variable as size. The language automatically decides which parameters are passed by value and which ones by reference. Although programmers can override this decision, for example, by using the `ref` keyword in C#.
- Objects are created using the keyword `new`, similar to dynamic objects in C++. An instance of a class without an explicit call to the constructor (using new) is, in fact, a pointer that is not initialized and should not be used.

```
MyClass m; //m is a pointer (reference) with no valid value
m = new MyCLass(); //m is initialized now
```

- Both languages use automatic garbage collection; objects that are created dynamically will be removed once they are no longer in use.
- Access modifiers such as public and private are added inline before each member of the class.

As a newer language, C# is considered by many a "simpler" and "improved" successor to Java. But C# has a much more limited user base as .NET framework was supported only on Windows-based systems. So, there was no standard browser support for it. The popularity of Java decreased when modern browsers started to remove support for it due to various performance reasons. On the other hand, it increased again thanks to its adoption as the native language in Android mobile OS.

12.2.1 Visual Effects and Image Processing

To demonstrate the use of dynamic polymorphism and virtual functions, let's consider another example. I introduced the Render class as a solution to modular scene rendering where a single object is responsible for all drawing operations. This allowed the rest of the program to not be concerned with the visual output. As long as `Render::Draw()` function is called with a `GameObject*`, the drawing will be done correctly. The `Render` class implements information hiding as it can achieve various ways of drawing without any need for the rest of the program to deal with it.

For example, the `Render::Draw()` can be modified to apply a global visibility to all objects, regardless of their own visibility value:

```
void Render::Draw(GameObject* obj)
{

    if (obj->visibility != 0)

    {

        ofSetColor(255, 255, 255, GLOBAL _ VISIBILITY*255/100);

        Draw(obj->currentShape, obj->x, obj->y);

        ofSetColor(255, 255, 255, 255);

    }

}
```

Another way that the `Render` class can apply styling effects to drawing is through **Image Processing**, i.e., pixel-level manipulation of images. Image processing is an essential part of computer science and computer engineering. Many applications rely on the computer's ability to process images, from computer vision (surveillance and object recognition) to photo and video editing. A particular example is the process of applying visual effects, which we commonly see in movies and photo editing. Changing the colors, adding highlights and borderlines, adjusting contrast and brightness, and sharpening and softening are examples of visual effects in photo editing.

In this section, our goal is to define modules that apply different visual effects to an image and then provide our render class with these modules, so any game object can be rendered with a different effect, as shown in Exhibit 12.1.

EXHIBIT 12.1   Using visual effects for rendering. Enemies are rendered with a "red" filter, even though they have the same shape (image file).

Before we get to the design of our rendering system with effects, let's see how an Effect module works.

12.2.2 Basic Image Processing

Manipulating images can take various forms, but they all boil down to a basic task: changing the pixels, which means changing the color of pixels or the RGB values. The basic operations for working with pixels are reading and writing the color of a pixel. In different graphics libraries, these are referred to as different names such as ReadPixel, GetPixel, WritePixel, PutPixel, or SetPixel. The information required for these operations includes the bitmap on which the operation is performed, the XY coordinates within the bitmap, and the color, which can be a user-defined type or a series of RGBA values (red, green, blue, and alpha).

For example, using SDL/SDLX, we had

```
void SDLX_GetPixel(SDLX_Bitmap* bmp, int x, int y, SDLX_
    Color* c);

void SDLX_PutPixel(SDLX_Bitmap* bmp, int x, int y, SDLX_
    Color* c);
```

In OpenFramework, these are methods of a class called ofPixels which itself is a member of ofImage and holds the pixel data. OpenFrameworks also has ofColor class that encapsulates all the data and operations needed for a single color. The following code, for example, sets the color of the top-left pixel of an image to red.

```
void ApplyEffect(ofImage* img)
{
        ofColor c;
        c.r = c.a = 255;
        c.g = c.b = 0;
        ofPixels pixels = img->getPixels();
        pixels.setColor(0, 0, c);
        img->setFromPixels(pixels);
}
```

ofPixels::setColor() is the equivalent of SDLX_PutPixel() in OF. Note that the getPixels() function returns a copy of pixel data, not a pointer to the original. Once a pixel value is changed, the pixel data has to be written back to the ofImage object using setFromPixels(). This makes the operation quite slow if we are doing it for just one pixel. As we can see shortly, we usually perform such operations on all pixels.

You can test this function in our OF-based game by adding the following lines to the Game::Update()[1]:

---

[1] Instead of OF_KEY_UP, we can use any other key code. OpenFrameworks has defined some of the key codes such as arrow keys. For those that are not defined, the code used in event handlers and our keyDown array is the ASCII code for the character. So, for example, for character "A" we can use 65, or for character "a" 97. Recall that we can use the single quote " " operator to get the ASCII code, so "A" is 97.

```
    if (keyDown[OF _ KEY _ UP] == true)
        ApplyEffect(player.currentShape);
```

Now, let's try to change all pixels in the first row to red. To do this, the line that calls set-Color() has to be repeated for all pixels at y=0:

```
void ApplyEffect(ofImage* img)
{
    ofColor c;
    c.r = c.a = 255;
    c.g = c.b = 0;
    ofPixels pixels = img->getPixels();
    int w = img->getWidth();
    for (int i = 0; i < w; i++)
        pixels.setColor(i, 0, c);
    img->setFromPixels(pixels);
}
```

If we want to do the same thing for all the pixels in the image, we should repeat the loop for all the rows, which means adding another loop:

```
void ApplyEffect(ofImage* img)
{
    ofColor c;
    c.r = c.a = 255;
    c.g = c.b = 0;
    ofPixels pixels = img->getPixels();
    int w = img->getWidth();
    int h = img->getHeight();
    for (int i = 0; i < w; i++)
        for (int j = 0; j < h; j++)
```

```
                        pixels.setColor(i, j, c);

        img->setFromPixels(pixels);

    }
```

The double loop in the above code is essential in image processing as it allows us to go over all pixels. The following code reads the pixel color of all pixels, sets the G and B parts to zero, and writes the new value back, which will be shades of red.

```
void ApplyEffect(ofImage* img)
{
    ofPixels pixels = img->getPixels();
    int w = img->getWidth();
    int h = img->getHeight();
    for (int i = 0; i < w; i++)
        for (int j = 0; j < h; j++)
        {
                ofColor c = pixels.getColor(i, j);
                c.g = c.b = 0;
                pixels.setColor(i, j, c);
        }
    img->setFromPixels(pixels);
}
```

☞ **Key Points**
- GetPixel() and SetPixel(), or similar names, are the most common operations in image processing that read and write pixel values at the given coordinates.
- A double loop that goes over both X and Y in a rectangular region is a very common code structure in image processing. It is used to read and process images or regions of them.

☑ **Practice Task:** Use the above sample code for ApplyEffect(), and perform other effects. For example, reduce the RGB values.
  Hint: Make sure RGB values stay within 0–255 range.

12.2.3 Sample Image Effects

Using the above code, let's consider a couple of other examples of operations (effects) that we can apply to images.

Here is the code to change the brightness by a certain amount (called CHANGE). Note how we change the RGB components, but make sure they stay within the valid range (0–255). This restriction is important if we don't want our colors to look all wrong when they go beyond the valid range.

```
void ChangeBrightness(ofImage* img)
{
      ofPixels pixels = img->getPixels();
      int w = img->getWidth();
      int h = img->getHeight();
      for (int i = 0; i < w; i++)
          for (int j = 0; j < h; j++)
          {
                ofColor c = pixels.getColor(i, j);
                c.g += CHANGE;
                if(c.g < 0)
                        c.g = 0;
                if(c.g > 255)
                        c.g = 255;
                //repeat for c.a and c.b
                pixels.setColor(i, j, c);
          }
      img->setFromPixels(pixels);
}
```

A more complicated example is for detecting the boundaries of a shape. This operation is commonly referred to as **Edge Detection** and assumes that a shape (an object in the scene) has a different color from its surrounding area (foreground vs. background), as illustrated in Exhibit 12.2.

Here is the code. It assumes we have a variable of type ofColor that is set to red (G and B set to zero and R set to 255) and is called RED. See how the operation uses two pixels: the current pixel that we are processing (i,j) and the next pixel (i+1,j). If these two pixels don't have the same color (R, G, or B are different), then we set the current pixel to red to identify an "edge" point.

```
void EdgeDetection(ofImage* img)
{
      ofPixels pixels = img->getPixels();
      int w = img->getWidth();
```

EXHIBIT 12.2   Edge detection. The edge pixels of an object (black lines above) can be used to recognize the shape and so the object.

```
    int h = img->getHeight();
    for (int i = 0; i < w; i++)
        for (int j = 0; j < h; j++)
    {
            ofColor c1 = pixels.getColor(i, j); //current pixel
            ofColor c2 = pixels.getColor(i+1, j); //next pixel
            if(c1.a!=c2.a || c1.g!=c2.g || c1.b!=c2.b)
                    pixels.setColor(i, j, RED);
        }
    img->setFromPixels(pixels);
}
```

The above code only compares the pixel to the "next one." This is quite limiting for edge detection, as a pixel may be on the right, left, top, bottom, or a diagonal edge. The full code has to compare the current pixel to all its neighbors: $(i-1, j)$, $(i+1, j)$,$(i, j-1)$, and $(i, j+1)$ are what we call the **4-Adjacency** neighbors that only includes left, right, top, and bottom. The 8-Adjacency neighbors include the diagonal ones as well such as $(i+1, j+1)$.

Another problem with this code is that once we set a pixel to red, we have changed the original image. So, when we move to the next pixel, we no longer have the original value of the pixel we processed. A better alternative is to make and use a copy when the original value is going to be needed.[2] In the following code, we have two copies of the pixel data, pixels1 and pixels2. While we do all the reading from pixels1, the changes are applied to pixels2, which is then used to set the image.

```
void EdgeDetection(ofImage* img)
{
    ofPixels pixels1 = img->getPixels();
    ofPixels pixels2 = img->getPixels();
    int w = img->getWidth();
    int h = img->getHeight();
    for (int i = 0; i < w; i++)
        for (int j = 0; j < h; j++)
        {
            ofColor c1 = pixels1.getColor(i, j);  //current pixel
            ofColor c2 = pixels1.getColor(i+1, j); //next pixel
            if(c1.a!=c2.a || c1.g!=c2.g || c1.b!=c2.b)
                    pixels2.setColor(i, j, RED);
        }
    img->setFromPixels(pixels2);
}
```

---

[2] This is not always the case, and we may not need to have a copy of original data, for example, when changing the brightness. The original value of pixels is usually needed when we are performing an operation that depends on the value of neighboring pixels, so a changed pixel value may be necessary later.

> ✋ **Reflective Question:** Does it make sense to you to have two copies of pixel data in the edge detection code, read from one and write to the other? Pay attention to how the pixel we are writing to will be the neighbor for another pixel in the next operations. If we write a new value, the old value will be lost and can't be used for future calculations. In image processing it is important to use this two-copy system.

> ☑ **Practice Task:** Write a full edge detection code that considers all directions.
> ☑ **Practice Task:** Write a simple image smoothing code by changing the RGB values of any pixel to the average of its neighbors.

12.2.4 Effect Classes

To achieve the design shown in Exhibit 12.1, we can have a base Effect class that sets the general structure of all visual effects and then derive new classes from it for specific effects. The alternative is to have various effects implemented in one class as different methods, ApplyEffect1(), ApplyEffect2(), etc. The problem with this approach is that it is not very flexible as we can't add new effects without modifying the existing code. Inheritance allows us to extend classes in a more reusable and manageable way, as each class is independent, and even without access to source code, we can extend a library class (as we are doing for ofBaseApp in OpenFrameworks).

Sample Code #2-c shows the implementation of a base Effect class and then a derived RGBEffect that performs the operation I had demonstrated using a global function ApplyEffect(). We can use this child class as an example and develop more effects.

The base class has three properties. One determines the effect type, such as EFFECT _ RGB. The others are reserved for parameters that effect may need. For example, the RGBEffect uses one parameter that determines which color components have to be removed. EFFECT _ RGB _ RED is a red effect and removes blue and green. Effect class has a constructor that doesn't really do anything but may be needed in future versions. It also has an Apply() method that is equivalent to the ApplyEffect() function I showed earlier. This function is the most important part of any filter, and we need to discuss it in more detail.

Sample Code #2-c

```
1. //Main.h

2. #define EFFECT _ RGB          0 //effect type

3. #define EFFECT _ RGB _ RED    0 //effect parameter values

4. #define EFFECT _ RGB _ BLUE   1

5. #define EFFECT _ RGB _ GREEN  2
```

```
6.

7.   //Effect.h

8.   #pragma once

9.   class Effect   //base class for all effects

10.  {

11.  public:

12.     Effect();

13.     ~Effect();

14.    void Apply(ofImage* img) = 0; //pure virtual function

15.     int type;  //all effects have a type

16.     int param1; //effects can have two parameters

17.     int param2;

18.  };

19.  //Effect.cpp

20.  #include "Main.h"

21.  Effect::Effect()

22.  {

23.  }

24.  void Effect::Apply(ofImage* img)

25.  {

26.     ofPixels pixels = img->getPixels();

27.     int w = img->getWidth();

28.     int h = img->getHeight();

29.     for (int i = 0; i < w; i++)

30.         for (int j = 0; j < h; j++)

31.         {

32.             // base effect doesn't really do anything

33.             ofColor c = pixels.getColor(i, j);

34.             pixels.setColor(i, j, c);
```

```
35.             }
36.     img->setFromPixels(pixels);
37. }
38. //RGBEffect.h
39. #pragma once
40. class RGBEffect :   public Effect
41. {
42. public:
43.     RGBEffect();
44.     ~RGBEffect();
45.
46. void Apply(ofImage* img);
47. };
48. //RGBEffect.cpp
49. #include "Main.h"
50. RGBEffect::RGBEffect()
51. {
52.     type = EFFECT_RGB;
53.     param1 = EFFECT_RGB_GREEN;
54. }
55. void RGBEffect::Apply(ofImage* img)
56. {
57.     ofPixels pixels = img->getPixels();
58.     int w = img->getWidth();
59.     int h = img->getHeight();
60.     for (int i = 0; i < w; i++)
61.         for (int j = 0; j < h; j++)
62.         {
63.             ofColor c = pixels.getColor(i, j);
```

```
64.                    if (param1 == EFFECT _ RGB _ RED)
65.                        c.g = c.b = 0;
66.                    if (param1 == EFFECT _ RGB _ GREEN)
67.                        c.r = c.b = 0;
68.                    if (param1 == EFFECT _ RGB _ BLUE)
69.                        c.g = c.r = 0;
70.                    pixels.setColor(i, j, c);
71.              }
72.         img->setFromPixels(pixels);
73. }
```

When using the `GameObject` and its child classes, it made sense to have a base class that can be used on its own to create objects that are "basic" and have no special features. In the case of effects, though, this may not be an appropriate design. What visual effect the base Effect class should implement? There is no default operation for effects as they are all different. What is common about them is that they all have certain common properties and have a standard function to apply their effect. It is important that this common function exists, so the rest of the program can use effects regardless of their type.

In cases like this, when there is no default operation to provide for one or more functions in the base class, there is no need to have that class implement those functions. It is expected that the child classes will have the implementation, and the child class version of the function is always executed. In other terms, we have virtual functions that don't even have a base version. These are called **Pure Virtual Functions** and are determined by =0 at the end of function declaration in the header file (line 14). A class with a pure virtual function is called an **Abstract Class**. It cannot have instances (we can't create objects of `Effect` type) and can only be used as the base for other classes. The main purpose of abstract classes is to establish a common structure and appearance for a group of classes. For example, all effect classes will have members `type`, `param1`, `param2`, and `Apply()`. This makes it easy for the classes to be reused and managed, as the rest of the program doesn't need to deal with how the effect is implemented and work. Effects can be replaced with minimum change in the code that uses them. In other terms, abstract classes define a common interface for all child classes. That is why in languages such as Java, they are called **Interface**, a class with no function implementation that is only used as a base in inheritance.

☞ **Key Point:** Pure virtual functions have no implementation in base class and have to be implemented in the child. A base class with a pure virtual function is called an abstract class and can only be used as base, and not to define instances.

12.2.5 Rendering with Effects

I started this section with the problem of adding visual effects to our game rendering. The `Effect` classes take care of implementing the visual effect, but integrating them with the rest of the code is a different task. It can be done in multiple ways:

1. The `Render` class can have a series of `Effect` objects of different types. The `Game` class then sets a property in each `GameObject` that identifies the effect they need to have, or passes a new parameter to `Render::Draw()` to choose the effect.

2. The Game class has a series of Effect objects of different types. It passes a pointer to these effects along with the object to `Render::Draw()`, and `Render` class uses the effect it has been given.

Both methods work, but the first one will require the code in `Render` class to know all effect classes, which is more restrictive. New effects cannot be developed without changing the Render class code. As such, I choose the second method. It requires adding a new version of `Render::Draw()` that has an `Effect` parameter, as shown in Sample Code #3-c.

Note that the effect changes the image, but we don't want our original `GameObject` to be modified. We only want it to be displayed differently. Later, we may need to display it with another effect or no effect at all. So, the original image data should not be affected. As such, the new `Render::Draw()` creates a temporary copy of the object's shape (line 20), applies the effect to it (line 21), attaches that image to the object (lines 22 and 23), and then draws the object using the old version of `Draw()` (line 24). Once done, it restores the original image. The temporary image will be released at the end of the function as it is a local variable.

In `Game::Draw()`, we can provide an effect when calling the `Render::Draw()` for objects. The effect can be a local variable, but it's better to have a class member in `Game` for each effect class we want to use.

```
RGBEffect effect;
effect.param1 = EFFECT_RGB_RED;
render.Draw(&enemy, &effect);
```

Sample Code #3-c

```
1. Render.h

2. #pragma once

3. class Render

4. {

5. public:

6.     Render();
```

```
7.

8.      int  scrollX;

9.      int  scrollY;

10

11.     void Draw(GameObject* obj);

12.     void Draw(GameObject* obj, Effect* effect);

13.     void Draw(ofImage* img, int x, int y);

14. };

15. //Render.cpp

16. void Render::Draw(GameObject* obj, Effect* effect)

17. {

18.     if (obj->visibility != 0)

19.     {

20.             ofImage tempImg(*obj->currentShape);

21.             effect->Apply(&tempImg);

22.             ofImage* saveShape = obj->currentShape;

23.             obj->currentShape = &tempImg;

24.             Draw(obj);

25.             obj->currentShape = saveShape;

26.     }

27. }

28. //the rest of Render functions
```

12.2.6 Using Dynamic Polymorphism in Render Class

The key part of Sample Code #3-c is the use of dynamic polymorphism in the new `Render::Draw()` function. To have a common interface and to keep rendering independent of effects, we chose to provide all Effect child classes to `Render::Draw()` as `Effect*`. In line 21, when we call the `Apply()` function, the compiler doesn't know which version of `Apply()` should be called. The base class has no implementation, and the compiler doesn't know what is the child type for the object being passed to `Render::Draw()`. So, no Early Binding can happen. It is at run-time that the type of effect object will be determined, and the proper function will be called.

12.2.7 Polymorphism in Python and Javascript

In both Python and Javascript, there is no object type explicitly defined, and no early compiling happens as they are interpreted languages (see Section 3.1). When we pass an object to a function, we do not specify its type. For example, a simplified equivalent Python code for the `Render::Draw()` would look like the following:

```
def Draw(self, obj, effect) :
    effect.Apply(obj.currentshape)
```

The Python run-time environment will check the objects `effect` and `obj` to see if they have members called `Apply` and `currentShape`. Classes can be extended through inheritance, as I showed in Chapter 11, and child classes can override existing parent members. The child class version will be automatically used, and it can call the base class version if needed (either by using the name of the base class or the `super()` function).

Sample Code #4-p demonstrates polymorphism in Python. The class `ChildItem` derives from `Item` and overrides the `Report()` function. When a `ChildItem` object is passed to a function, and its `Report()` is called, the child version will be used. Note that the base class version can be called in two ways:

1. Using the base class name (line 13) which needs the `self` parameter

2. Using the `super()` function which needs no extra parameter.

Sample Code #4-p

```
1.  class Item :
2.      def − init − (self):
3.          self.data1 = 1
4.      def Report(self) :
5.          print("=====")
6.      print(self.data1)
7.

8.  class ChildItem(Item) :
9.      def − init − (self):
10.         Item. − init − (self)
11.         self.data2 = 2
12.     def Report(self) :
13.         Item.Report(self)    #needs self as parameter
```

```
14.            #alternative:
15.               #super().Report()      #doesn't need self as
parameter            print(self.data2)
16.
17.  def Test(obj) :
18.       obj.Report()
19.
20.  child = ChildItem()
21.  Test(child)
```

☞ **Key Point:** Dynamic polymorphism and virtual functions allow code to use base class while the actual objects are instances of child classes that maintain their identity.

✍ **Reflective Question:** Do you see the role of dynamic polymorphism in creating class libraries? A library can have a class like Render that only knows about GameObject. Programs using this library can develop their own child classes based on GameObject that have overridden functions. Instances of these classes can be passed to Render and still have their own overridden function used, even though the Render class has no idea about these new classes and functions. Without this feature, developing class libraries would be a lot harder.

12.3 PHYSICS CLASS REVISITED

To wrap up the discussion of polymorphism and OOP, and to include all the features we have reviewed in our sample game, let's try to add the Physics class (Section 10.1) to the OF-based project with an extra feature.

In Section 10.1, we assumed the ground is made of a series of flat levels. As such, we modeled the ground using two arrays that show the X location of bumps (changes in level) and the new Y (ground level) after that X value. The function GroundLevel() was in charge of returning the ground level at any given X based on these two arrays. Other assumptions about Physics included

- A gravity-based accelerated movement as controlled by FallJump() function

- A certain algorithm for controlling the movement as implemented in Move()

- Collision detection based on linear distance using Collision() and Distance().

Sample Code #5-c shows the OF-based implementation of this class. To use it, we have to add a Physics object to the Game class and pass the objects to its Move() function instead of calling the object's Move().

Sample Code #5-c

```
1.   //Physics.h
2.   #pragma once
3.
4.   class Physics
5.   {
6.       int xLevels[MAX_GROUND_LEVELS];
7.       int yLevels[MAX_GROUND_LEVELS];
8.       int nLevels;
9.   public:
10.      Physics();
11.      ~Physics();
12.
13.      void Move(GameObject* obj);
14.
15.      virtual int GroundLevel(int x);
16.
17.      bool Collision(GameObject* obj1, GameObject* obj2);
18.      float Distance(GameObject* obj1, GameObject* obj2);
19.      float collisionDistance;
20.  };
```

As you should expect by now, a child class of Physics can extend its functionality by adding new types of physics-related features or changing the existing ones. Any of the member functions can be modified by a child class to extend the class. For example, the collision can be detected not based on distance but based on any corner of one object being inside the boundary of the other object, or the distance can be between centers and not the top-left corners.

EXHIBIT 12.3   Ground level can change as steps, ramps, or ram shapes.

Not only we can have multiple child Physics classes, but also we can use them at the same time, in the same game. One object may follow a gravity-based fall/jump, while another uses a different system, or one level of the game may have a platform-based ground, while another uses a randomly shaped one, as shown in Exhibit 12.3.

Dynamic polymorphism allows us to create different Physics classes but use all of them similarly in the Game class, through calling the Move() function. Using virtual functions guaranties that the child version of the functions will be executed regardless of how the function is called.

12.3.1 Heightmaps

To demonstrate the use of dynamic polymorphism, I revise the sample game with the following features:

- A new child class based on Physics that uses a heightmap image to represent the ground level. This class will modify the Physics::GroundLevel() function and adds a new method to load and process a heightmap image.

- A Physics* member in the Game class instead of an actual Physics object. This pointer can be set at any time (including constructor) to control which Physics the game uses.

Sample Code #5-c shows the new child HMapPhysics (Heightmap-based Physics) class and its use in the game project. Our design process for this new class followed these steps:

- Determine which new information we have to add new data members

  - Instead of two xLevels and yLevels arrays, we need to have the Y value for every possible X (from zero to the width of the image). This means we will need an array with dynamically defined length to hold Y values, and X values are the index to this array. No separate X array is needed as we include all X values. Since the array length (image width) is determined dynamically, we also need a variable to hold that information.

    ```
    int* groundLevels;

    int heightMapWidth;
    ```

- Determine which new operations need to be performed on these data members to add new member functions

  - We will need to read an image and identify the Y values for the ground level. This should be done in anew function.

  ```
  Void ReadHeightMap(ofImage* hmap);
  ```

- Determine which existing functions have to be modified to incorporate new members

  - GroundLevel() function has to change to make use of the new array.

The constructor sets the array pointer to NULL and width to zero, and the destructor frees the array if not NULL. The creation of array happens in the new ReadHeightMap() function (line 26). The code for this function is based on the image processing examples I discussed in Section 12.2. The code

- Identifies the width and height of the image and allocates memory for the array (line 32)

- Starts a double loop for X and Y to go over all pixles

- For each X, starts from the top and looks for a black pixel. When it is found (line 42), the value is saved in the array for that X, and the rest of the inner loop (Y) is skipped (line 45).

The Game class has a new constructor that receives a Physics* (line 59) and uses it to move the objects (line 66). The main() function creates an HMapPhysics object (line 72) and passes it to the Game (line 76).

Different types of Physics can be created and passed to the Game class. The Game::Update() treats them all as Physics*, but since GroundLevel() is virtual, the child version will always be called and correct ground values will be used.

Sample Code #6-c

```
1. //HMapPhysics.h

2. #pragma once

3. class HMapPhysics : public Physics

4. {

5.     int* groundLevels;

6.     int heightMapWidth;

7. public:

8.    HMapPhysics();

9.    ~HMapPhysics();
```

```
10.    void ReadHeightMap(ofImage* hmap);

11.    int GroundLevel(int x);

12. };

13. //HMapPhysics.cpp

14. #include "Main.h"

15.

16. HMapPhysics::HMapPhysics()

17. {

18.    groundLevels = NULL;

19.    heightMapWidth = 0;

20. }

21.    HMapPhysics::~HMapPhysics()

22. {

23.    if(groundLevels != NULL)

24.        free(groundLevels);

25. }

26. void HMapPhysics::ReadHeightMap(ofImage* hmap)

27. {

28.    if (hmap == NULL)

29.        return;

30.

31..    heightMapWidth = hmap->getWidth();

32..    groundLevels = (int*)malloc(heightMapWidth *
            sizeof(int));

33..    ofPixels pixels = hmap->getPixels();

34..    int w = heightMapWidth;

35..    int h = hmap->getHeight();

36..    for (int i = 0; i < w; i++)

37..    {
```

```
38..          groundLevels[i] = h - 1;
39.           for (int j = 0; j < h; j++) //going down
40.           {
41.               ofColor c = pixels.getColor(i, j);
42.               if (c.r == 0 && c.g == 0 && c.b == 0)
43.               {
44.                   groundLevels[i] = j;
45.                   break; //go to next i
46.               }
47.           }
48.      }
49. }
50.
51. int HMapPhysics::GroundLevel(int x)
52. {
53.      if (x>0 && x<heightMapWidth)
54.           return groundLevels[x];
55.      else
56.           return -1;
57. }
58. //Game.cpp
59. Game::Game(Physics* p)
60. {
61.      physics = p;
62. }
63.   //in Game::Update()
64.   //player.Move();
65.      int oldx = player.x;
66.      physics->Move(&player);
```

```
67.            render.scrollX += (player.x-oldx);
```

68. //Main.cpp

```
69. int main( )
70.    {
71.            ofSetupOpenGL(SCREEN_WIDTH,    SCREEN_HEIGHT,OF
               _WINDOW);
72.            HMapPhysics p;
73.            ofImage hmap;
74.            hmap.load("images/hmap.bmp");
75.            p.ReadHeightMap(&hmap);
76.            ofRunApp(new Game(&p));
77.    }
```

12.3.2 Static Members

Even though the example in this section uses only one Physics object, it is possible to have multiple and switch between them as many times as needed, as we did for effects. In such cases, it is possible to have members in Physics class that are always the same for all instances of that class. The collisionDistance value is an example. It determines the minimum distance that is considered a collision. Any class can have such members that are shared among all instances. The keyword static identifies such data members, as shown below. When used for a member function, it means that the function uses only static members and, as such, can also be shared among instances of the class.

The following code shows how to modify the Physics class to use static members:

```
//class members in Physics.h
static bool Collision(GameObject* obj1, GameObject* obj2);
static float Distance(GameObject* obj1, GameObject* obj2);
static float collisionDistance;
```

If a data member is static, we need to define it as a global variable separately, outside the class:

```
//global variable in Physics.cpp
float Physics::collisionDistance = COLL_DISTANCE;[3]
```

---

[3] If a global variable needs to be used in other source files, you may use the keyword extern.

Static member functions can be called with class name instead of object name because they no longer belong to a single object but to the whole class (all instances). The Game::Update() class needs to be modified when using the Collision() function:

```
if (Physics::Collision(&player, &prize) && prize.visibility>0)

prize.visibility --;
```

The advantage of having such static functions is that we no longer need to have an instance of a class to access its functions. This feature is very helpful when designing classes with a series of functions and no data (for example, a collection of Math helper functions).

☞ **Key Points**
- Static data members are shared between all instances of a class.
- Static member functions can only use static data. They are called through the class name, not the object.

HIGHLIGHTS

- Polymorphism is the ability of objects to behave in different forms, as instances of both the parent and the child class.

- Binding is the act of associating a name to a content, for example, deciding which version of a function needs to be called. Early Binding happens at compile-time while Late Binding at run-time.

- Instances of the child class can be passed to functions that need a parent object.

- When a child object is passed to a function that needs a parent, the object will be treated as parent within that function. If the function accesses one of the methods, the parent version will be used.

- In the above example, if the method is virtual, then the child version will be used.

- Virtual functions are expected to be overridden, so the child version is always used.

- GetPixel() and SetPixel(), or similar names, are the most common operations in image processing that read and write pixel values at the given coordinates.

- A double loop that goes over both X and Y in a rectangular region is a very common code structure in image processing. It is used to read and process images or regions of them.

- Pure virtual functions have no implementation in the base class and have to be implemented in the child. A base class with a pure virtual function is called an abstract class and can only be used as a base, and not to define instances.

- Dynamic polymorphism and virtual functions allow functions to treat parameters as the base class, while the actual objects are instances of the child classes that maintain their identity.

- Static data members are shared between all instances of a class.

- Static member functions can only use static data. They are called through the class name, not the object.

END-OF-CHAPTER NOTES

A. Things I Should Mention

- The term "polymorphism" is also used in biology and zoology. It refers to possible genetic variations that exist in a population.

- Function Overloading (having multiple functions with the same name but different parameters) is sometimes called functional polymorphism.

B. Self-Test Questions

- What is static polymorphism?

- What is dynamic polymorphism?

- What are virtual functions?

- What is a pure virtual function?

- What is the use of an abstract class?

- What are static members?

C. Things You Should Do

- Learn about 3D heightmaps. The concept is the same, but instead of an array, we will have a 2D image.

- Check out these web resources:

  - https://learnopengl.com/

  - https://openframeworks.cc/documentation/3d/

  - http://www.cplusplus.com/doc/tutorial/polymorphism/

- Take a look at these books:

  - *Computer Graphics: Principles and Practice in C*, by James Foley et al. (for 3D graphics)

  - *C++ Programming Language* by Bjarne Stroustrup (for OOP, in general, and inheritance and polymorphism, in particular).

D. Reflect on the Experience of Reading This Chapter

- What did you expect from this chapter before reading it?

- What was it about, and what did you learn?

- What tasks did you perform, and what difficulties did you face?

- How did you feel about the material and tasks presented in this chapter?

- How can you improve your learning experience?

- How do you see this topic in relation to the goal of learning to develop programs?

- Do you feel overwhelmed with the notions of polymorphism and virtual functions? Feel free to review the text and make sure you follow along with the sample codes and tasks.

- Do you see how polymorphism helps with a generic Render class while using child GameObject classes?

PART 6

Moving Forward

The best thing about a Boolean is even if you are wrong, you are only off by a bit.

GOAL

In the previous parts of this book, I reviewed some of the major concepts in programming through the lens of modularization. We saw how Structured and Object-Oriented Programming use various forms of modules to create complicated software programs by assembling simpler elements into hierarchies of modules.

While we covered significant topics that should make you ready to tackle difficult software tasks, there are many more things to learn about. In this last part of the book, and through three short chapters, I introduce some of them to help you move forward.

Software Design

Topics
- Design patterns
- Data structures
- Software Architecture

At the end of this chapter, you should be able to:
- Have a general idea about the reviewed topics
- Plan further studies and learning activities on these topics

OVERVIEW

At the start of this book, I discussed the software development process involving requirement analysis, design, implementation, and testing. The terms "programming" and "coding" throughout this book are used to certain parts of this process. Some people apply these terms to the specific act of "producing the lines of code," and as such, talk about software designers vs. software programmers. While this practice is not uncommon, and we all should be aware of it, I prefer to reserve the term "implementation" for that purpose. I do appreciate, and sometimes use, the terms "software designer" vs. "software programmer" as parts of the programming team, but at least for the purpose of this book, the terms "programming" and "coding" include acts of both design and implementation. Even though these two acts are separate phases of the software development process, the boundaries are not very clear for multiple reasons:

- Almost all software development projects involve going through all phases iteratively to refine the software, add features, and remove defects.

- Small projects cannot afford to have different people working on different tasks, so one person frequently does all or many of the activities individually.

In previous chapters, I discussed some fundamental concepts of software design. As a starting programmer, you are likely to work on relatively big projects where more experienced people do the high-level design. But still, you need to do some low-level design work to get a module or algorithm working. You may also work on smaller projects where you have to design the whole software. Finally, I hope by now, you are keen on learning more. So, in this chapter, I will briefly review some more advanced concepts in software design, including design patterns, data structures, and software architectures.

13.1 DESIGN PATTERNS

If you look back at many of my examples in this book, you will notice that a few general organizations were frequently used in many of them. The two most common ones were the command processor and Game Loop. Both of these cases demonstrate a reusable organization of software that provides an acceptable solution to a common design problem:

- For command processor, the problem was to respond to a series of events or commands. The solution was a loop that consists of (1) receiving the command and (2) dispatching the command to a series of modules, each corresponding to one command.

- For Game Loop, the problem was updating the game world and redrawing it every frame. The solution was again a loop that consists of (1) receiving user input and other information to update the variables and (2) perform all drawings for a new frame based on the new values of the game variables.

Another example is the double buffering (discussed in Section 6.3), which is commonly used in graphics applications to create smooth video frame transitions. Such solutions are generic and reusable because they are not for a particular case and deal with common software problems from different applications. We refer to these solutions as **Software Design Patterns**. The concept of using patterns in software programs has been around since the 1960s, but the specific term was popularized through the 1994 book, *Design Patterns: Elements of Reusable Object-Oriented Software*, by Erich Gamma, Richard Helm, Ralph Johnson, and John Vlissides, the so-called *Gang of Four*. Others have applied the concept to specific domains such as Robert Nystrom's 2014 book, *Game Programming Patterns*.

Two other frequently used and helpful design patterns, particularly for multimedia applications such as games, are

- **Flyweight**: This pattern involves using the same resource for multiple similar objects to increase the efficiency of the program. For example, a series of enemies or particles can all have the same shape. There is no reason to load and create hundreds of copies of the same image if our game uses hundreds of enemy or particle objects. For this reason, our GameObject class in both SDL and OF versions included a pointer to shape, and not an actual SDLX _ Bitmap or ofImage object. Any class that owns the objects (Game or ParticleSystem, for example) can load one copy of the image and assign it to all objects:

//Game.cpp (Section 10.1)

```
prize_shape = SDLX_LoadBitmap("Prize.bmp");

for (int i = 0; i < NUM_PRIZES; i++)

{

prize[i].shape = prize_shape;

prize[i].SetVisibility(100);

prize[i].y = 380;

}
```

//ParticleSystem.cpp (Section 10.2)

```
GameObject* p = new GameObject();

p->x = sourceX;

p->y = sourceY;

p->SetVisibility(100);

p->shape = pshape;
```

Another way of achieving this result is to use static members, as I did for Physics class in Section 12.3. Static class members are only created once and are shared among all instances of that class.

- **State:** The behavior of many software systems can be described as a series of specific situations (certain values for their variables) and events that cause the transition from one of these to another. Each one of these situations is called a State, and a system described in terms of pre-defined states is commonly referred to as a **Finite State Machine (FSM)**. For example, a guard character in a game can have Idle, Watch, and Attack states.

  - Idle involves doing nothing. The state may change if an enemy is approaching, and the distance is less than a certain value. This distance defines the event that causes the transition.

  - Watch involves constantly rotating to always face the enemy when it comes close. So the action in this state starts by detecting where the enemy is and then changing orientation.

  - Attack involves moving towards the enemy. It happens when the enemy is getting closer to what the guard is protecting.

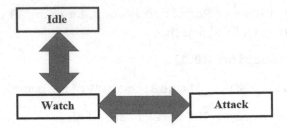

EXHIBIT 13.1   Finite state machine for a guard character. Based on the enemy's distance from the guarding point, the guard moves between three states.

Exhibit 13.1 illustrates the states in the guard example. State Machines are another common design pattern where the behavior is defined through a series of state modules, each including an Action and a Transition function.

Refer to many printed and online references for more on design patterns and how to use them. Good starting points are the books I mentioned earlier:

- *Design Patterns: Elements of Reusable Object-Oriented Software* by Erich Gamma, Richard Helm, Ralph Johnson, and John Vlissides

- *Game Programming Patterns*, by Robert Nystrom

- *Head First Design Patterns: A Brain-Friendly Guide* by Eric Freeman, Elisabeth Robson, Bert Bates, and Kathy Sierra.

13.2 DATA STRUCTURES

The design patterns mentioned in previous sections were mainly conceptualized based on the operations that the program performs. On the other hand, there are also common patterns in organizing data. While they can be considered a design pattern, such organizations of data are commonly referred to as **Data Structure**. Abstract Data Types (Section 9.1) defines the logical relation of data elements and their operations. Data structures define the specific organization of data for ADTs.

Some common data structures are introduced as follows:

- **Arrays** are a collection of related elements with the same type and role.

- **Records** are a collection of related elements with the same or different types and roles. Structures and classes can implement a record. Each item in a Record is called a Field.

- **Tables** are the most common way to represent a database. Rows of a table usually include all the data for one item (for example, information or a student). Rows can correspond to a Record, and the Table can be an array of Records.

- **Stacks** are arrays with a specific order for the reading and writing. Adding a new member to the stack (Push), always places the new data at the top. Removing an existing member (Pop) always takes the member at the top. So, Stacks are Last-In-First-Out (LIFO) arrays.

- **Queues** are similar to stack but use a First-In-First-Out (FIFO) method. The new members are added to one side, and existing members are extracted from the other side.

- **Lists** are arrays with specific size and order, using index values to point to members.

- **Linked Lists** are lists where each member points to the next or previous (Single Linked) or both (Double Linked). Linked List members cannot be accessed using an index value (a disadvantage), but the list can grow dynamically, and new members can be inserted anywhere with only modifying the pointers in previous or next member.

```
class Item

{

public:

        Item();


        //for linked list

        Item* next;

        Item* previous;

};

//imagine currentItem is pointer to an item

//inserting a new item to the list after currentItem

Item* newIitem = new Item();

Item* nextItem = currentItem ->next;

nextItem->previous = newItem;

currentItem ->next = newItem;

newItem->next = nextItem;

newItem->previous = currentItem;
```

- **Trees** are Linked Lists where each item can point to more than one as the next. Trees form hierarchical and one-directional organizations that start from one item and grow at each level.

- **Graphs** are similar to Trees, but there is no single direction, which means any item can point to any other item.

Understanding data structures is essential in managing data in any large application. The proper data structure allows the program to efficiently present and process the data. There are many online and printed references to learn about data structures. Most of these references also cover algorithm design since data structures are tightly related to algorithms for processing them. Some good examples are

- *Data Structures and Algorithms Made Easy* by Narasimha Karumanchi
- *A Common-Sense Guide to Data Structures and Algorithms* by Jay Wengrow
- *Data Structures Using C* by Reema Thareja
- *Algorithms and Data Structures in C++* by Alan Parker.

13.3 SOFTWARE ARCHITECTURE

Modularization is the key thread that connected almost all the concepts discussed in this book. It is a basic design concept, applicable from LEGO® models to software programs, that allows us to create reusable and manageable modules and then build bigger modules using the collections of smaller ones. We saw how a loop is a small module that repeats a few lines of code, or an array is a module that combines similar data elements. I discussed functions and objects as a more complex example of modules, and we used aggregation and inheritance to build more complex objects.

Large systems (software o other types) are generally divided into **sub-systems** or **components**, each made of simple and reusable modules, building a hierarchical structure. As we moved forwards in the book, it was clear how different modules could be considered parts of a bigger module. For example, in a game engine, Render and Effect classes could be parts of a sub-system that manages visual output. Similarly, we could have sub-systems in charge of network communication, user interaction, physics simulation, and other tasks. Together, these sub-systems or components define a large, complicated system. While the design of smaller modules like functions and objects is associated with choosing data items and algorithms to process them, the design of large systems involves choosing the right components and defining their relationships. The organization of a large system (with various tasks and modules) is usually referred to as **System Architecture**. The task of architectural design is essential in big software production projects and frequently involves using well-established architectural patterns.[1]

Among common software architecture, I can mention the followings:

- **Input-Process-Output (IPO)** is a three-section architecture that breaks down the software into three components (each made up of various modules, as illustrated in Exhibit 13.2):
  - Input, to receive, manage, package, and distribute various input information.

[1] As if software programmer vs. software designer terms did not cause enough confusion, some people use the term "software architect" to refer to high-level designers.

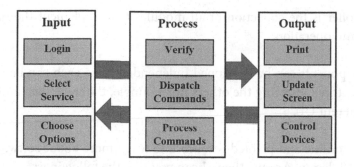

EXHIBIT 13.2   IPO.

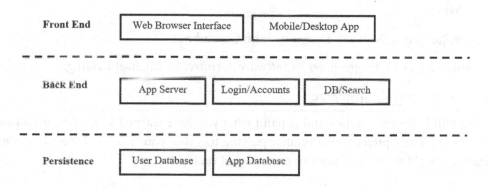

FXHIBIT 13.3   Layered architecture.

- Process, to perform core operations on the data

- Output, to provide processed data to users and other systems.

- **Layered Architectures** provide different levels of abstraction that usually start with user interface (both input and output). Exhibit 13.3 shows examples of layered systems commonly used in web applications and networking.

- **Client–Server Systems** follow an architecture that includes one or more main components that provide data and services (servers), and a large and potentially infinite number of simpler components that use those data and services (clients). Most Internet services such as web, email, and file transfer use a client–server architecture.

- **Model-View-Controller (MVC)** is another common architecture that divides a software program into three components:

  - Model is the collection of all modules that hold and manage the program's core information.

  - View is the collection of all modules that interact with the user and other systems.

- Controller is the collection of all modules that implement the program's main logic and operation.

These components can be developed independently, so each can be replaced with a new one without affecting the others (for example, the same data and operation but with a different UI).

There are many online and printed resources to learn more about software architecture and architectural design. Among them, I can mention the followings:

- *Clean Architecture: A Craftsman's Guide to Software Structure and Design* by Robert C. Martin

- *A Philosophy of Software Design* by John Ousterhout

- *Design It!: From Programmer to Software Architect* by Michael Keeling.

13.4 SUGGESTED PROJECTS

Nothing will help you practice and expand what you have learned so far better and more than defining some projects that require putting together your knowledge and search for some new information. Here are two examples of such projects:

- An Image Editor that performs common editing tasks on images such as resizing, changing colors or brightness, and applying filters

- Game AI to have more complicated enemies in your game.

13.4.1 Image Editor

Write a program with the following requirements:

- A toolbar at the top of the screen as the main user interface

  - A toolbar has two main components: a series of icons (small mages) and support for mouse click on them. You can have a single image for the whole toolbar or multiple small icons displayed.

  - Toolbar includes New, Open, Save, a series of effects (a.k.a. filters), Zoom In, Zoom Out, and Exit commands.

  - For effects, include (at least) converting to grayscale, changing brightness, and smoothing.

  - Zooming in and out involves creating new images with double or half size. Think about what the pixel colors of the new images should be.

- A main area where the image is shown.

13.4.2 Game AI

Write a 2D game program with the following requirements:

- A player that moves up, down, left, and right.

  - Imagine a top view.

- Enemies that attack the player with different algorithms.

  - One group randomly moves.

  - One group goes towards the enemy.

  - One group protects a certain point by moving in front of it and following the player only if it is close to the protected point.

- Use finite state machines to define the behavior of the third group.

Software Projects

OVERVIEW

There was a time that software development was an ad-hoc process, jumping in front of a computer and hacking your way through code. It didn't take much time after the production of larger software products that people in the software industry realized that producing software has a lot in common with other types of production; it needs proper production lines, plans, and management. The use of the term **Software Engineering** started in the 60s as a response to what was considered a software crisis, the growth of software products without proper management. Dealing with producing and maintaining these software products in a more systematic way and through proper production and operation standards and practices is the subject of software engineering as a discipline that brings together computer science, computer engineering, management, and business fields.

In this chapter, I briefly review some of the most important concepts in software engineering. Project management and version control are among subjects that anyone involved in the software industry should be familiar with.

14.1 PROJECT MANAGEMENT

Managing a software project starts by understanding different activities (tasks) that are involved in it. As I mentioned in the first chapter, the most common tasks are requirement analysis, design, implementation, testing, and maintenance. A linear arrangement of these tasks into separate phases results in what is called the **Waterfall Process Model of Software Production**. While easy to understand, the Waterfall Model is not very practical as almost no software project is so linear and one-directional. Repeating phases and overlapping them commonly happens in small and big projects due to new features or new understandings.

More modern **Process (or Lifecycle) Models** for software development are iterative and cyclical. Among them are **Incremental (or Evolutionary)**, **Spiral**, and **Agile** models. Fast-paced mobile and web-based software projects are particularly suited to the Agile model that allows dealing with vague or changing requirements and technologies and has a light-weight methodology. It is based on short **Sprints** that add new features to the software and involves constant collaboration between team members, as illustrated in Exhibit 14.1.

Once the proper process model is accepted, the project management involves multiple activities. These activities and the way they are performed depend on the management framework and method. For example, two common Agile-based methods are **Extreme Programming** and **Scrum**. Regardless of methods, there are a series of tasks that are performed, in different ways, in all project management approaches:

- **Task Identification**, sometimes called **Work Breakdown Structure**, is to define smaller tasks that together form the projects. For example, UI design and asset development can be tasks in a game production, themselves divided into smaller ones.

- **Scheduling** is assigning timing requirements (start date, end date, deadlines, etc.) for tasks. It can also include dependencies that show which task has to end before another can start.

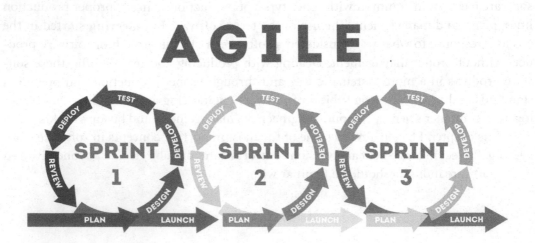

EXHIBIT 14.1   Agile process.

- **Resource Allocation** is to identify which team members or tools are used for each task.

- **Risk management** is to identify risk items (things that can go wrong), plan what to do to prevent them or what to do if they happen, and monitor the risk items.

- **Quality Control** involves testing and evaluating the product to see if it matches the design (**Verification**) and the original requirements (**Validation**).

- **Change Control** (sometimes called **Configuration Management**) is the process of requesting changes (due to found bugs or added features) and tracking all the changes made as to the result of those requests. An important part of Change Control is **Version Control**.

Software Project Management shared many concepts and tools with general project management, although there are specifics as well. A great starting point to learn about various aspects of managing software projects is the book *Software Engineering* by Ian Summerville. Software Engineering Institute (SEI, https://www.sei.cmu.edu/) is a research organization that provides various resources on the subject. Remember that even small projects need to be managed properly.

14.2 VERSION CONTROL

Any software programmer faces one or more of the following cases at least once[1]:

- Losing our files (or changes) when you have no proper backup.

- Changing part of the code and needing to temporarily undo to check something.

- Changing part of the code, only to realize later that they need to get back to what they had earlier.[2]

- Changing part of the code, only to realize later that you don't remember why and how you made that change and now can't do any more changes.

- Making different versions of your code and not being able to track the order or included changes.

- Sharing your code with a collaborator and not knowing how to combine changes that you both have made.

All the above cases, and many more, frequently happen during software development projects and illustrate the need for version control. Managing different versions of your software (including not just code but also all data files and documents associated with the

---

[1] It will be much more than once if they don't learn a lesson from that first case.

[2] Sometimes, this happens minutes before an important presentation. You do a tweak to make things look better, but it all crashes.

software) is one of the most critical parts of project management. There are many tools available for this purpose, and it is essential that you and your team agree on the version control system you are using for your project.

Common approaches for version control are

- **Manual**: You create different versions of your files, name them in a clear way (such as v1, v2, etc., or May 2020, June 2020). You will then save these versions in a backup computer for future use.

  - A common naming convention for versions includes major (release), minor (update), and build (small update or developmental) numbers. For example, v1.2.300 means the first major release, second minor release, and build number 300.

- **Dedicated (Centralized) Server**: You install dedicated software for version control that can be local or on a server managed by and available to your team/organization. A popular example of this approach is using **Concurrent Versions System (CVS)** and its successor, **Apache Subversion (SVN)**.

- **General-Purpose (Distributed) Server**: You use a publicly available server that is installed and hosted by service providers such as **GitHub** (https://github.com/). The advantage is that you don't need to worry about managing the version control server. The drawback is that your organization or team will lose full control on storage and features.

Both SVN and GitHub (and many other alternatives) have client programs that can run on your computer and help with version control, in addition to web-based interfaces. Visual Studio and many other IDE's support both centralized version control and distributed version control through connection to Git or other service providers (this may require add-on extensions).

CHAPTER 15

Concluding Notes

> **Topics**
> - Summary of some key points discussed in the book
> - Some practical advice on software development

> **At the end of this chapter, you should be able to:**
> - Move forward with a plan and confidence in your abilities as a starting software developer

OVERVIEW

I covered some fundamental programming subjects throughout this book. I started by understanding what a program is and how it is made of code and data. Then, I introduced the concept of modularization as a thread that connected most of the topics in this book: arrays, and user-defined types as modules of data and functions as modules of code. Objects came after those to encapsulated code and data, and provide features such as inheritance and polymorphism.

In this concluding chapter, I summarize some of the key points and provide a few practical advices that can help you as you move forward in your programming journey.

15.1 SUMMARY OF KEY POINTS

- Programming is more a way of thinking than a way of writing instructions. It is a logical and creative way to solve problems.

- Each program consists of two main things: data (information) and code (operations on the data).

- Programming is heavily based on the notion of modularization, defining modules of data and code in a hierarchical way.

- Data is presented through variables. The First Golden Rule of Programming states: "For every piece of information that the program creates and/or needs, define a variable."

- Code is primarily organized through functions.

- Arrays and user-defined types group data items together.

- Programs follow a sequential execution model by default.

- Structured Programming uses selections, iterations, and functions to control the execution of the program.

- Object-Oriented Programming (OOP) is based on the notion of objects (and classes), modules that group together related data and code.

- The pillars of OOP are encapsulation, inheritance, and polymorphism.

- Graphics operations allow programmers pixel-level access to the visual output.

- Many languages such as C/C++ and Python don't have built-in support for graphics. They require the use of third-party libraries.

- Games are good examples to learn programming as they incorporate almost all essential features.

- The common structure of a computer game program includes an initialization part and a main loop, itself divided into Update and Draw sections.

15.2 SOME PRACTICAL ADVICE

- You learn programming by writing code (a lot of it) and thinking about your code. Think about why things happened the way they did and imagine what happens if you change things. Then, try to explain the results. Sense-making is as important as the work itself.

- Add detailed comments to your program. It helps others who need to use your code and yourself when you come back to it later.

- Use naming and coding conventions that make sense to you and make your code more readable.

- Work with others. Debug their code and try to explain yours to them. Practice Pair Programming.

- Pay attention to annoying syntax and logic rules, such as

  - == is not the same as = (but is frequently mistaken).

- Comment symbols are tricky: for example, /* vs. */
- Most programming languages are case sensitive.
- Most programming languages need a symbol (commonly;) at the end of statements. Not Python!
- Most programming languages don't care about white space and indents. Not Python!
- You need to have matching { } pairs.
- In typed languages like C/C++, operations on integer variables are done in integer mode. So, 3 divided by 2 is 1!
- Variables (especially pointers) need to be initialized before use.
- Dividing by zero can crash the program. Check before any division.
- Arrays start from index 0.
- Don't get scared by a long list of compile errors. In many cases, fixing one thing takes care of so many errors.
- Save and back up your files regularly.
- Use a version control system.
- **Never think you know enough. Keep learning.**
- **Programming can be fun. Enjoy!**

Bibliography

Adams, E. (2014). Fundamentals of Game Design. Pearson Education.

Agarwal, A., & Lang, J. (2005). *Foundations of Analog and Digital Electronic Circuits*. Elsevier.

Atencio, L. (2016). Functional Programming in JavaScript. Manning.

Bansal, A. K. (2013). *Introduction to Programming Languages*. CRC Press.

Boshernitsan, M., & Downes, M. S. (2004). *Visual Programming Languages: A Survey*. Computer Science Division, University of California.

Ceruzzi, P. E., Paul, E., & Aspray, W. (2003). *A History of Modern Computing*. MIT press.

Cormen, T. H., Leiserson, C. E., Rivest, R. L., & Stein, C. (2009). *Introduction to Algorithms*. MIT Press.

Dix, A., Dix, A. J., Finlay, J., Abowd, G. D., & Beale, R. (2003). *Human-Computer Interaction*. Pearson Education.

Foley, J. D., Dam, A. V., Feiner, S. K., Hughes, J. F., & Carter, M. P. (1997). Computer graphics: Principles and practice, in C. *Color Research and Application, 22*(1), 65–65.

Freeman, E., Robson, E., Bates, B., & Sierra, K. (2008). *Head First Design Patterns*. "O'Reilly Media, Inc."

Gamma, E., Helm, R., Johnson, R., & Vlissides, J. (2009). *Design Patterns Elements of Reusable Object-oriented Software*. Addison Wesley.

Gries, P., Campbell, J., & Montojo, J. (2017). *Practical Programming: An Introduction to Computer Science Using Python 3.6*. Pragmatic Bookshelf.

Huntley, J., & Brady, H. (2017). *Game Programming for Artists*. CRC Press.

Karumanchi, N. (2011). *Data Structures and Algorithms Made Easy: 700 Data Structure and Algorithmic Puzzles*. CreateSpace.

Keeling, M. (2017). *Design It!: From Programmer to Software Architect*. Pragmatic Bookshelf.

Kernighan, B. W., & Ritchie, D. M. (1978). *The C Programming Language*. Englewood Cliffs, NJ: Prentice Hall.

Kodicek, D. (2005). *Mathematics and Physics for Programmers*. Nelson Education.

Krug, S. (2013). *Don't Make Me Think, Revisited: A Common Sense Approach to Web Usability*. New Riders.

Marschner, S., & Shirley, P. (2015). *Fundamentals of Computer Graphics*. CRC Press.

Martin, R. C. (2018). *Clean Architecture: A Craftsman's Guide to Software Structure and Design*. Prentice Hall.

McGonigal, J. (2011). *Reality Is Broken: Why Games Make Us Better and How They Can Change the World*. Penguin.

Murray, J. H., & Murray, J. H. (1998). *Hamlet on the Holodeck: The Future of Narrative in Cyberspace*. MIT press.

Nardon, L. (2017). *Working in a Multicultural World: A Guide to Developing Intercultural Competence*. University of Toronto Press.

Norman, D. (2013). *The Design of Everyday Things: Revised and Expanded Edition*. Basic books.

Nystrom, R. (2014). *Game Programming Patterns*. Genever Benning.

O'reilly, T. (2009). *What Is Web 2.0*. "O'Reilly Media, Inc."

Ousterhout, J. (2018). *A Philosophy of Software Design*. Yaknyam Press.

Parker, A. (1993). *Algorithms and Data Structures in C++* (Vol. 5). CRC press.

Patterson, D. A., & Hennessy, J. L. (2020). *Computer Organization and Design, MIPS EDITION: The Hardware/Software Interface*. Morgan Kaufmann Publisher.

Prensky, M. (2007). *Digital Game-based Learning*. Paragon House.

Raymond, E. S. (2003). *The Art of Unix Programming*. Addison-Wesley Professional.

Reese, R. M. (2013). *Understanding and Using C Pointers: Core Techniques for Memory Management*. "O'Reilly Media, Inc.".

Sax, D. (2016). *The Revenge of Analog: Real Things and Why They Matter*. Public Affairs.

Schaller, R. R. (1997). Moore's law: Past, present and future. *IEEE Spectrum*, *34*(6), 52–59.

Silberschatz, A., Gagne, G., & Galvin, P. B. (2018). *Operating System Concepts*. Wiley.

Sommerville, I. (2015). *Software Engineering 9th Edition*. Pearson.

Spraul, V. A. (2012). *Think Like a Programmer: An Introduction to Creative Problem Solving*. No Starch Press.

Sternberg, R. J. (Ed.). (1994). *Thinking and Problem Solving*. Academic Press.

Stroustrup, B. (1985). *The C++ Programming Language*. Addison-Wesley.

Tanenbaum, A. S., & Bos, H. (2015). *Modern Operating Systems*. Pearson.

Thareja, R. (2011). *Data Structures Using C*. Oxford University Press, Inc.

Toal, R., Rivera, R., Schneider, A., & Choe, E. (2016). *Programming Language Explorations*. CRC Press.

Tselikis, G. S., & Tselikas, N. D. (2017). *C: From Theory to Practice*. CRC Press.

Waldrop, M. M. (2016). The chips are down for Moore's law. *Nature News*, *530*(7589), 144.

Ware, C. (2010). *Visual Thinking for Design*. Elsevier.

Watt, D. A. (2004). *Programming Language Design Concepts*. John Wiley & Sons.

Weinberg, G. M. (1998). *The Psychology of Computer Programming* (Silver Anniversary). Dorset House.

Wengrow, J. (2017). *A Common-Sense Guide to Data Structures and Algorithms: Level Up Your Core Programming Skills*. Pragmatic Bookshelf.

Index

Note: **Bold** page numbers refer to tables; *italic* page numbers refer to figures and page numbers followed by "n" denote endnotes.

Printed in the United States
By Bookmasters

Printed in the United States
By Bookmasters